The Evil Within Us

Evolution of Social Systems

&

The Ideal State

Ideal Publishing Inc.

Published by Ideal Publishing, Inc.

16133 Chester Mill Ter., Silver Spring, MD, 20906

Printed in the United States of America

ISBN: 978-1-7354591-1-0

Dedication

This book is dedicated to the memory of those who spent their lives working for, and suffered for, a better future for mankind and those who laid down their lives, or lost the lives of their loved ones in this struggle: Abraham Lincoln, Albert Einstein, Charles Darwin, Deng Xiaoping, Joe Slovo, Karl Kautsky, Karl Marx, Leon Trotsky, Mao Zedong, Mikhail Gorbachev, Mohammad Ali Jinnah, Mohandas Gandhi, Nelson Mandela, Patrice Lumumba, Rosa Luxemburg, Salvador Allende, Sigmund Freud, Socrates, Vladimir Lenin and Zulfiqar Ali Bhutto

Contents

List of Illustrations

List of Maps

List of Tables

List of Photographs

Preface

This book is a review of human evolution and development of human society and the "state" as a powerful social group. The purpose is to determine the direction in which humanity and human institutions have moved and are moving at this time and, especially, to look at what lies ahead in the future. With this in mind, I have tried to work under the constraints of a concise and simple presentation for the average college student. My intention is to determine and describe, based on historical human experience, what structure and policies define an ideal state. Political scientists disagree on the exact definition of a state. To me, a state is a social group with a central authority, or government, that exercises sovereignty over the territory of the social group, managing affairs of the group, by means of institutions it develops and controls. The "United States", "United Kingdom" and the "Russian Federation" are examples of present-day states, but tribes and nations of the earlier times of our history were also states.

It was quite clear from the start that, if the material is to be understandable to the average undergraduate, some new terminology had to be developed and old terms had to be made more descriptive, or abandoned. In this context *socialism, capitalism and communism* are the most difficult terms to deal with since, due to the strong emotions they have aroused in the political discourse of the recent past, each of them has assumed numerous shades of meaning. Thus, socialism can mean quite different things, depending on whether Adolf Hitler is the speaker, or Vladimir Lenin, or Zulfiqar Ali Bhutto of Pakistan. Capitalism means one thing to Marx and quite another to President Reagan. Similarly, communism means quite difference things, depending on whether you are listening to Marx, President Nixon, or a member of the communist party of China. So, I have avoided the use of these terms to the extent possible and have focused on the fundamental concepts and visions of human society and the relevant state structures, which are implied by these terms. I have coined terms that are more conceptually descriptive, even if they appear to be blunt to some of the readers of this book and, maybe, crude to others. The most difficult term to avoid in this context was "socialism". We are social beings and our development as families, tribes and nations, etc., has been due to increasing socialization of our social groups. Thus, in a way, "socialism" has been with us throughout our existence. At present, this term has developed a specific meaning – "solution of our social problems by our collective efforts, as equals, by developing social institutions". This is the meaning with which I intend to use this term, whenever it is unavoidable. This, I hope, would help in an objective understanding of history and the concepts which are dealt with in this book.

The concept of nationhood is also very confusing if you are listening to politicians of this age. I have kept the traditional concept of a *nation*, which has, till recently, been used widely. According to this concept, a nation is a human social group with four characteristics of its members in common - race, land, language and religion. These characteristics are important in this sequence - race, or genetic heritage, being the most important. Common historical experience also increases the national feelings of a group. Language, in the broad sense of the word, i.e., self-expression, and religion indirectly cover, virtually, the whole culture of a group. Also, a group is a nation, by this concept, even if all the characteristics are not uniformly shared by the whole population of a group. Thus, the Irish are a nation whether they live in Northern or the Southern part of Ireland, or in the United States. Similarly, Koreans are a nation, whether they live in North or South Korea. Also, Arabs are a nation, even though they are split up into many states. Pakistan is a state, but Pakistanis are not one nation. There are ten nations in present-day Pakistan – five are quite large in terms of population, but there are five other smaller ones. Pakistan, according to this concept, is a multi-national state. Pakistan, also, has communities of immigrants, mainly belonging to the territory of present-day India. They and, even some members of the nations inhabiting Pakistan, have become attached to Pakistan more than their respective nations. I refer to such a population as a community that has developed strong *"state loyalties"*, and *nationalism* has become of secondary importance to such communities. This process is in progress in all states of the world, but needs to be identified as such for clarity of discussion. Also, by this concept of nationhood, United States is not a nation, but is a multi-national state. In fact, members of practically all nations of the world reside in it, or are its citizens. United States is also, basically, a European colony, but has become diverse with time. However, its mass media do refer to it, and sometimes even to its territory, as a nation. Britain and India are other multi-national states. Germany, Japan and France are national states and the respective populations are nations. The term, "nation", is used in this context in what follows.

Here, I feel I should describe my attitude towards my national and state origin and my feelings about the United States, of which I became a citizen in 1990. It was a very pleasant surprise to find how my fellow Americans dealt with me. I was always treated with kindness and respect. I was especially impressed by the level of honesty in the new society I found myself in. I have never seen such a high level of honesty and consideration anywhere - and I have been to most of the important countries of the world. I was always frustrated about how the governments in Pakistan managed their state. Starting with the 1956 invasion of Egypt by British, French and Israeli forces, I was forced to try to understand what religion is and what human conflicts are really about. I was twelve years old then, a student of sixth grade in school. As time passed and I had contact with other human beings all over the world, I lost my class, religious and national prides and prejudices and became *just a human being*. When I chose to become a US citizen, my approach to problems of humanity was already a global one. I am a citizen of the US, but I have not lost my feelings for the rest of humanity by the change in my citizenship. As the reader would discover, I do criticize the US and Pakistan, among other states, for actions these states have taken in the global political arena. My attitude towards the US and my fellow countrymen is simple – we want a better US and, if possible, we want it to become an ideal for the rest of mankind. Unfortunately, this is not what has happened in the US during its existence as an independent country, despite all the claims of our rulers. However, I am sure the US would become a better state for its people and for the rest of mankind in the future. It has, already, faced up to its shortcomings and mistakes and has overcome many of its weaknesses and, I am sure, it would continue to do so in the future. *A great America is what we, the people of the US, would continue to dream about – an America that is sensitive to the needs of its children, the elderly and the disabled, an America that truly ensures equality and full freedom to all its citizens in all spheres of life and an America that inspires admiration of the rest of mankind, not its hostility or fear.*

As is well understood now, no matter what stage of history we are looking at, human society has always split up, broadly, into two camps - economic abusers and those who get abused. Different thinkers have given different names to these two classes at different stages of history, as the basic mode of economic production has evolved and the characteristics of the ruling and the ruled class have changed accordingly. The ruling classes are referred to as "tribal chiefs", "feudal landlords", "Aristocrats", "Nobles", "Bourgeoisie", etc., and the abused classes have been referred to as "serfs", "slaves", "Commoners" and "proletarians", etc. In this process that has always split humanity since the beginning of civilization, the basic economic behavior and characteristics of the two classes have always been the same. What Socrates faced two and a half thousand years ago, is what we face even today. Accusers of Socrates alleged that he was corrupting the young by teaching them to study things in the sky and below the earth and by not believing in "Zeus" and other "gods" of Athens. Today, young men and women not only study "things in the sky and below the earth", but also things they cannot directly see with their own eyes. They have all become students of Socrates. And where is Zeus, where are the other "gods" of Athens? We look at their statues and laugh at stories of their petty quarrels. Are they not dead? Is Socrates not alive in all of us?

…. "do not be offended at my telling you the truth: for the truth is, that no man who goes [against] you or any other [crowd], honestly striving against the many lawless and unrighteous deeds which are done in a state, will [survive]; he who [really fights for justice, must lead a private life, not a public one].

…. Me you have killed because you wanted to escape the accuser, and not to give an account of your lives… there will be more accusers of you than there are now….. If you think that by killing men you can prevent [them] from censuring your evil lives, you are mistaken "….

Plato (1896), *Apology.* Translated by Benjamin, Jowett. Urbana, Illinois: Project Gutenberg. Retrieved October 15, 2019, from www.gutenberg.org/ebooks/1656.

Note: "Apology" here means "defense". This is part of what Socrates said in his defense, according to his student - Plato. Accusers of Socrates alleged that he was corrupting the young by teaching them to "study things in the sky and below the earth" and by not believing in "Zeus" and other "gods" of Athens.

I have chosen the South Asian word *"looteras"*, as a generic term, to denote the abusers, or members of parasitic classes, at any stage of history. Rulers like Genghis Khan, Hitler, Pol Pot and killers of natives of the Americas, were not just exploiters, though – they were savage mass murderers also. This term, and what it denotes, according to my understanding, appropriately describes the ruling classes of the past and present, their conditions of existence and their "morality". Thus, according to this concept, human society has always split up into one or another type of *"exploiter class"*, and the *"people"* as the exploited and abused class. It does not, however, mean that all the members of the ruling class are aware of their role, but may think of themselves to be entitled to greater wealth and their main role in ruling the people, as a matter of right. The term however, looks extreme to most people and somewhat extreme it is, despite the fact that it denotes a historically extreme and internationally condemned phenomenon. However, I was at a loss to find a different term which would better describe the ruling classes throughout human history. Many intellectuals use the term "elite/elites" to describe these classes, but this term has interpretations that tend to glorify them - this, certainly, is not my intention! If the readers can suggest an alternative term, I would certainly consider it!

The division into the two classes is dynamic since members of a class are constantly struggling against their fellow members and the members of the opposite class, some managing to move up or being pushed down into the other class. As a rule, the ruling class exploits the people and grows rich at their cost, without working in proportion to what it grabs. As a result, the people work, but do not receive what they deserve on the basis of their efforts. The division of the population of a state, purely on the basis of income, is not the objective here - since the population of a state can be categorized into numerous "classes" based on arbitrary ranges of income - such terminology is in widespread use today and is quite confusing. Most Americans, for example, believe that they belong to the "middle class" and not to the "poor" class, although their standard of living may correspond to either of the two basic classes! They may be poor but they do not want to acknowledge that, or they may be super-rich but do not wish to brag about the wealth in their possession.

Ever since I gained consciousness in a small town in Pakistan, I have always been puzzled by the world around me, and especially, by the conditions under which the people of Pakistan live. The challenge to understand them and their conditions of life, in time, became the challenge to understand life and the world that I found in existence. I hope this effort of mine, as summarized here, would serve to motivate the reader to deeper thought and analysis. My fundamental realization is that mankind has developed two opposite sets of survival skills and behavioral patterns. I refer to one of these sets as *individual survival skills* or skills that embody and promote personal and selfish needs. The other set is referred to as *group survival skills*, i.e., those skills that embody and promote group needs, and individual needs indirectly – sympathy, compassion, civility, consideration, etc. These group skills are the basis of our civilization and are the reason why we are the dominant species on this planet. When we refer to good and evil, we are in reality referring to these opposite skills and qualities within all of us.

It is shocking to realize that the history of mankind is, basically, a history of violence, like that of all other species. Our ancestors have killed and robbed each other, besides killing and eating other species. They have attacked families and abducted other men's women and children. They have stolen from each other what was not theirs. They have tricked and deceived each other. They have enslaved and abused other human beings. They have glorified conquest and abuse of other human communities. Our ancestors have done all this, while desiring to live peaceful lives, while wanting to raise their children in peace and happiness. Today, we can ask any human being how he or she wishes to live and, chances are, he or she would express a desire to live in peace with his or her family and friends and wish others to do the same. This is the fundamental contradiction of our existence, which is validated by the law of evolution and by an understanding of the structure of the human mind. We have been adapting and competing - that is what we have been doing: and violence, loot and plunder have been our history. In other words, our history has been a history of *"lootera-ism"*, but, still, we have had an ever-growing desire for this to come to an end.

The subject matter at hand is based on a brief review of history. The first chapter describes the beginnings of the evolution of mankind in the age of savagery. Chapters-2 deals with feudalism and a very long period of development of our feudal societies. Chapter-3 summarizes how the world affects us and how we change it - the fundamental logic of historical change, as I see it. Chapters 4 and 5 describe the fundamental concepts of economics. Evolutionary history of mankind in the period of civilization, beyond the stage of feudalism, is reviewed in Chapters 6-12. Economics of each stage and the accompanying conflicts are dealt with as the need arises. Chapter-13 describes the semi-global empire that has developed since the end of the Second World War. After describing two current existential threats to mankind, in Chapter-14, historical stages of social, economic and political development of mankind are summarized in Chapter-15. After a brief description of the human personality and its development, sources of human conflicts have been described in Chapter-16. Fundamental political issues covered by the current global political discourse are described in Chapters 17-19. Chapter-20, then, covers the structure of an ideal state, based on the lessons learned from history of mankind and its evolution. Then, the possible future of mankind, as I visualize it at this time, is depicted in Chapter-21.

Throughout the text, I have drawn on the work of numerous thinkers and geniuses who spent their whole lives, and in some cases sacrificed their lives, while searching and investigating one or another aspect of human existence, with the scientific knowledge that was available to them. Others spent their lives, or lost them; in leading humanity toward those goals that they had come to believe would create a better world. The quotations from, or references to, their works indicate what they were able to see in their times, regarding what we see as objective reality now - when we have the luxury of drawing on the enormous growth in scientific knowledge available to us. The quotations are a tribute to the work of those heroes of humanity, but do not imply any judgment on the sum total of their intellectual contributions. For a better and deeper understanding of those philosophers, scientists and political leaders, the reader would have to refer to their original works. I have, however, reviewed, and quoted from, some of those writings that I consider to be important to a thorough understanding of the subject matter of this book. I have also quoted some political leaders when they were dead wrong - in order to indicate what disasters their misconceptions ultimately led to.

This book is not written as a scholarly work with references to works of other scholars to justify each and every conclusion I have drawn. That would have been an impossible task, considering that it covers the whole evolutionary history of mankind in less than four hundred pages. Such a review of history was also unavoidable. I have, thus, referred to quite a few books and articles mentioned in notes at the end. These could get the reader started in finding the answers to questions which are bound to arise in his or her mind. The referred books and articles, in turn, provide numerous further references for the reader, to understand the specific issues they have dealt with.

A bibliography at the end of this book lists those books which, to my mind, provide the background information required for an in-depth understanding of the issues dealt with in this book. Due to the dramatic development of the internet, a lot of information about the history of mankind is available on-line. I like to recommend Wikipedia for this purpose. Also, many video recordings of lectures, book reviews and historical events are available on YouTube. I like to recommend videos of talks by Professors Noam Chomsky, David Harvey and Richard Wolff. Also, talks by Tariq Ali about a wide range of political issues, especially about the Soviet Union and Pakistan are very informative. Martin Jacques, Kevin Rudd - the ex-Prime Minister of Australia, and Professor Peter Nolan of Cambridge University are experts on China. Also, Yukon Huang and Louis Gave have specialized in different aspects of the Chinese economy. I would like to encourage the reader to look into videos of their talks, to get a better understanding of the subject matter of this book.

Modern social sciences have provided the tools for my analysis and it would be a challenge for many readers to verify and understand the validity of my conclusions. I, thus, expect a lot of disagreement and would welcome all criticism. Criticism of recent historical events, i.e., events of the last four hundred years, is especially important, since I have not always described them in sequence and have omitted many events that I considered irrelevant or relatively unimportant for my purposes. I have, also, deliberately repeated descriptions of many events - sometimes to stress their importance and, sometimes, just to summarize what has been dealt with before, to develop my arguments further. I would like to know about such errors in terms of sequence of events, or otherwise, if they are noticed by the readers - so that those may be corrected in a future edition.

Tayyib A Tayyib,

March 25, 2020

Chapter – 1

Life on Earth

Life's Basic Questions

What is life? Why am I alive? Where am I really? What would happen to me in the future, especially after my death? These are questions which bother us all the time, after we gain consciousness? We never get a satisfactory answer. Some religious people may give us some explanations but those explanations raise more questions than they answer. The science of Biology has developed as our science of life. It is based on all that we have learned about ourselves and other life forms on our planet, earth. We have come to know a lot about ourselves and other living beings but life's basic questions still continue to be puzzles we have not been able to solve completely.

Whatever its reasons for coming into being, a few characteristics are very clear about life and living beings. An individual, of any form of life, tends to grow and tends to reproduce himself, i.e., each life form tends to grow in numbers. Plants live on nutrients in the soil, use the sun and water. Some plants feed on each other too and are unable to reproduce without the help of insects for pollination. All forms of life use natural, existing, non-living physical sources, but in addition to those, they basically consume other living beings for survival and growth. As human beings we need the sun, the air, and water, and some other nonliving elements of the earth too. However, we cannot survive without consuming other forms of life – vegetable and animal.[1]

Basic Laws of Biology

Two basic laws/theories form the foundation of Biology. Virtually all living beings are composed of cells, which reproduce themselves. Also, all forms of life evolve with time. These are the two basic laws/theories of Biology.

Cell Structure of Living Beings

Virtually all living beings are composed of one or more cells. Viruses are the only exception. They consist of DNA or RNA with a protein cover. They cannot reproduce themselves. Instead, they use cells, which they infect, to reproduce them. Bacteria are unicellular organisms, while animals and plants are multi-cellular. All cells arise from preexisting cells. Thus, bacteria, being composed of one cell each, reproduce themselves by division into two cells. Our bodies are composed of many different types of cells. Those cells reproduce themselves to repair damage done to our bodies or just to cause growth of our bodies. Cells in the reproductive systems of humans, reproduce human beings. Cells thus, are, the fundamental units of life and the cellular structure is a basic theory of life.[2]

Law of Evolution

All forms of life compete with each other. As environment changes, living beings try to adapt to those changes to survive. The DNA of a living being determines its structure and basic behavior. There are differences in the DNA of members of a population. Those differences are inherited by the offspring of a living being. The inheritance is basically random, resulting in wide variations of inherited characteristics. The process of natural selection ensures that individuals with certain forms and combinations of DNA survive and reproduce. What it means is that certain off-springs of a living being have a better chance of survival and reproduction in competition with others. This process of natural selection ensures that those individuals with better abilities in terms of competition and adaptation are able to survive to reproduce themselves. That is the conclusion that an understanding of evolution necessarily leads us to. Evolution is the basic law of living beings. Some people refer to it as a theory, but it has stood the tests of time and scientific evaluation, and has been accepted by biologists as the most important law of that science. Facts that have come to our knowledge since it was initially proposed, especially the discovery of the structure of DNA and the recent understanding of its role in evolution, have only tended to strengthen the premises on which it is based. Hence, we would refer to it as the *"Law of Evolution"*.[3]

Characteristics of Life

Competition and Survival

In the processes of growth and competition, one instinct common to all living forms is the instinct of self-preservation. The reaction to any danger is anger and defensive aggression. The organism under threat either attacks the threatening individual, or runs away from danger if it can, or, what is relatively rare, it may compromise with it, so the two may coexist while competing at the same time. The competition is for habitat and resources that are essential for survival. Survival of the fittest is the fundamental principle, and this is referred to as "natural selection". The "fittest" doesn't mean the strongest, or necessarily the cleverest species or individuals, but those who have skills for long-term survival and reproduction.

Competition and Adaptation

There is continuous competition among all forms of life that exist at any time. As the environment changes, the individuals of all species learn to adapt to it to survive and reproduce. The adaptation causes mental and physical changes. Organs needed for competition grow, while those that are not essential tend to wither away. Those individuals who are able to devise new tools and techniques that help them in survival and reproduction tend to succeed in competition while others don't. This is as if the competition had started with the purpose of one species winning over all the others and that is, in fact, what has happened, with mankind becoming the dominant species.

Gender Bifurcation

The struggle for survival causes some species to split up into two genders. This has generally happened to animals who have to move to survive. This is one form of adaptation and is a survival skill. It has not generally happened with plants. Most plants have both male and female flowers, at this time, and the female flowers get fertilized by pollination by insects. Some plants, also, reproduce themselves by giving birth to baby plants from their roots. In animals, two genders are the norm. The reason is that the animals have to move or fly during their search for food. If an animal is carrying a baby, this creates a "security risk" if the animal has to defend itself from predators also. Thus, most animals have split up into two genders in the evolutionary process. The females carry and give birth to their offspring and the males are responsible for providing food and protecting the females and the young. Thus, this splitting up into two genders helps the organism in survival. But the bifurcation cannot happen suddenly. Since it is based on evolutionary experience, it happens in stages. We know the existence of some organisms which seem to be in transition from one to two genders. At this stage they have developed both the male and female sexual organs in the same body and are able to reproduce themselves. At a later stage these organisms may gradually split up into two genders. Now, it is important to note that both the new genders are likely to have some quality of the previous generations which were unisexual and so some genes do get carried over to some individuals of both the genders. This variation is part of the evolutionary process. Hence, the presence of homosexuality in some individuals, even after long periods of time, should not be a surprising characteristic.

Honey bees are unique in the way they have evolved and they can exist only as families and their individual existence, over a considerable period of time, has become impossible. Thus honey bees only exist as families, or colonies. There are seven known species of honey bees, of which six are native to Southeast Asia, which is considered to be the origin of this group of insects. The seventh species, known as the "European bee", is found in Europe, Africa, North and South America and Australia. Four of the seven species, build their nests in the open, on trees and rocks. The other three have learned to build their nests in hollow spaces, for better protection from predators and the weather.

A colony of bees has a fully developed female, or "queen", which is the mother of all the bees in it. After mating, it has the ability to lay fertilized or unfertilized eggs at will. The other bees are, mainly, worker bees which do all the work of gathering nectar, raising the young and protecting the colony. Worker bees are pseudo-females, since they can lay eggs but cannot mate and cannot lay fertilized eggs. Only males, or drones, emerge from their eggs. Males only exist in the colony during the mating season, i.e., in the spring season. Their only purpose in life is to mate! They cannot even eat on their own. They have to be fed. Thus, drones in a colony have a mother, no father, but they do have a grandfather! Thus, this insect, perhaps, due to its small size, has not bifurcated into two genders. In fact, there are somewhat like two and half genders of bees! The bees in the colony communicate by chemical signals, which mainly originate from the glands of the queen. They also communicate by making different sounds and by "dancing". If the queen dies, and the worker bees are unable to replace her with another queen raised from a fertilized egg of the previous queen, the colony dies. The colony has bees of different ages performing specialized jobs based on their age, including house-keeping and cleaning, feeding the young and defending the colony against attack by predators. In spring, the colony raises male bees, or drones, for mating, and then produces several queen bees. After mating, each of the queen bees leaves the colony with a group of other bees of the colony. Thus, a colony of bees produces several colonies of bees and this is how honey bee colonies reproduce. Thus, the family or colony of bees, which may consist of more than a hundred thousand individuals at certain time of the year, is really an individual! And honey bees, effectively, have a socialized existence only.

Social Group Formation

One important characteristic of competition between species, especially animal species, is that groups within each species also compete with each other, and that *forming a group is a very important and basic survival skill*. So, there is competition for survival of a species, i.e., competition among species, competition among social groups within each species, and within sub-groups of those social groups - the smallest sub-group being a family where two genders exist, and competition for survival of individuals within each social group at the lowest level and within all individuals of a species. This group structure exists within many species, and very clearly exists within mankind. It is the most important technique, in fact, of survival. It helps in hunting and gathering, in defense and survival, in search and movement over long distances. Birds of many species form groups at certain times, e.g., during migration over long distances, or for breeding. Such groups help the birds, which normally exist as pairs, to more easily navigate over long distances with much less chances of error and to be more secure against predators. The same applies to schools of fish during migration or otherwise. Penguins near Antarctica, gather in one spot of the continent to breed and they form pairs only for breeding purposes. In the harsh climate of Antarctica, they could not do so, even as pairs. Bison, deer, wildebeests, elephants and zebras form groups for security during grazing and breeding. Tolerance of each other is the common characteristic of these social groups. Wolves and wild dogs also form groups for hunting. They have an alpha-male hierarchy in the group, which gives sexual and other privileges to some males in the group. But the group is not based on compassion and "civil behavior". Lions and Cheetahs also live as family groups, but are unable to go beyond family formation, due to their extreme individuality and intolerance of each other. We can see that the deer have been able to populate virtually all continents, but the lions, which are their predators, are confined to only a few places in Africa and Asia, and are, naturally, facing extinction.

Origin of Life and Species

Looking at all forms of life on our planet in terms of our current scientific knowledge, the best guess is that this phenomenon came into being in a very elementary form, billions of years ago. Some unknown process caused it to happen and it started as a unicellular life form with a simple structure and became complex with passage of time. Hundreds of million years old fossils show that life forms have existed on this planet for a long time. Many life forms have gone extinct and we have only their fossil remains.

"The main conclusion here arrived at, and now held by many naturalists who are well competent to form a sound judgment, is that man is descended from some less highly organized form. The grounds upon which this conclusion rests will never be shaken, for the close similarity between man and the lower animals in embryonic development, as well as in innumerable points of structure and

4

constitution, both of high and of the most trifling importance- the rudiments which he retains, and the abnormal reversions to which he is occasionally liable- are facts which cannot be disputed. They have long been known, but until recently they told us nothing with respect to the origin of man. Now when viewed by the light of our knowledge of the whole organic world, their meaning is unmistakable. The great principle of evolution stands up clear and firm, when these groups of facts are considered in connection with others, such as the mutual affinities of the members of the same group, their geographical distribution in past and present times, and their geological succession. It is incredible that all these facts should speak falsely. He who is not content to look like a savage, at the phenomenon of nature as disconnected, cannot any longer believe that man is the work of a separate act of creation. He will be forced to admit that the close resemblance of the embryo of man to that, for instance, of a dog - the construction of his skull, limbs, and whole frame on the same plan with that of other mammals, independently of the uses to which the parts may be put - the occasional reappearance of various structures, for instance of several muscles, which man does not normally possess, but which are common to the Quadrumana - and a crowd of analogous facts - all point in the plainest manner to the conclusion that man is the co-descendant with other mammals of a common progenitor."

Charles Darwin (1871), *Descent of Man and Selection in Relation to Sex*, Chapter 21: General Summary and Conclusion: Project Gutenberg. Retrieved October 15, 2019, from www.gutenberg.org/ebooks/2300.

Since the earth had no oxygen in its atmosphere when it came into existence billions of years ago, various species evolved from unicellular organisms in water, to bigger organisms within water, to those which could survive on land and water together after an atmosphere containing oxygen had developed. Then organisms evolved which could survive only on the land, some of whom developed the ability to move on land and fly in the air. So, all species have evolved that way, from living beings in the water, basically, to living beings in the air, the journey from the sea to the air required the greatest set of skills in survival. Since we have not acquired the skills to fly and we cannot live in water any more like our ancestors, we normally live in the air on land as human beings. We have evolved basically by competing for the land, with other living beings and have taken over most of the land and its resources, destroying many species, and our distant ancestors destroyed others. So, a lot of the species are extinct, and the rest are going extinct at this time, mostly because of our ability to expand at their cost. Also, new ones are coming into being.[4] Thus, there is a hierarchical tree-like relationship between all living beings - from the earliest, most primitive organisms, to living beings like plants in water, to those beings which could move in the water, and other species which moved onto the ground, to flying species, and so on. Although we cannot fly or stay underwater for too long, as we have adapted to life on land, we have acquired the skills to survive in the other environments. We can be a fish when we want to and a bird when we want to, but we are land-based animals most of the time where we are most comfortable. Thus, we can compete in every environment - the sea, the land, and the air. Now, we are even trying to compete with each other in space where, it seems, no other species has been. But this is how we are related to other life forms, especially other apes. Thus, strictly speaking, all forms of life are connected to each other by their genetic heritage. We are all part of one big phenomenon and all forms of life are our relatives. Some of them are not aware of what is happening to them. We seem to be more aware, but even we don't know what is happening, in its entirety.[5]

Thus, it is very important to know and understand the Law of Evolution and its implications. Any doubts on this matter have to be put to rest before one can understand the world. Not to understand ourselves, not to understand our origin, not to understand the conditions which shape us, makes it impossible to understand life and the behavior of mankind in today's world.

Origin of Mankind

We look and behave very much like the other apes. Of all the species, we are closest to Bonobo and other apes in terms of genetic heritage. Other apes, chimpanzees, orangutans and gorillas, have a lot of behavior which is similar to ours. If we look at the family tree of the apes, and examine the evolution of the apes and the emergence of mankind as Homo sapiens, we can see the close connection between mankind and the rest of the species. But as time has passed, it seems, the distance between us and our closest relatives, has grown dramatically, because our knowledge and our understanding of our surroundings is so much more now than of any other species, that no other species can compete with us. It seems that the branching off of the Homo sapiens from the rest of their relatives happened one to two hundred thousand years ago. It looks like a long time, but it is a very short period on an evolutionary scale. If we consider the two billion years of existence of life on this planet as one day, then we have existed for only the last 12 seconds of the day! A lot had happened on this planet before we even evolved into Homo sapiens, and certainly all the other living beings were not created by us or for us. We evolved as one species of apes and have only existed somewhat in our present form for a relatively short period of time.[6]

Dispersal of Homo-sapiens

Somewhere in a part of Africa, now called Ethiopia and the Great Rift Valley, the Homo sapiens developed. Their ancestors had migrated to new lands in search of new sources of livelihood. They did the same. The struggle originally was in Africa, and it seems that when it had reached a level that the human species had gone ahead of the other animals, then humans were able to move to new areas within Africa and to the Middle East and Central Asia, through the only land corridor there was - that of the Sinai Peninsula. Initially, settlement of Europe was not possible, because Europe was covered with ice during the ice age. So, the earliest arrivals into Asia stayed in South and Central Asia till the ice began to melt. Then, humans moved into Europe and Northeast Asia. Only about 30,000 years ago, Northeast Asians moved over the Bering Sea into what is now called Alaska and from there to the rest of North, Central and South America. During this period, mankind also moved down into the southern tip of Africa and human beings in South Asia managed to reach Australia. This is how the world originally got peopled. Populations developed different racial and other characteristics because of isolation and different challenges in their environment.[7]

One factor that we can identify in this process of peopling of the world is that shortage of, or deficiencies in, the habitat forced people to move. Even in relatively recent history, those areas which cannot support their people have been the source of people who, hardened by competition and compelled by shortage of habitat, moved out in violent confrontation with other tribes and people and captured new areas of the world. The people of the Arabian Peninsula went out all over the Middle East and even Central Asia and North Africa. The people of Mongolia captured parts of Europe, the Middle East and China. The Europeans stayed where they were, content and happy with plenty of food and other necessities of their existence. But, later on, when the population of Europe grew and competition became severe, Europe exploded onto the rest of the world with a genocidal frenzy never before seen in human history.

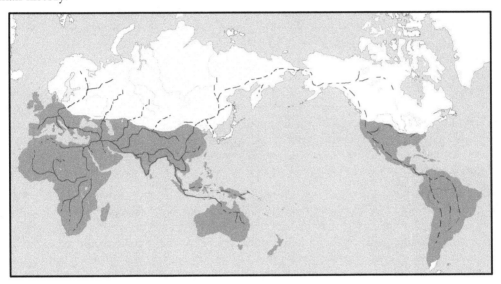

Figure 1-01: Spread of Homo-sapiens

(Map based on world map at: https://www.d-maps.com/carte.php?num_car=3229&lang=en)

Human Social Groups in the Age of Savagery
Family

The bifurcation of the human species happened at an early stage of evolution as it has happened to other species of animals. At that stage, human beings were living just like any other animal species - killing each other, eating each other, and eating everything they could find, fighting over habitat, trying to protect themselves. But the need for procreation was very much present in the genetic heritage of both genders. Thus, it is reasonable to assume, that the interaction between the two genders must have occurred as it happens between those of other animals at this time. Family as a long-term phenomenon may not have existed at that

7

time. But at least temporary cooperation between the two genders must have existed for successful procreation to have taken place.

"In the several great classes of the animal kingdom,….the differences between the sexes follow nearly the same rules…the males are almost always the wooers….The female, on the other hand, with the rarest exceptions, is less eager than the male….she is coy, and may often be seen [endeavoring] for a long time to escape from the male….The exertion of some choice on the part of the female seems a law almost as general as the eagerness of the male."

Charles Darwin (1871), *Descent of Man* and *Selection in Relation to Sex:* Urbana, Illinois: Project Gutenberg. Retrieved October 15, 2019, from www.gutenberg.org/ebooks/2300.

Darwin's observation, about the males being the wooers and the females being less eager, is an important observation, because the evolutionary logic behind it underlies the main psychological differences between men and women, even at this time. The implications of this observation are far reaching. What he is saying is that in species after species, it is the female who is "coy". What that leads to is that at earlier time, i.e., in the age of savagery, there was a very big difference in the situation of a human male and a female. The female always knew that the child that was going to be born to her was her child, and the male never knew and could never be sure. So, if we are looking at this process on the evolutionary scale, then, since there were no social controls related to the family as a unit, the males had to be very uncertain in this regard. Thus, the whole process of evolution has instilled in them this uncertainty, which has become genetic.

The uncertainty of males has two effects. First of all, the female, since she is sure that the child that she would carry would be hers, would be looking for a man who would make a commitment, who would raise the children with her, who would be around to impart all survival skills, that he has, to his children. On the other hand, the male would be eager to impregnate as many females as he can, so that he can be sure to pass his genes to the next generation. He would also be very possessive of the females he has in his control, and would likely violently keep them away from other males, to make sure that he would be able to pass on his genes to the next generation. Thus, the male was pulled in two directions - by the female or females to stick around and to protect his family and by his genetic *uncertainty* to move on. Thus, the human females developed this lack of trust of the males due to their experiences during the age of savagery and this "coyness" ultimately helped create the human family. But this family did not have to be and by all indications was not a family of one man and one woman. Rather, a few females with one man had to be the norm, especially at that stage. In fact, we know that polygamy has been practiced by men, in some parts of the world, till now. In time, the family unit emerged in mankind, as it now has in many other animals. In this family the male of the species was dominant because of the need for security. The behavior of the male was the same as the behavior of the males of many other species at this time, and like any other species of apes, the male of the human species tended to collect as many females as he could and protect "his" females and offspring. At the same time, the "coyness" of the females, or sexual selection, had an effect on the ability of the human species to form a larger family, because "loyal" and "reliable" males got more opportunity to mate and have offspring. The same qualities could help in the creation of extended families, or groups of families at a later stage.

> **"The family then may be called the first model of political societies: the ruler corresponds to the father, and the people to the children; and all, being born free and equal, alienate their liberty only for their own advantage. The whole difference is that, in the family, the love of the father for his children repays him for the care he takes of them, while, in the State, the pleasure of commanding takes the place of the love which the chief cannot have for the peoples under him."**

> Jean Jacques Rousseau (1920), *The Social Contract & Discourses*. Urbana, Illinois: Project Gutenberg. Retrieved October 14, 2019, from www.gutenberg.org/ebooks/46333.

This era was an era of intense competition with other species also. The development of the five fingers on the hand, especially the thumb, meant that man could make and shape tools for his convenience and for his protection. Man had developed the brain and the intelligence to make tools and move with his family from one place to another, providing security to his family with the weapons he had begun to learn to make. Thus, "hunting and gathering" was the mode of production at this stage. In the hunter-gatherer stage of development, economic activities like hunting of animals, fishing and gathering of fruit, were engaged in by a man and his children, or perhaps sometimes, by females. The results of these activities were shared by the whole family. This was the mode of production and distribution during this early stage of development of human beings. There was an obvious need for more males at this stage, so that the family could engage in hunting of bigger animals, or defend itself from them. This need must have given rise to the desire to retain more male members as part of the family.

It should also be noted that, even at this early stage of the development of a family, the balance of a family had to change with time. This change happens even now. When a child is born, it is more dependent on the mother through its early stages of life for its physical needs, for its basic training. As time goes by, the child learns its basic skills, but then, to learn more advanced skills, it depends more and more on the father, and less and less on the mother. And as that happens, there is a continuing support and conflict between the parents and the child, because the balance of power is changing, because the relationship of the child and other children is changing. So, although the family was, and is, about reproduction, it already had in itself an element of competition and conflict. This is the situation even in families of today, no matter what is the stage of development of the society around them.

At this stage, the primitive condition of mankind meant that no sense of just and unjust, good and evil had yet developed. Cannibalism and murder, even mass murder of whole families can be visualized at this stage. This age encompasses a very long period of tens of thousands of years, of which there is no record. So, the habits learned and changes that occurred in mankind were absorbed into the genetic makeup of mankind, but these continue to control our behavior even at this time. The human mind had continued to absorb the lessons of human existence. In short, the human family had become an embryonic "state", the man being its head and the "ruling class" and the women and children being the "people". It could not be a very abusive state though, because of the genetic relationship of the "population" of the state, i.e., the individuals in the family. The statement made by some political thinkers that "the state started as a city-state" is an oversimplified misconception. States started when the first human families developed. As tribes evolved from families in the hunter-gatherer societies, these social groups became more and more powerful and organized and at a certain stage of their development, we begin to refer to them as states, although even human families had the same characteristics - a piece of territory over which the family exercised "sovereignty" and had specialized functions for its members including relations with other families.

Extended Family

With time, the individual family existence began to change into a group existence, like a group of chimpanzees or a group of some other apes or a group of wolves. Human beings, very early, found the power of being in a group. So, they developed extended families, and continued to hunt and gather whatever they wanted, as extended families or bands. For families, there was a pressing need to change the mode of production to more hunting than gathering, as hunting satisfied more family needs than only gathering of fruit from trees. Larger families could also defend themselves better. So, security also required not only better hunting tools but more males in the family. The mode of production, thus, began to change towards more hunting as extended families became possible. The extended families, of course, continued to compete with each other, and killed each other, and ate each other and destroyed each other. After tens of thousands of years of existence as families, it is reasonable to assume, that man had accumulated skills based on his experience - the brain had continued to grow and human intelligence increased, and the human mind developed to include a social-self, based on the family's existence and the rules of the extended family. This implies the coming into being and growth of the Ego and the Superego, in the human personality at this stage. What this means is that the qualities of human sympathy and compassion, of civil behavior as dictated by life of an individual in a group, continued to grow beyond the level attained as a family only. Thus, the process of development of group survival skills, and hence of civilization, had to grow to include more members of a family for a longer time than before.

The process of growth beyond a family, however, met a natural barrier - that of incest. Existence as families implied that children left the family when they became adults, as they could not be tolerated by their parents as competitors - sons as competitors to each other or to their fathers; and daughters as competitors to each other or to their mothers, or wives/mates of their fathers. If the offspring could continue to live with the family of their parents, they had to do so according to the rules imposed on them by their parents. Thus, they either had to give up all incestuous relationships or the group had to become like a pack of wolves, with its dominance hierarchy. The mind and the social self of the individuals having been developed to understand and control their behavior in accordance with the requirements of the developed love, compassion and civility within the family, it seems individuals could submit to the requirements of the family and avoid incestuous relationships within the family.

If a family succeeded in keeping its offspring as adults within the family, it meant that the family would grow beyond the elementary family due to two factors - the adult offspring would add to the numbers and their mates from elsewhere would also add to the size of the extended family - however, this would be more likely for adult male offspring than adult females. The family would thus grow fast. Also, offspring of such adults would be hybrids. This would mean that the vigor and genetic survivability of the individuals, and hence the extended family, would also grow. On the other hand, the families, which allowed the offspring to stay on with the family without any control over incest, could not grow much without causing conflict and could not develop the vigor and genetic survivability that comes with hybrids coming into existence within the family. Thus, the extended families with incestuous relationships would tend to perish in competition with the non-incestuous families which would be much larger in numbers. Since, incest has developed as a taboo in all of mankind, it is quite clear that the taboo started its development at this stage of evolution, since the successful extended families, and hence tribes, would not result without it. Thus, mankind was able to cross this hurdle in development of civilization, while other species failed, as we can see from their behavior even at this time. To create a group, some skills are required in individuals – these *include love, sympathy and respect for rights and needs of others, cooperation with others and loyalty to the group. Without these qualities in the individuals, a group could not survive for even a short time. Creation of a group is the most important survival skill, since an individual becomes much stronger with the help of a group when needed. We refer to sympathy, cooperation and loyalty, etc., as group survival skills.*

Some other animals, like some species of chimpanzees, or wolves, have learned to form extended families or packs. Wolves, for example, live in packs and hunt together. Chimpanzees may form social groups larger than families. Such animals tend to create dominance hierarchies to deal with loss of members when the offspring become sexually active. Such groups contain alpha, beta and omega males. Some have alpha females too. This causes the extended family to become larger, but the conflict between sexually mature individuals continues in one form or another and a big tribe cannot come into being.

The development of the extended family had a big effect on another aspect of human existence. Since the number of males in the extended family could grow, it became possible for some males not to worry about their security and to focus on thinking and experimenting to make new tools and weapons and find new ways of improving the habitation of the extended family and its hunting and defense tactics and skills. Thus, the increase in the size of the group, under conditions of civil behavior and compassion, i.e., active group survival skills, increased the possibilities of further innovation for the extended family as a whole.

Nomadic Tribes

The extended families evolved into mini-tribes and larger nomadic tribes, *as their group survival skills developed further*, and continued the original hunter-gatherer existence along with shepherding and "farming" of other domesticated animals. The increase in the size of the extended family, or tribe, and the changes in what was available for hunting and gathering, with annual changes in weather, can naturally be expected to put pressure on a tribe to move in search of resources. Thus, the mode of production started shifting from hunting and gathering to "mobile" farming. When the extended families developed into mini-tribes, mankind began to advance from the hunter-gatherer stage and production for the whole mini-tribe began to take over from production for a family only. This had to be accompanied by specialization into fishing and hunting of different animals by small bands within the tribe. Members of a tribe, otherwise, hunted together, engaged in fishing, or raised sheep, horses and other animals together. The products from these activities were shared by the whole tribe. This was the fundamental mode of production during the early nomadic stage of tribalism. This change in the "mode of production" presented new problems to mini-tribes and necessitated larger tribes. The development of nomadic tribes happened due to increase in the size and concentration of tribes necessitating grazing of animals over different pieces of land at different times of a year. This could only happen with further development of the social relationships and tolerance. Thus, the process had to be gradual and, in fact, was a transition from savage existence of extended families, to tribes settled on land. Nomadic tribes can be found even at this time, in parts of Central Asia, like Afghanistan. They can also be found in the deserts of the Arabian Peninsula, and in Africa. This stage of human evolution is considered to be the "beginning of civilization", but as we can see, the processes involved at this stage, had to be present, in an embryonic form, in the previous stages of human existence also.

A tribe consists of a number of genetically related families, but there is always competition between these families over the limited resources that the tribe has. The same way, there is competition between a tribe and other tribes, over land and other resources. But at the same time, there is unity, because one tribe and another may be genetically related. They are formed because one tribe breaks up into two or more tribes as it grows. The new sub-tribes may have chosen different hunting grounds. They are genetically related and, relatively closely situated, so that they are always competing at the same time. But, then, an economic relationship may also develop between the tribes. One tribe may produce something because of the resources it has access to and another may have different resources available and may produce something else. The two tribes may exchange their products. Native American tribes, or the so-called "Red Indian" tribes, used to exchange their produce. They would bring in gifts to the other tribes. Things that one tribe had gathered, like animal skins, might have been in excess of its needs. The tribe could exchange them for something that it needed. Generally, this was mutually beneficial trade without any money involved, i.e., barter-trade. So, exchange comes into being between tribes historically. But everything produced by a tribe, was actually produced by a family, so it was actually exchange between families of different tribes. Tribal societies do not have much industry to speak of, especially, when they are always moving. That's why they survive as tribes. The relationship, between one tribe and another, is basically economic, especially if they are nomadic. That's the strongest bond they can have, although genetically too they may also be related. Economics tends to bring them together, and tribalism, or tribal identity, separates them. When they face a common enemy, they tend to come together, although there are conflicts between them. The conflicts are, basically, over grazing rights over land and over habitat, but there may also be conflicts related to inter-tribal mating. Most primitive tribes don't allow their members to get married to outsiders, but this does happen. Economic relationships tend to unite tribes, by making them interdependent. The fierce battles between tribes are over habitat, because one tribe may grow bigger and, then, it may want the habitat, or the land, belonging to other tribes, leading to genocidal conflicts. For example, North American native "red Indian" tribes had land claims about what was their "own" hunting territory. They did not tolerate other's intrusion into their territory.

Generally, the early stages of the evolutionary history of mankind, from the hunter-gatherer stage to emergence of nomadic tribes, are all referred to as the age of savagery. This era was spread over one to two hundred thousand years. In fact, we have gone through three clearly definable stages of our social development during this period, as were described in this chapter. This was a very long period with the processes of development of *group survival skills* slowly but clearly changing the ways mankind existed. At each stage of development, the mode of production was the leading cause of social change. Competition, among families and tribes, meant that there was continuing conflict and violence over natural resources and habitat.

CIVILIZATION AND NATURAL SELECTION

...... "It must not be forgotten that although a high standard of morality gives but a slight or no advantage to each individual man and his children over the other men of the same tribe, yet that an increase in the number of well endowed men and an advancement in the standard of morality will certainly give an immense advantage to one tribe over another. A tribe including many members who, from possessing in a high degree the spirit of patriotism, fidelity, obedience, courage, and sympathy, were always ready to aid one another, and to sacrifice themselves for the common good, would be victorious over most other tribes; and this would be natural selection. At all times throughout the world tribes have supplanted other tribes; and as morality is one important element in their success, the standard of morality and the number of well-endowed men will thus everywhere tend to rise and increase.

...... Now, if some one man in a tribe, more sagacious than the others, invented a new snare or weapon, or other means of attack or defense, the plainest self-interest, without the assistance of much reasoning power, would prompt the other members to imitate him; and all would thus profit. The habitual practice of each new art must likewise in some slight degree strengthen the intellect. If the new invention was an important one, the tribe would increase in number, spread, and supplant other tribes.

...... All this implies some degree of sympathy, fidelity, and courage. Such social qualities, the paramount importance of which to the lower animals is disputed by no one, were no doubt acquired by the progenitors of man in a similar manner, namely, through natural selection, aided by inherited habit. When two tribes of primeval man, living in the same country, came into competition, if (other circumstances being equal) the one tribe included a great number of courageous, sympathetic, and faithful members, who were always ready to warn each other of danger, to aid and defend each other, this tribe would succeed better and conquer the other.

...... He who was ready to sacrifice his life, as many a savage has been, rather than betray his comrades, would often leave no offspring to inherit his noble nature. The bravest men, who were always willing to come to the front in war, and who freely risked their lives for others, would on an average perish in larger numbers than other men. Therefore it hardly seems probable that the number of men gifted with such virtues, or that the standard of their excellence, could be increased through natural selection, that is, by the survival of the fittest; for we are not here speaking of one tribe being victorious over another.

Although the circumstances leading to an increase in the number of those thus endowed within the same tribe are too complex to be clearly followed out, we can trace some of the probable steps. In the first place, as the reasoning powers and foresight of the members became improved, each man would soon learn that if he aided his fellow-men, he would commonly receive aid in return. From this low motive he might acquire the habit of aiding his fellows; and the habit of performing benevolent actions certainly strengthens the feeling of sympathy which gives the first impulse to benevolent actions. Habits, moreover, followed during many generations, probably tend to be inherited."

Charles Darwin (1871), *Descent of Man and Selection in Relation to Sex, Chapter V: On Development of Intellectual and Moral Faculties:* Project Gutenberg. Retrieved October 15, 2019, from www.gutenberg.org/ebooks/2300.

SEXUAL SELECTION

...... "Sexual selection depends on the success of certain individuals over others of the same sex, in relation to the propagation of the species; while natural selection depends on the success of both sexes, at all ages, in relation to the general conditions of life. The sexual struggle is of two kinds: in the one it is between the individuals of the same sex, generally the males, in order to drive away or kill their rivals, the females remaining passive; while in the other, the struggle is likewise between the individuals of the same sex, in order to excite or charm those of the opposite sex, generally the females, which no longer remain passive, but select the more agreeable partners.

The laws of inheritance determine whether characters gained through sexual selection by either sex shall be transmitted to the same sex, or to both, as well as the age at which they shall be developed. It appears that variations arising late in life are commonly transmitted to one and the same sex. Variability is the necessary basis for the action of selection, and is wholly independent of it. It follows from this, that variations of the same general nature have often been taken advantage of and accumulated through sexual selection in relation to the propagation of the species, as well as through natural selection in relation to the general purposes of life. Hence secondary sexual characters, when equally transmitted to both sexes, can be distinguished from ordinary specific characters only by the light of analogy. The modifications acquired through sexual selection are often so strongly pronounced that the two sexes have frequently been ranked as distinct species, or even as distinct genera. Such strongly marked differences must be in some manner highly important; and we know that they have been acquired in some instances at the cost not only of inconvenience, but of exposure to actual danger.

..... The belief in the power of sexual selection rests chiefly on the following considerations: Certain characters are confined to one sex; and this alone renders it probable that in most cases they are connected with the act of reproduction. In innumerable instances these characters are fully developed only at maturity, and often during only a part of the year, which is always the breeding season. The males (passing over a few exceptional cases) are the more active in courtship; they are the better armed, and are rendered the more attractive in various ways. It is to be especially observed that the males display their attractions with elaborate care in the presence of the females, and that they rarely or never display them excepting during the season of love. It is incredible that all this should be purposeless.

...... He who admits the principle of sexual selection will be led to the remarkable conclusion that the nervous system not only regulates most of the existing functions of the body, but has indirectly influenced the progressive development of various bodily structures and of certain mental qualities. Courage, pugnacity, perseverance, strength and size of body, weapons of all kinds, musical organs, both vocal and instrumental, bright colors and ornamental appendages, have all been indirectly gained by the one sex or the other, through the exertion of choice, the influence of love and jealousy, and the appreciation of the beautiful in sound, color, or form; and these powers of the mind manifestly depend on the development of the brain."

Charles Darwin (1871), *Descent of Man and Selection in Relation to Sex, Chapter 21: General Summary and Conclusion:* Project Gutenberg. Retrieved October 15, 2019, from www.gutenberg.org/ebooks/2300.

Chapter - 2

Feudalism

Tribalism Based on Agriculture

When it was discovered by nomadic tribes that crops could be planted and reaped and that this activity could involve less effort than hunting, gathering or even nomadic shepherding, the tribes began to settle on fertile lands. This fundamental change in the mode of production brought a dramatic change in human social existence. It began to occur about six thousand year ago and it is considered by many to be the beginning of civilization of mankind, although there is no fine line between this stage of human existence and the age of savagery. Nomadic way of life has continued till now to some extent, in some parts of the world. Agriculture brought the need to plough the land and plant seeds of certain crops. Crops required protection and care and had to be reaped at a certain time. All this required constant presence of those engaged in this activity. So, tribal settlements began to arise. Also, there was a lot of work in which animals could help. So, the need for domestication of more species of animals arose in the course of development of this mode of production. Some animals began to be protected because they were needed for their meat, milk or for their work. This began to affect the evolution of certain species of animals and mankind became the breeder of some species. Mankind was not fighting other species of animals for survival anymore, because this species had gone far ahead of others. It was now fighting those which were a nuisance, and protecting those which were useful to it.

As tribes settled down on land, the struggle between them became a struggle for possession of land and protection of the crops on the land. Land, thus, became central to the social system and, with passage of time, attachment of tribes to "their" land grew. Tribes at this stage were frequently engaged in fighting and in trying to reach accommodation with neighboring tribes. This process of competition and cooperation also involved exchange of produce and participation in other social activities. Since the most fertile land was close to rivers, and this land could also be easily used for agriculture, many tribes settled on the river banks. Thus, started the oldest of the well-known civilizations - the Mesopotamian civilization developed on the banks of the Tigris and Euphrates rivers in what is now called Iraq. The Egyptian civilization developed on the banks of the Nile. The Indus valley civilization developed on the river Indus and its tributaries and the Chinese and other civilizations developed on river banks in the Eastern regions of Asia. These were the oldest civilizations of what came to be known as the age of feudalism. They developed about six thousand years ago and lasted for several thousands of years as developing and changing centers of feudalism. Even now, some of these river valleys have not really come out of the feudal state of social organization.

Dynamics of Tribal Conflict

Intra-tribal conflict is basically between the ruling class and the people of the tribe. The ruling class in a tribe has been the embodiment of *higher intelligence, knowledge, skill, selfishness and aggressive tendencies. It consists of individuals with highly developed individual survival skills.* The other members, or the "people" in the tribe, are likely to be accommodating, cooperative and of more civilized nature, in general. They are the ones who keep the tribe together with their group survival skills, i.e., *sympathy, love, courtesy, consideration and compassion for each other.* However, the extra sexual access available to the ruling class would tend to lower the level of civilization of the tribe, but the "coyness" of the females and the larger number of individuals in the ruled class would tend to increase the level of civilization. Thus, it is difficult to decide what would happen to the level of civilization as a result of the inherent conflict between the two classes of a tribe. However, it is quite clear that in reality tribes did grow into bigger tribes and the smaller tribes could only lose to the bigger ones in competition. Only the bigger tribes could carry the genes of a higher level of civilization embodying more developed *group survival skills, because development of these skills was the necessary requirement for continuation and enlargement of such a group.* Also, the bigger tribes did merge, or fuse, into advanced non-tribal feudal societies. Examining the dynamics of inter-tribal competition, conflict and violence, one can see that the competition involved several processes, which we describe as follows:

Expansion:

This is the normal growth of a tribe. The tribes originally consisted of several families genetically related to each other. With time, each of the families spawned new families, which acquired new members from intra-tribal or inter-tribal marriages. Thus, individuals from other tribes were also absorbed into a tribe, which could create new sources of conflict within the tribe or with other tribes, but could potentially affect the cohesion of the tribe itself. One would expect difficult new members to be thrown out if they affected the cohesion of the tribe in a negative way, but those who helped the tribe in becoming more cohesive would be retained by it. On the other hand, difficult members would be better accommodated by a tribe with a high level of civilization. Thus, this process of acquisition of new members can only be seen as a process of improvement in the level of civilization of the tribe or its power. Thus, tribes tended to grow and break up as a function of their cohesion, or level of civilization. The size of the tribe was critical in its ability to compete and survive.

Contraction:

A tribe would contract without division due to disease or accidents, or due to loss of members in conflicts with other tribes. The loss of members could be viewed as increase in its level of civilization, due to loss of its more aggressive members in combat.

Division:

Tribes could split because of internal conflicts mainly in their ruling classes or just because of a low level of civilization in general. This could only weaken the tribe. In fact, both the new tribes would be weak and vulnerable to attack by other tribes. Thus, conflict in the ruling class of a tribe would increase chances of its annihilation, by creating smaller, less competitive, tribes.

Annihilation:

Conflict between two tribes could annihilate one tribe and may weaken the surviving tribe. In this process some members of the losing tribe, with a high level of civilization, may survive the conflict and may be able to survive as members of the surviving tribe, but the individuals with a low level of civilization would not be able to survive, due to their inability to adjust to the resulting new conditions. Often, tribes were taken into physical and sexual slavery – one tribe would, thus, rule another and make it work as a set of slaves. This was the beginning of Imperialism – enslavement and rule of one tribe over another, for extraction of labor as "tribute".

Coexistence:

This happens between two, or more, tribes which have achieved a high level of civilization. This was not happening in the early stages of settlement of tribes, but is widely at work in tribes at this time. Such tribes, ultimately, fused into a bigger tribe, or a nation.

Fusion:

This could not happen in the early days of tribalism. But even during that period, two tribes could fuse into one tribe, if they were originally created by a division of one tribe due to conflict in the ruling class, if the conflict had later disappeared. In recent times, many tribes have been able to fuse together into one nation.

Conflicts within and among tribes combined with the low level of development of technology, caused the population growth of the relevant territories to remain low. The ruling class of a tribe was very small at that time and the standard of living of its ordinary members was very low. Even, at present, wherever tribalism dominates, these conditions prevail.

For nomadic tribes and tribes settled on land, one can confidently conclude that the inter-tribal and intra-tribal conflicts could only result in, mainly, the genes of a higher level of civilization being transmitted to the next generation. Thus, the sizes of tribes have consistently increased as a historical tendency and civilization of mankind has risen to higher and higher levels, throughout its evolutionary history and the social groups at this higher level of civilization have an overwhelming advantage at our current state of existence.

Religion and Feudalism

As tribes learned to coexist with each other and to trade with each other, the tribal identities of their members began to melt away and new ways were found by individuals to control, use and abuse each other. The most important development was the emergence of a folk lore and mythology to strengthen and use the feelings of common destiny and sympathy among individuals. Myths had the effect of giving a super-human character to the stories of the time. Story tellers appeared and the power of the story, the power of the myth and the power of deception and control of the human mind was discovered. Rituals developed and had the desired effect of reinforcing the mythology and the feelings of civility, compassion and togetherness felt by communities of the times.

Religious belief systems bound those who worked on the land to the owners of the land, and controlled all those in an area with many tribes, without the use of physical violence. *Religion was based on the feelings of love, sympathy, courtesy, consideration and compassion for each other that human beings had developed over the long period of their evolution and also the fear and helplessness of mankind to explain its existence.* Such belief systems developed in all the initial centers of civilization - the religion of the Greeks and the Romans, religion of the pharaohs and Judaism in Egypt, religion based on fire worship in Mesopotamia, the Hindu religion in the Indus Valley, the Buddhist religion and its off-shoots elsewhere in Eastern Asia, Christianity and Islam as off-shoots of the Judaic religion. Thus, the breakdown of tribal ruling classes, within an area, led to the development of a new feudal ruling class.

"The fear of ancestors and their power, the consciousness of owing debts to them, necessarily increases, according to this kind of logic, in the exact proportion that the race itself increases, that the race itself becomes more victorious, more independent, more honoured, more feared. This, and not the contrary, is the fact. Each step towards race decay, all disastrous events, all symptoms of degeneration, of approaching disintegration, always diminish the fear of the founders' spirit, and whittle away the idea of his sagacity, providence, and potent presence. Conceive this crude kind of logic carried to its climax: it follows that the ancestors of the most powerful races must, through the growing fear that they exercise on the imaginations, grow themselves into monstrous dimensions, and become relegated to the gloom of a divine mystery that transcends imagination—the ancestor becomes at last necessarily transfigured into a god. Perhaps this is the very origin of the gods, that is, an origin from fear! And those who feel bound to add, "but from piety also," will have difficulty in maintaining this theory, with regard to the primeval and longest period of the human race. And of course this is even more the case as regards the middle period, the formative period of the aristocratic races—the aristocratic races which have given back with interest to their founders, the ancestors (heroes, gods), all those qualities which in the meanwhile have appeared in themselves, that is, the aristocratic qualities."

Friedrich Nietzsche (1887), *On the Genealogy of Morals, Second Essay: "Guilt, Bad Conscience, and the Like"*. Urbana, Illinois: Project Gutenberg. Retrieved October 16, 2019, from www.gutenberg.org/ebooks/52319.

All the exotic qualities which ultimately began to be called godly qualities were ascribed to the new ruling class and especially to its leaders. So, kings arose, who were also considered gods. Ways were found to propagate these beliefs and to entrench them into society, by rituals that encompassed all activities of social life at that time. It was the duty of the peasant, for example, to serve his master, and then not to leave him, because it was "immoral" to leave the master! The peasant had to serve his masters, and he or she had to accept all the rules that the masters passed down to him, or her, in the name of religion, or otherwise. Thus, religious belief systems developed into powerful means of control of inter-group and intra-group behavior, as the ruling class developed an unholy alliance between its two wings - one consisting of kings, nobles and powerful military men and the other consisting of religious clerics with their hierarchical orders. The concept of a "god-king", thus symbolized the unity of the two wings. Control by the sword was exercised by the former wing and control by resort to myths and rituals was exercised by the other. This gave the ruling classes much greater power to control the people.

Religion can, thus, be described as the ideology of feudalism, i.e., its base has always been the feelings of an individual being part of the human community, feelings of compassion for all of mankind, or simply civilization developed over a very long period of more than a hundred thousand years of our evolution as homo sapiens, but its superstructure is a set of myths and stories of conflicts of tribal existence of mankind, and rituals developed in all aspects of life in the era of feudalism – in more recent times.

In the process of its development, religion also defined a social group which could reach across nations, or include many nations with the same belief system. In this feudal stage of history, there were no boundaries, or well-defined borders, between tribes or nations. There were no methods of control, like the police, which all modern states have. Modern states maintain armies, have police services, secret services, and taxation and other departments of social control. In the old feudal societies, those means of control were missing. Religion played the role of a means of control. Simultaneously, landlords and their chief, the king, emerged in such a society. The king became the source of all military power also. The king was, thus, assigned divine power, in addition to control of all military power. So, we can look at religion as an ideology of primitive feudalism, and also as a control-mechanism during feudal times when social groups were beginning to grow and even later when nations developed. With time, as nations have multiplied and expanded, so has religion. In today's world, Christianity is the religion with the biggest number of believers, followed by Islam, Hinduism and Buddhism. Basically, there are two belief systems at this time. One belief system covers Judaism, Christianity and Islam. Hinduism, Buddhism and Shinto-ism form another belief system. These two belief systems are quite distinct from each other - in the way they visualize a Creator of mankind, human birth and death. There are many sects in each of these religions. Voodoo and belief in magic form another belief system, though it does not have many followers.

In today's world, there are more than two billion Christians, about one and a half billion Muslims, one billion, or so, Hindus and, maybe, one to two billion Buddhists of various sects. Also, there are large, and growing, numbers of people who no longer profess religious beliefs. These religions go across boundaries of nationhood, and even political boundaries of states. Religion plays a unifying, and at the same time, a divisive role. Within a state, it can divide nations. In multiple states, it may unify populations. It all depends on which segment of which population in which state believes in what. Religion is, thus, one source of conflicts within and among states.

Structure of a Feudal Kingdom

When we refer to Greek city-states we are, basically, referring to tribal-states. Athens can be called a state because one very small group of people, who were very close in race and culture, lived within well-defined boundaries. It was a tribal society which developed to a highly organized level in terms of agricultural production, military capability and civilized behavior. Thus, when a number of families came together and organized themselves as a tribe in a city, a city-state came into existence.

To try to understand a kingdom, as a state in feudal times, one could examine feudal society at the time and its organization from bottom up. As feudal society developed, most human beings were reduced to working on land as peasants, or as slaves, while land was owned by the landlords. Some of the peasants worked as servants, or slaves, for the landlords, and they also maintained their households. At this stage, the landlords maintained small armies of soldiers and the soldiers were used to control the people working on lands "owned" by the landlords. In general, the landlords took whatever they could from the people working on the land. A variety of relationships developed between the landlords and the working people in different regions of the world, during this stage of human evolution. The biggest landlord in an area would declare himself to be the "king" of the area and the others had to plead allegiance to him or face war. Thus, the king was the landlord of the landlords, and generally claimed to be the landlord of the whole kingdom. Since religion had become part of the system of control, the king would maintain a religious chief also, who would be declared to be a great respected figure. The king was also the military chief and commanded his army, to enforce his will on the kingdom. Without the army, the landlords would refuse to pay "tribute" to him. Thus, the local landlord was taking most of the produce from the people actually engaged in agriculture, based on his claim of ownership of the land. "Ownership" itself was achieved by violent military conquest, however. Part of the produce, collected this way by a landlord, would be sent to the king as "tribute", i.e., the king's share of the loot. The more primitive the feudal society, the more abusive the relationships were. Slavery existed as part of feudalism for thousands of years. Rebellions by peasants and slaves happened frequently. Also, the landlords would rise against the king whenever they could and would try to become kings themselves. Thus, wars during feudal times were wars over "tribute", or loot, and how it was to be shared. In time, the church began to own land also and became somewhat independent of the landlords and the king, effectively becoming a landlord itself, but the "unholy alliance" between it and the king, to subjugate the people, continued throughout this period.

Nations and National States

Tribes, which developed in the same territory, tended to develop similar languages and were also genetically close, since they were the product of division and fusion of the same original set of tribes. When large tribes had stayed on the same land for a long time, had developed strong, tribal attachment to the land and had grown strong in numbers, they began to recognize the rights of other tribes to their traditional tribal lands and settled down to a life of coexistence with them. In the meantime, the mode of production also began to change, because of the increased size of such tribes. A class of artisans developed in each tribe, to

23

service agricultural equipment and to meet other needs of the tribe not directly connected with agriculture. Thus, ironsmiths, shoemakers, weavers, developed. These people had developed new skills that did not require ownership of land. As this small-scale "industry" developed, its products began to be exchanged between tribes. This further helped that process of coexistence. The threats of subjugation by distant tribes further forced tribes with a similar genetic heritage, language and religion, to merge together and to develop a national identity, instead of separate tribal identities. However, the process of formation of *advanced feudal societies*, or nations, was a long evolutionary process, basically precipitated by the change in the mode of production and exchange.

Thus, because of these economic and social pressures, tribal society tended to become one huge tribe, a tribe of tribes. The new larger social group maintained the common language of its components, had a common genetic heritage and "national" territory, comprised of the tribal territories of all the tribes involved, and may have had a common belief system, or religion. Thus, basically, common genetic heritage, land, language and religion define a nation, or a "national state". These are the primary variables in development of *advanced feudalism* and the evolution of a nation. Culture is dependent on language, in the wider sense of individual self-expression and group-expression, and religion. It also depends on the history of this social group and its national territory. It is, thus, a secondary or dependent variable.

Dynamics of National Conflicts

Religious difference between tribes may hinder the process of integration into one nation, but are generally not able to stop it. Thus, a nation may continue to have more than one religion. It does have one national territory, though, since the contiguous territories of tribes have been merged into one. Such a non-tribal nation may also continue to have families identifying themselves with their tribal origins, but the national identity predominates. Nations continued to compete with other nations and tribes on adjoining territories. Merger of tribes could not happen without merger of their ruling classes. *Thus, the basic conflict within a nation continued to be the conflict between its ruling class and its people.* Further, conflicts between a nation and other nations and tribes, were driven by the desire of the ruling classes for more tribute. *Thus, such conflicts, also, were really conflicts between their ruling classes.* This inter-national competition, generally, involved the following processes:

Expansion:

This may happen due to the normal growth of a nation's population, or due to inter-marriages with individuals of adjoining nations. The territory of a nation may also expand because more territory is grabbed from another nation, but this can only happen after a large-scale conflict. In general, nations grow in proportion to their level of civilization, by merger of national tribes, and become more and more cohesive.

Contraction:

Generally, this would not happen to a nation unless there is a very large-scale invasion. In recent history, this has happened to the native nations and tribes of North and South America and Australia, as a result of the *Great Genocide* unleashed on them by European nations and empires.

Division:

Nations resist being divided. In recent history, this has happened to the Vietnamese, German, Korean and the Arab nations. The German and Vietnamese nations have re-united. The Arab nation has so far failed to unite, because of its tribal character. However, the Yemeni Arabs have re-united into one state. Koreans continue to resist their political division into two states.

Annihilation:

A nation, generally, has a large population. Annihilation is, thus, highly unlikely. This, however, has happened during the *Great Genocide* of the native people of North and South America, Africa and Australia.

Fusion:

This does not normally happen if nations exist as independent nations or states. In multi-national states, this process occurs more or less peacefully, depending on how well the state is managed.

Coexistence:

This happens between two, or more nations, or national states, which have achieved a high level of civilization. This process is widely at work between national states at this time.

Thus, we can conclude that the co-existence of national states tends to create a multi-national state, based on respect for the rights of all nations. Inter-national conflicts within multi-national states, also, tend to lead towards fusion rather than division, if tolerance is employed for the management of such a state. Thus, if the rights of all nations are guaranteed, then more multi-national states are likely to emerge. Existing multi-national states can also break up if these rights are not guaranteed. This has recently happened to the Soviet Union, Yugoslavia, Pakistan and Czechoslovakia. Since the level of civilization of mankind has been rising consistently, the emergence of more multi-national states is a certainty.

Structure of a Feudal Empire

As feudalism continued to develop, tribes grew in terms of population, or merged into *advanced feudal societies*. At some point, the "needs" of the kings and their aristocratic or "noble" companions could not be met by the people under their control.

Then, the kings began to invade neighboring tribes and peoples to increase the "tribute" extracted by them. For example, the English kings invaded and occupied Scotland and Wales and subjected tribes in these two areas to economic and sexual abuse, thus increasing the "tribute" extracted by the English ruling class. This was a feudal empire which later on became the British multi-national state of today. To look at this process in more detail we can examine the Roman Empire. The head of the Empire was called an Emperor, or Caesar, who was a king of kings. Caesar ruled with the help of a coterie of "senators", or landlords and property owners, the rich and powerful in Rome. The people of Rome were divided into citizens and slaves, but it was not enough for Caesar and his senators to abuse and exploit only these people.

"It is primarily involved in this hypothesis of the origin of the bad conscience, that that alteration was no gradual and no voluntary alteration, and that it did not manifest itself as an organic adaptation to new conditions, but as a break, a jump, a necessity, an inevitable fate, against which there was no resistance and never a spark of resentment. And secondarily, that the fitting of a hitherto unchecked and amorphous population into a fixed form, starting as it had done in an act of violence, could only be accomplished by acts of violence and nothing else—that the oldest "State" appeared consequently as a ghastly tyranny, a grinding ruthless piece of machinery, which went on working, till this raw material of a semi-animal populace was not only thoroughly kneaded and elastic, but also moulded. I used the word "State": my meaning is self-evident, namely, a herd of blonde beasts of prey, a race of conquerors and masters, which with all its warlike organisation and all its organising power pounces with its terrible claws on a population, in numbers possibly tremendously superior, but as yet formless, as yet nomad. Such is the origin of the "State." That fantastic theory that makes it begin with a contract is, I think, disposed of. He who can command, he who is a master by "nature," he who comes on the scene forceful in deed and gesture—what has he to do with contracts?"

Friedrich Nietzsche (1887), *On the Genealogy of Morals, Second Essay: "Guilt, Bad Conscience, and the Like"*. Urbana, Illinois: Project Gutenberg. Retrieved October 16, 2019, from www.gutenberg.org/ebooks/52319.

Roman armies captured many territories, like Egypt, Palestine, Greece, Tunisia and Spain, for "the greater glory of Rome". The way "the greater glory of Rome" was implemented was very simple. For example, Palestine was subjugated and the local landlords were ordered to pay a certain amount of "tribute" to Rome (for its "greater glory", of course). The Ruler of Egypt was also ordered to do the same. A contingent of the Roman army was stationed in Egypt to make sure that the ruler did not break this "agreement". Similarly, kings or landlords of Spain, Tunisia, etc., were required to pay a tribute specified by the Emperor. Thus, the Caesar's Rome received tribute from several kingdoms and territories, including territory within its direct control. The local kings and landlords of those lands had to collect this tribute, as a tax on the producers. Thus, the landlords and kings and Caesar met the "needs" of their fiefdoms, kingdoms and the empire. This is how a feudal state converted itself from a kingdom to an empire. In principle, the mode of loot and plunder remained the same; it was only implemented on a grander scale. This is what made the Roman Empire great, i.e., a great evil, like all other empires in history. It and its fragments lasted for about fifteen centuries. An empire is, thus, a natural stage in the growth of social groups and competition between them. It is the result of the growth and better organization of a social group, making it possible for its ruling class, driven by greed, to project its power further due to the development of technology – specially that related to weapons of war. Some feudal empires became very

26

powerful, lasted for a long time and, ultimately, evolved into multi-national states, when sub-sets of their tribes evolved into nations. Britain is one such example. Isolation of three sets of peoples on one island helped the English ruling class in this process.

Evolution of Feudalism

Throughout the age of feudalism, every tribe and nation had a basic conflict - a conflict between its ruling class and its people. Those individuals, who were more clever, selfish and self-centered, were able to become dominant in these groups because of their ability to use their *individual survival skills* against other individuals of their own group. They became the nobles, military leaders and kings of their tribes or nations. This class tended to usurp the material wealth of the tribe, or nation, depriving the people of the social group. It also tried to grab the possessions of other tribes and nations, leading to inter-tribal conflicts. Thus, every conflict between tribes or nations was really a conflict between the ruling classes of the two social groups involved.

In the early stages of conflicts between tribes, the total destruction of one tribe by another was the norm. Later, loot and plunder of other tribes and nations became the main motivation for large tribes or nations that had already acquired sufficient land and other resources for their basic needs. The main mechanism of this loot and plunder was the *"tribute"* extorted from the people of the relevant subjugated tribes, or nations. Thus, a high level of violence persisted throughout the age of feudalism. The age of savagery had been even more violent. *Thus, the history of mankind is, basically, a history of violence.* Human beings have killed each other, kidnapped or raped the women of other men or tribes, robbed each other, stolen from each other, deceived each other and enslaved others if they could, throughout human history. Thus, hordes of Genghis Khan; Huns from the Caucasian mountains; the Mughal armies invading India; The Arabs invading what is now Pakistan; the Mongols attacking China or what is now Russia; the Crusaders attacking Jerusalem; Arabs invading Central Asia, North Africa and Spain; Greeks, under Alexander "the Great", invading what is now Iran, Afghanistan, Pakistan, or Iraq - all behaved like savages that they were [1]. The same applies to the tribes of North and South America and Australia and New Zealand, before *the Great Genocide* was unleashed on them, by Europeans about four centuries ago. During advanced feudalism, the degree of inter-tribal and inter-national violence had decreased with passage of time. Genocide of whole tribes became slaughter of old men and women and enslavement of young men, women and children. Later, slavery also decreased as a practice. However, this dying practice was revived by Europeans during the colonization of Africa and the Americas. At this time, genocidal slaughter of whole tribes is a rare occurrence. Also, openly practiced slavery has come to an end.

From the beginning of civilization, till the eighteenth century, feudalism reigned supreme everywhere, including Europe. Kingdoms were in power everywhere, including Europe, Asia and Africa. Aristocracy became the ruling class all over Europe. Feudalism in Europe became the most advanced social system, as innovation progressed in terms of creation of agricultural implements, boats and weapons of war. Armies grew in size and organization. Competition and conflicts among kingdoms grew in intensity as wars and invasions became more violent in scale, intensity and frequency. European states developed the most advanced feudal

27

societies, as tribes merged into nations. European population had been growing slowly, but steadily, due to inter-tribal conflicts. As feudalism evolved into non-tribal feudalism, it begun to grow faster. Agriculture and fishing were the main economic activities. Under-employment and unemployment on agricultural land began to increase. Peasants, serfs and share-croppers worked on the land and the land owners received income from the land without doing much work. The families of the agricultural workers were very poor, but they were able to survive because food was never too far from them. Adults could feed on it or the children could steal it whenever they felt the need. Even though the poor agricultural workers had no proper homes, lacked clothing and shoes and had no proper medical care, they were still able to feed themselves. Thus, if major epidemics did not break out, their population continued to grow steadily. *Primitive feudalism*, in which discrete tribes existed, claiming discrete pieces of land as their tribal territories, came to an end. Instead, *advanced feudalism emerged* in several regions of the world, especially in Europe. Under these conditions, the standard of living of the people did not improve. Further impoverishment of the people resulted. The fast increasing population was displaced towards cities by modernization of large farms and their expansion by violent takeovers of lands owned in common and lands owned by serfs and small land owners, resulting in migration of dispossessed farmers towards cities.[2] At present, we can see this process in progress in South and South-East Asia and in some states of the Middle East and non-Arab Africa. In these areas, although industry is growing, changing the social systems, but society basically remains at the advanced feudal stage.

With the emergence of *advanced feudalism*, the desire for conquest of more territories also grew. Europe exploded into an imperial frenzy. Several European states established empires during fifteenth to nineteenth centuries. New routes were discovered to reach India and the East, which was considered to have great riches to offer to the conquerors - gold and spices. Discovery of America and Australia opened up whole new continents for colonization and economic exploitation. In the seventeenth to nineteenth centuries, the process accelerated further and imperial powers frequently got into wars with each other, while grabbing more and more territories. Soon, virtually, the people of the whole world were being ruled from Europe, Japan and the US.

Some empires had become so huge that their territories could not be kept fully under the control of the imperial powers and conflicts developed between the newly created colonies and their creators, especially in the Americas. Spain and Portugal lost control of their colonies in North and South America. Britain lost control of its thirteen colonies in North America and the United States was born.

Chapter - 3

Dynamics of Social Change

Individual and Society

There are two aspects to a study of the relationship between individuals and their social group, or society in general, to which they belong. We need to look at how society affects an individual and how individuals affect their society. First, we look at how society affects an individual's development. There are two parts that make up a person - what you were born with and what the environment does to you. Your body is what you were born with - your face, your arms and legs, your skin, etc., including the genes that are passed on to you by your parents. Thus, at birth your body is given to you along with its innate abilities and your brain is given to you along with its innate abilities. The gene pool, or DNA, passed on to you determines how you would grow and what form and shape you would take, physically, as you grow up, and how sharp you would be in absorbing information that is fed to you by your environment and how intelligent you would be in understanding and using it. In short, at birth, you are given not only the little body you have, but also the way it would grow. In other words, your future growth and performance is also partly predetermined for you, at birth. The instinct to survive is born with a person - in fact with any living organism. The sexual instinct is also born with an individual, but it is inactive at birth. Later on, it is awakened at a certain age. So, a person is born with certain instincts, the instinct of self-preservation and the sexual instinct are the two basic instincts, which govern the behavior of any organism, including a human child. We do not have any control over what kind of body, or brain, or instincts we are born with. We have no control over who our parents would be and what kind of genetic heritage they would pass on to us. In fact, the parents too have no control over the kind of a child they would have. They only know they want a child, and the sexual instinct to reproduce controls their behavior.

The other factor, that controls the development of an individual, is the environment. "Environment", is an abstract reference to the non-living and living objects that we come in contact with during our life time. It means the food we eat, the air we breathe, the water we drink, the trees and animals, the sunshine and the moonlight we come in contact with, and other chemicals we take in, knowingly or unknowingly. It also means our parents and what they do to us, and for us - how they bring us up and what they teach us. It also means, how others - relatives, teachers, religious leaders, members of a tribe, our nation and the state affect us. It also means the system that we are educated in and the other media that influence our thinking and beliefs. In short, we do have individual physical existence, but we have virtually no control over our genetic heritage and our environment. At a certain stage in our lives, we do make choices, as mature individuals, which seem to be independent choices, but in the ultimate analysis, even these choices are dictated to us by our genes and our environment. This raises the question as to what relationship we have to the society we are born in, and grow up in, and what part of what we do and achieve in our lifetime truly belongs to us, and we can lay claim to. In other words what achievements in our lifetime are purely our achievements and credit for those does not go to others?

Kids have a very close connection with their mother first of all, and later on with the father too, and they instinctively feel secure near their parents. This is where they find love and security. When a child begins to grow, he doesn't want to go away from the father or the mother. There's an area after which he feels insecure and comes back. He wants to play around you. It seems that we are born with a blank mind, because, as all of us know, we remember nothing of that stage of our lives later on. A baby is born with all the five senses. It moves its head towards wherever light is coming from. The light affects its eyes and it reacts by looking in that direction. If we play music at one spot, the baby turns its head around to figure out the source. It reacts to any sound like that. Later on, it is affected by taste and smell also, and then it tries to touch, taste and eat everything. Then the stage comes when the baby begins to talk and is able to communicate. Communication is non-verbal in the beginning, all it does is cry, and make noises, and then somebody else has to figure out why it is uncomfortable. Maybe it is hungry, maybe it is wet. Suppose a baby is near a heater. The heater emits heat and light. A baby cannot distinguish between the light and the heat, so it is the parents' job to scare him so he would not touch the heater and get burned. Parents have to raise their voices so that the baby does not get hurt. It cannot communicate and so doesn't understand instructions. The parents have to shout and, ultimately, the baby gets conditioned to avoid the heater - since it learns that every time it approaches the heater, something unpleasant happens.

This is how a child learns what is light and what is heat and what are the names of different colors. The child's parents teach him these names - red, green, yellow, the names of colors are given to a child and the child learns to identify different colors. If a parent tells a child that a certain color is red, the child learns to call that color red. The parent may experience one sensation and the child a different sensation, but both call it by the same name - red. There is no way to know if the two sensations are the same. Names of different objects and persons are taught to a child, different sensations are described to him and the child is taught to distinguish between different states of the same object. For example, a door may be closed or open. A ball may go up or down, etc. In short, a child is taught language skills and slowly is able to communicate in the parents' language. As language is learned, the ability to think starts developing. The more he knows, the more his ability to process new information and new ideas develops. So, the parents lay a foundation. On that foundation a building begins to rise - the whole structure of logical thinking. Everything at every stage of our life is built on what we were programmed with before. Intelligence implies the innate ability to think, and the training that the environment has put onto us. Slowly, a person's intelligence and his ability to analyze events and processes grows - as determined by a person's genetic heritage and training.

When we are born, our first and most basic instinct is to survive, we have no control over having that instinct, but what really helps us to grow is a sense of trust in the environment that we are in. That trust is developed primarily with our parents and at first with the mother of a child. That trust is based on a feeling of security that a child has. And as we grow, we begin to gain confidence in our ability to navigate and explore our environment. We tend to mimic our parents' behavior, as it is, to us, a trustworthy way to meet our needs for survival. Our parents have also mimicked their own parents, learning the same survival skills, learning what works. We may even inherit their ideas. Because our trust has built up to such a level, we tend not to even question their ideas. We tend to even adopt their behavior. We, thus, tend not only to look like them but to think and behave like them.

Thus, the most important elements of the environment for a child are his parents. Later on, in his life, his brothers and sisters and other relatives play a significant part in bringing a child up. Then, teachers, religious elders, books, television, newspapers have their influence on his growth and development. Depending on the child's social group, tribal, national, state influences are also internalized by a child, from the environment, as discussed later in Chapter-16, and become part of his personality. As the mind grows, he has more and more complex challenges. Being a human being, there are many more intellectual challenges in later parts of one's life, because his thinking is now, sometimes, going beyond his parents and elders. Thus, at any stage an individual's development, the previous stage is the prerequisite. When we do research work, we clearly stand on the shoulders of those who have researched the same area or similar issues before us. We try to understand what they came across and where they failed to grasp the processes that were hidden from them because they did not have the insight to grasp them. With our new knowledge that we have acquired over time, we may be able to understand the same processes and make new discoveries that our predecessors missed.

Without the work of the predecessors, what the research worker has done is impossible to conceive. But in the process, he would have done something new. He is the vehicle from which this thing is being created. He cannot distinguish between what he has achieved and what the environment has done through him. Society is evolving and knowledge is being passed from generation to generation. So, ultimately, the society is responsible for his achievement.

The law of evolution tells us that the whole species is developing. It may be growing unevenly, but it is growing and the individuals in it are not growing completely separately. Human beings are social animals and our social existence, or socialization, has grown in all directions over time. It has affected our innovative and religious behavior. Our economic institution have become more social, our scientific endeavors have also become more social and our religious beliefs are giving way to our social urges – For example, the class structure of Hinduism is dying, so are the irrational religious myths being challenged by our scientific training, while the basic evolutionary morality is continuing to survive and grow. Not only the law of natural selection is at work, but the law of sexual selection also has its influence.[1]Each individual is not disconnected from other individuals in society; each individual has a social existence. Each individual exists like a bee in a beehive. It just doesn't look like a bee in the beehive, because the bee has much more intricate chemical connections with other bees in a hive and cannot exist for long without the others. Human beings seem to be independent but really are not, because the connections between humans are mainly mental and are not clearly visible. Since they are not visible, they are easily confused, because we look at the person's physical self, and we say "this is him". But that's not him we are talking about, because a human being is a lot more than his physical self. So, it is in a society where individuals develop and knowledge is being created and being passed on to the generations to come. In the same way, the genetic make-up of a society is being developed. Now if a person becomes a great scientist, it is the society's achievement and if he becomes a great criminal, it is also the society's responsibility, because he didn't make himself into a criminal. If he wants to murder people, then, he has been trained by his environment to murder. The process by which an individual is raised has gone wrong, we can say, but it is still a process over which he has no control, and could not be held responsible for, entirely.

Clear understanding of the relationship of the individual and society has far-reaching implications. If in a state, e.g., the United States, some individuals make great scientific discoveries, then, this is an achievement of the United States, not of the individuals really. And if the United States produces a lot of criminals, then it is also its responsibility. There is something wrong in the society which is creating a lot of criminals. All the criminals may be killed, or confined to jails, but twice that many may be created by the society, if conditions remain the same. Millions are in jail today in the United States, and the crime rate is among the highest in the world. Something has gone wrong with the American society! Politicians are trying to fix the product of society, the criminals. We cannot hope to fix the product of society and hope to fix a societal problem. The problem is to be found in the society of the United States.

Similarly, the basic ideas in organized religions go somewhat like this - A Superior Being, or some Superior Beings, gave you life and made you superior to other species. So, you have to go by the commands of those Superior Beings (i.e., whatever the religious authority in society tells you). If you do, you will be rewarded. If you don't, you would be punished - right now, or in a later life. For a person to become a criminal, i.e., someone who breaks a society's rules of behavior, and the person's personality to get that distorted, we have to look in the environment to find what made it so. Nobody is born a criminal. Nobody is born a murderer, or a thief - a child does not even know what murder or theft is. The environment teaches him what is theft, what is bribery, how to steal, everything is taught to him. He commits a crime, initially at least, only if society creates unacceptable conditions for his existence. Crime is not in his genes, but the instinct to survive is! So, the whole structure of religious beliefs falls, as regards crime and punishment, if this truth is acknowledged. If we understand this, then, we cannot justify Paradise, we cannot justify Hell, we cannot justify people being punished for what they do - even rewarded for what they do. A society in which criminals are being punished by putting them into jails, only isolates society from the criminals, and, in a way, protects society from those criminals, but it doesn't stop the process which creates criminals.

So far, we have analyzed how the personality of an individual is created by the society in which the individual is born and grows up. But, when an individual has grown up from a child to an adult, he becomes a full member of that society. He is then able to affect the development of other children who are growing up in that society at that time. Thus, the individual, at a certain time, becomes part of the society that is creating the personalities of other individuals. In fact, a number of individuals are developing in a society, their number depending on the size of the population, and these individuals are being added to that society. As they grow up, they and their ideas and behaviors have a growing influence on the society as a whole. Thus, the society is being continuously changed by new individuals who join it and grow up in it. Their arrival brings a change in society and the new ideas and innovations they bring, change society also, and its beliefs and social and economic relationships. Thus, the relationship of an individual and society is a two-way street - Society creates individuals and individuals constitute society and change it by their entry into social life and becoming members.

The Logic of Change

When we study the evolution of species on our planet, especially evolution of mankind, we can understand how change occurs in society and how it has always occurred. It is obvious that nothing remains constant on this planet, or in life. Things are constantly changing. Every object changes with time. Trees grow, rivers flow, birds move from one place to another, old ones die and new ones are born, tide rises and ebbs, sun rises and sets. We move and work and then we go to sleep. New buildings are built and old ones fall or are torn down. Weather changes constantly and seasons change bringing changes in all vegetation and natural life forms. Our creations, like buses, trains and airplanes move constantly, moving goods and people from one place to another, over short and long distances. In short, every passing second brings a new "world order", i.e., a new order, or set of objects and relationships, in which the objects throughout the universe are arranged and related to each other.

What is not always noticed in the way the world changes, is that every change is opposed by the status quo, i.e., the way things are and have been before. One of the laws of physics is that *every action has an equal and opposite reaction*. This is true of all physical phenomenon. What needs to be realized is that this is also true of social, sociological and political changes. New ideas are always received with skepticism and always face opposition. If an individual comes up with a new idea, the first reaction in society ranges from doubt and incredulous laughter to open hostility. If the individual persists, then the idea may be examined more thoroughly to challenge its validity, to degrade its value, or to render it useless in comparison to existing ideas. When and only when the new idea passes these tests, and the resistance they entail, the process of acceptance and implementation may start. The new idea may be a new concept of political organization, or simply a concept of a new kind of plough, or a weapon like a sword or a bomb. If the idea is accepted by society, the opposition to its use does not stop. There may be individuals who feel they can propose a better form of political organization, a better plough, sword or bomb. Their set of ideas continues its opposition till it is either totally abandoned, or a compromise emerges. If we think of the original set of ideas for creation of a plough or a weapon, as a thesis and the opposite set of ideas as the anti-thesis, then we can say that as a rule this conflict between the thesis and the anti-thesis continues till there is a synthesis between the two, i.e., some aspects of the two sets of ideas may be combined, i.e., a compromise would emerge, and a new kind of plough, or a weapon, would be the result. Then a set of ideas would emerge which oppose the synthesis, i.e., the synthesis would become a new thesis and a new anti-thesis would emerge in opposition. This is how human ideas develop in society and bring about change. This is, basically, how the German philosopher, Hegel, understood the process.

Similarly, all movements of objects, both living and non-living, in the universe, face opposition, but the movement occurs if the driving force behind the changes is greater than the force of resistance. For example, if a new plant starts growing at some spot on the earth, the first resistance it faces is that of the soil itself. It has to make its way out of its seed and spread its roots into the soil, despite its physical resistance, and has to push its stem up through the soil to emerge above it. Throughout this state of its growth, the young seedling is helped by the water, air and the nutrients in the soil. As it continues to grow, it faces opposition of insects and other animals who may eat it up and of the wind which may blow it away, or micro-organisms which may cause disease. But if it succeeds in facing all these dangers and resistances, then it continues to grow. It may grow up to be full-fledged tree and may bear fruit. The seeds in the fruit may cause new baby-plants to emerge. As this tree grows and stands up to all the opposing forces of its life, in time it grows old and weak and may fall to the ground in a wind-storm, with its roots broken up and pulled out of the ground. This brings the tree's life to an end and so brings the conflict with the existing forces to an end. But, then, a new conflict is created by the existence of a dead tree lying on the ground. The dead tree may ultimately become dust, fighting with the wind and insects which try to end its existence, and ultimately succeed. This would also end the series of conflicts starting with the initial conflict which started with a little plant emerging from a seed. But in this whole set of conflicts, there are two sets of forces. One set helps the plant or tree in growing and the other is opposing it.

If we look at the details of the growth of this plant from a seed, we can see that in the original set of objects - the seed, water, air, nutrients, soil and birds, some help and some oppose the growth of the seedling. The seed, water, air and nutrients would help and may be called the "thesis". The soil and birds would oppose and these may be called the "anti-thesis". The conflict between this thesis and anti-thesis may result in a "synthesis" that allows the seedling to emerge and grow, but the synthesis is then opposed by a new anti-thesis consisting of a new set of forces, e.g., the wind and new micro-organisms and animals. If the plant survives this conflict then a new synthesis would emerge in the form of an adult tree.

This new thesis, in turn, would be opposed by a new set of forces which would ultimately succeed in knocking the tree down. But the downed tree again has a thesis, and an anti-thesis resisting its existence as a dead tree, and so on. Thus, the conflict between a thesis and anti-thesis, both consisting of sets of objects, ultimately results in its resolution, creating a new conflict. If the sets of objects include human beings, then any "situation" or thesis has two sets of forces acting on it, including the ideas in human minds that incorporate and reflect the situation. The result is that the situation is changed as a compromise and a new situation then results, which incorporates the resultant of human ideas and power. This is how Karl Marx looked at, and understood, change in human society and the action and interaction of objects, including human beings and their ideas.

Examining the two ways of looking at the process of change in the world, especially in human society, the question does arise as to which one is more valid. New ideas affect human beings and those human beings bring about change in society. The change in a society also affects thinking human beings and gives rise to ideas in their minds. Thus, ideas bring about physical changes and physical changes create new ideas. To determine which process comes first is somewhat *like a chicken and egg paradox*. Chicken is born first, or the egg?

Since, the process of change in human society is continuous, both processes are at work all the time. If importance of one or the other is to be determined then, Marx's concept seems to be more valid. To understand this requires understanding evolution better, since what we are discussing are micro processes in the evolution of mankind and hence of the world. If we think backwards, we can say that a physical process was responsible for the birth of each individual and the personality of each individual was also created, or developed by, a set of physical processes between the environment and the individual. These processes consisted of development of language, thinking, etc., and imparting of data related to numerous processes, in short knowledge and ideas. Each individual up the ladder, i.e., in the previous generation, was created this way. If we keep going like this up the evolutionary ladder, we reach the point where life was created on the planet. At this time, there was only one physical process that created the first living form - a unicellular organism. Since it was the first living being on the planet, no ideas or skills could be conveyed to it by the non-living environment of the planet. There was something in the set of objects on the planet, constituting the environment that resulted in a "living", or potentially "thinking" being. Ever since then, the living and non-living environment has continued to affect all living beings on the planet and living beings have reacted to the changes in the environment and tried to survive by adopting new ideas and behaviors.

If we think of the living beings of today and the environment at the time of the creation of the first life form, then *all that has happened, in the intervening period, is that the environment, and the set of objects it consisted of, has produced today's living beings and their ideas and not vice versa.* Since the total environment of the planet, its climate and its physical movements including rain, wind and sunshine remain outside our control even now, the situation has not changed much.

Let us look at development of some military aircraft. The Soviet Union produced an aircraft which was identified as Mig-17. Later on, Mig-19, Mig-21, Mig-23, Mig-29 and Mig-31, etc., were developed by the same state. Similarly, the United States developed aircraft that were identified as F-10, F-14, F-15, F-16, F-21 and F-22, etc. Teams of scientists and engineers, who could understand the sciences of the air and the earth and who had the knowledge of electronics, created the earliest version, Mig-17 or F-10, of these aircraft. There were two sets of pressures on both the teams - one was the pressure of competition with the other state's team and the other was the conflict within those in the same team, but who had different ideas as to how to improve, and make more potent, the existing model of an aircraft. We can think of the two sets of ideas of the same teams as the thesis and the anti-thesis, dealing with the capabilities of their own aircraft and those of the competing aircraft. The result, for each team, was a synthesis that resulted in the new version of each state's aircraft. What we have discussed is, of course, the *evolution* of American and Soviet fighter aircraft, from F-10 to F-22 and from Mig-17 to Mig-31, respectively.

Evolution of mankind has been a very slow process and the logic of change described above has been at work throughout. Many challenges to human survival emerged and we developed survival skills at each stage, including the ability to form groups and work in cooperation. *Human beings developed the skills to collect and acquire products and also to exchange them. This is a skill unique to the human species.* These economic survival skills ultimately became dominant in the development of our social groups. As time passed, evolution became more and more the evolution of economic activities. Some people, at each stage, took advantage of these evolutionary changes to become dominant in their social groups and, thus, formed the ruling class. When the fundamental economic system of production changed, it changed the relevant social group and the way it was organized. *Thus, the changes in the mode of production had the most profound effect on development of human social groups and of mankind in general.*

Chapter - 4

Basics of Economics

As we look at evolution of economic activities in humans, we see that in the hunter-gatherer stage of development, economic activities, like hunting of animals, fishing and gathering of fruits, were engaged in by a man and his children, or when the extended families developed, by the male members of the extended family. This was the early state of socialization of economic activity and the mode of production consisted of progressively developed hunting and gathering activities. When the extended families developed into mini-tribes, large nomadic tribes, land-based tribes, nations and national states, the mode of production advanced from the hunter-gatherer stage to production for the whole tribe to production, and exchange of goods, with many tribes over a large feudal population. In this process, when the social groups became large, like a nation or a state, not only was production of goods a bigger activity, but their distribution also became a big task. In other words, goods were distributed in such a way that they reached all the individuals in the social group and were then consumed. Thus, large-scale production, distribution, and consumption of goods occurred as civilization advanced. Study of these three aspects of economic processes is undertaken by modern economics.

Normally, it is the effort of the producers to produce more than what is needed for consumption. What is not consumed can, thus, be saved for the future, or can be exchanged for products that may improve or expand the operation of the social group, like a family, a farm, or a company. In a simple case, a family working on its farm may produce more food-grain than it needs. Some of the food-grain may be saved to meet unforeseen needs. The rest may be sold to obtain other goods that the family needs. In addition, some of the surplus may be sold to buy agricultural implements, or, in modern times, to buy a tractor. The addition of new implements on the farm increases the efficiency of the farm and may result in higher production in the next farming season.

As social groups increased in size, the process of choosing what to produce and what not to produce also progressed and specialization into one activity or another naturally followed. In later stages of development of tribes, it became possible for a class of individual to forego agricultural production completely and to focus on acquiring and using skills needed by those engaged in agricultural production. Some of those in the ruling classes of tribes, also, became traders, engaging in exchange of products between individuals and tribes. Also, groups of artisans specializing in making swords, bows and arrows and repairing them, spinning of yarn and cloth, blacksmiths engaged in making ploughs and repairing them, those engaged in making and repairing shoes, etc., developed in tribes and their villages. Thus, specialization grew as economic skills progressed in an atmosphere of increasing socialization in every field of human activity. At present, we have reached a very high level of socialization of all activities and the most recent development, which occurred only a few hundred years, has been the development of the factory as a production unit. Production in a factory is socialized. Workers produce one or another part of a machine, or consumable product. They work together so that no one can claim that he produced an item of the output of a factory, e.g., a car. A car is the product of all the workers of the factory and their separate contributions cannot really be separated or quantified. The same happens in a mine. This is socialized production at its best and is the consequence of socialization of all other activities. Even this production in factories has continued to increase in size of production units and specialization of human effort, giving rise to huge societies and states, based on use and abuse of this process of human economic evolution. To understand the processes of modern economic activities, we need to look at some basic concepts. This should give us some insight into why the world is the way it is at present.

Natural Resources

Our planet has its naturally occurring landscape – Its mountains, plains, rivers and oceans. We exist on this planet along with a huge number of other species – plants and animals, etc., on land, in lakes, rivers and in the oceans. There are minerals that exist near the surface of the earth, or deep underground. The air on this planet and the sunshine are essential to our survival. All these living and non-living resources are referred to as "natural resources" to be used by us. Since these natural resources require no human effort or labor in coming into being, their value, in terms of human labor, is zero, i.e., their coming into existence costs us nothing.

Production

Our planet has been changed by living beings which are always on the move. The greatest change has been brought about by us humans. We are constantly using the natural resources of the planet and are creating new objects, or products - roads, railway tracks, airports, residential and commercial buildings, factories, farm houses, machines, food grains, etc. The main natural resources used are the land, plants and the animals on the planet and its mineral resources and we use them in different ways to create different products. Minerals may be processed to obtain metals. Forests are cut down to obtain wood which may be used for building houses. Crops we need are planted and the plants and their fruits are used by us in different ways, including as fodder for animals we rear in our animal farms. We protect the species we need and we destroy the ones we do not need, or those which cause us harm. Air, light of the sun and water flowing in rivers may be used for generation of electricity. We have thus transformed this planet by our economic activities. Our production results in two basic kinds of products and services. *Consumer goods and services* are directly consumed by us. Clothes and home appliances are consumer goods. Restaurants provide food and the services of waiters who serve that food to customers. Consumer goods and services also include weapons and soldiers who use them. Soldiers provide their services for our protection or for destruction of humans or individuals of other species. *Producer goods and services* are meant for production of other producer or consumer goods, but are not directly consumed by us. For example, machine tools and services of engineers may be used for production of cars which we use directly, or these may be used for production of cranes, which can be used for construction of houses for us. Producer goods also include skilled labor produced by universities and training centers, such as engineers, doctors and scientists, who provide their services for production of goods, or to provide services like teaching and research. Many services produced and provided by a society can be used for both consumption and production. Transport, telephone and communication systems and services of engineers and doctors are such dual-use services.

Production may take place at different kinds of production and service centers, but, the basic process involved is the same - application of human labor to natural resources in their virgin or modified form. A production center may be a farm, a factory, a restaurant, a bank, an insurance company, a university or a training center. The products of these production centers may be physical products like food grain, machines, chickens, milk, meat, butter, trained engineers, technicians, doctors and scientists, or they may be services provided by a bank, insurance company, or a transport company, etc. The output of a production center may be both physical products and services - as provided by a restaurant, like burgers being provided by it and services provided by its waiters.

Capital

Capital is actually what has been produced but not consumed by a certain society - it generally refers to unconsumed production over a long period of time, or title to it in financial form, i.e., in the form of money. It is just the sum total of unconsumed products of different forms - each of the products being the result of labor. Since all products require labor for their production, and incorporate that labor, capital is, thus, just unconsumed labor objectified in the form of various unconsumed products, which may be used in the production process, as tools or facilities for production. These can, thus, be referred to as products incorporating objectified labor. Some capital accumulated by a society may be non-productive capital – like pyramids, tombs, monuments and mansions which may be of some symbolic or emotional value for individuals, or a community, but which cannot be used in any process of production. They are, however, generally included in the total "wealth", or total accumulated capital of a state. Productive capital may take different forms as follows:

- **Fixed Capital** - land prepared for a factory or agricultural farm, buildings, machinery, electronic systems of communication used by a factory or mine, etc.

- **Facilitating Capital** - Electric power supply, gas supply and water supply, etc., used by a farm, factory or mine. Oil and lubricants used for maintenance of buildings and machinery.

- **Unfinished Input Capital** - Raw materials, unfinished products and components input to a production center. This form of capital is processed and transformed into the output products or services of a production or service center.

- **Human Capital** – low skilled, medium skilled or high-skilled human beings which provide their efforts or labor for production. Living labor of these human workers gets incorporated into the goods and services provided by a production or service center. There can be no production without human labor. There is no such thing as an "unskilled worker". Every human being has some ability and skill to produce or to help in production of goods or services. Totally disabled individuals are the only unskilled human beings, who are not able to participate in most production or service activities. Like all other forms of capital, *human capital* is the result of human labor or efforts by individuals incorporated into themselves by study and practice and the labor of their parents, teachers and trainers also, incorporated into them. They also consume food and other consumable products, containing objectified labor, during the period of their development into fully trained individuals. Thus, those individuals become low skilled workers like janitors or cleaners, medium skilled technicians like electrician, plumbers and masons, or highly skilled engineers, doctors and scientists, etc.

All forms of capital have a useful life. Facilitating capital has a short life span in a production center. It gets consumed in the production process. Unfinished input capital is transformed during the production process, into products or services produced and, thus, has a short life span in its original form. Fixed capital has a relatively long useful life. Buildings may last for thirty or fifty years. Machines may last for five to ten years in a factory. Land is a natural resource and lasts forever, but the improvements on land, in a mine, agricultural farm, or factory may not last for a very long period and may require renewal. Human capital, or human workers, have a useful life of thirty to fifty years, depending on the kind of skill they have and use.

Since all other forms of capital, besides human capital, have human labor incorporated, or objectified, into them as non-living labor, we refer to human capital use, or labor provided by living human beings, as "living labor".

Factors of Production

Human beings produce all kinds of products and services. Thus, living labor or use of human capital is essential for production. At the earliest stage of our existence, our ancestors only knew how to gather fruit from trees and bushes and to catch other animals by hand if they could. "Production" at that stage was the result of two factors:

- Natural resources

- Living labor.

As our ancestors developed the skills to create and use tools like bows, arrows and spears, they developed their hunting skills. Since the hunting and other tools required living labor to be used, they incorporated living labor and that labor became objectified into the form of tools. Thus, production at that stage required three factors:

- Natural resources
- Labor objectified into tools
- Living labor

When human beings learned to domesticate and use other animals, for meat, milk, hides and eggs, etc., the basic factors of production remained the same, although the fundamental mode of production had changed. Similarly, when tribes became nomadic, use of animals for transport services changed the mode of production fundamentally, especially with the invention of the wheel, but the basic factors of production remained the same. Only more living labor went into making wheels and carts when they came into use. When tribes settled down to engage in agriculture, a few thousand years ago, production did become quite complex with tools like ploughs and water-wheels coming into use. Throughout this period of thousands of years, production required the same three basic factors of production. After passage of several more thousands of years, the tools for production of goods and services have become much more complex. Also, we have learned to use more natural resources, in more ways, to create many more new products and services for use in our societies and states. Further, higher and higher skills are needed for operation of those complex tools and machines which have been developed by us. Thus, we produce more and more highly skilled workers for production and the human capital in our societies and states has grown accordingly. The basic factors of production can, therefore, be identified as follows:[1]

1. *Natural resources in their original or modified form.* In other words, we need capital for production of goods and services. The basic components of capital have been described before. These are:

   • Fixed capital

   • Facilitating capital

   • Unfinished Input Capital.

   All three components of capital are generally needed in a production or service center. Each component of capital consists of numerous tools, machines, buildings and materials. The requirements of each production or service center are different, depending on what is being produced.

2. *Living labor. It is essential for production of goods and services.* All *socially necessary work* in a mine, factory, or bank, or any other form of a production or service center is considered living labor. Efforts made by an unskilled worker, an electrician, a plumber or a driver of a truck, the work of a scientist, a research worker, a doctor in a hospital, an engineer designing a new machine or building, or the management work done by a factory manager – these are all forms of labor which result in production of a product, or provision of a service to society.

Some economists tend to classify all kinds of tools, machines, buildings, land and materials in natural or modified form as basic factors of production. Capital and labor are also specified as factors of production, as if the other forms of consumer and producer goods and services were not included, or objectified, in them. This tends to confuse students of economics.[2]

Means of Production

When we refer to means of production, we could be referring to implements and tools used to produce products and services. Implements and tools, so used, were the means of production in primitive stages of our social development, as described before. At present, if we are analyzing the economy of the United States, a European state, or China, we are referring to not just simple tools and implements but, among others, to:

- Coal mines, or iron, aluminum and copper mines producing minerals for copper and aluminum smelting factories and steel mills, etc. Naturally occurring coal, iron ore and copper ore are the natural resources accessed by such mines and coal, iron ore and copper ore are their products. The steel mills and smelting factories convert the ores into copper, aluminum and steel ingots.

- Whole factories producing railroad engines and other equipment for railways, materials for construction of buildings, military equipment and equipment for complete cement factories and sugar mills and machinery of various kinds, including cars and home appliances.

- Farms or agrobusinesses producing food grains, milk, kettle, cotton, oil-seeds and a range of other agricultural products.

- Sugar mills processing sugar cane, cotton mills producing cotton yarn, textile factories, etc.

- Television and radio stations, railways and bus-transport companies, banks and insurance companies, universities and other educational institutions, providing services of different kinds. Universities, for example, produce engineers, doctors, lawyers and other trained individuals with various professions or expertise. Human beings form the "raw materials" for a university, or any other educational institution. The output of such an institution is trained manpower. All an educational institution does is to train the individuals who enter it for education. It provides a service. So, a university is a "means of production". Similarly, a radio or television station is a means of production, since it produces, or provides, news and entertainment programs, or information in general, to a population. Bus-transport and railway companies provide transportation services for goods and people.

So, by "means of production", we mean agriculture farms, mines, factories and other institutions which produce products and services. These huge sophisticated institutions and organizations are the present day "tools" of production, requiring thousands of workers for their operation!

Characteristics of Production
Diminishing Returns

In a feudal society, when agriculture has not been mechanized, one person can efficiently cultivate only a small piece of land, say, five acres. If two persons work on the land, they may be able to produce more grain on that piece of land. But if the number of persons working on the same piece of land keeps on increasing, then the output of the land does not grow proportionally. In fact, after a point, each added person adds a decreasing amount of output, or marginal output. This is because too many workers do not help each other and start getting in each other's way. This phenomenon is referred to as the law of diminishing returns. It was noticed, first, by a British clergyman and economist named Thomas Malthus. Malthus, based on his observations during his lifetime, i.e., 1766-1834, thought that the human race was doomed because population was growing geometrically and doubling every 25 years, while agricultural production was growing arithmetically, i.e., a few percentage points per year. He thought these processes would drive wages down to subsistence levels, because of the huge increases in population. However, much more land has been brought under cultivation and the mechanization of agriculture has dramatically increased global agricultural output, while the population has tended to move to other occupations in new industries, or new areas of the world. Despite these changes, the wages of workers did go down to subsistence levels, in fact to below-subsistence levels. This, brought death to large numbers of people, till the population stabilized. However, the increasing population was not the only reason for subsistence-level wages. We would investigate this issue in later chapters.

The law of diminishing returns does not apply to production on a farm only. It applies to all other forms of production centers - factories, mines, banks, etc. Every production center has its optimal labor force for its level of automation. If the number of people, is increased or decreased from that level, production per person tends to go down. In other words, the cost of production tends to go down initially as manpower is added but, after an optimal point, it begins to go up.

Specialization

In a village, or a tribe, if people divide their everyday work among themselves and every person does only one kind of work, then this can help everybody, if they exchange the output of their work. In fact, they can produce more this way. When this was realized, specialized occupations of blacksmiths, carpenters, shoe-makers, hair-cutters arose in the early evolution of mankind. In agricultural communities, production also became specialized. A family would produce corn and wheat, while another would produce sugarcane and fodder. Yet another would raise farm animals and produce milk and meat. Such families would exchange their products within a village and all would benefit because of this specialization.

When the factory system came into being, the process of specialization was accelerated by the owners of those factories and mines, for the same reasons. Thus, workers began to specialize in their occupations - machine operators, welders, carpenters, mechanics, etc. Even the management became specialized and occupations of foremen, shift supervisors, etc., came into being. The development of industries moved towards activities that required more and more mental work and the occupations also became more and more demanding in mental abilities and training. Now, it is quite common to think about specialized surgeons, doctors, engineers and software developers and further specialization of occupations and production activities continues, with companies engaged in only mining, oil production, production of computers and cell phones and those producing computer software, etc.

Productivity

Productivity refers to how much is produced by a worker in a certain amount of time. Some workers are more efficient than others in terms of their use of time and energy. Thus, productivity is measured in terms of output per worker. Productivity, however, depends on the production process. If the process is modified by increasing the use of machines, productivity increases. It does not necessarily mean that a person becomes more efficient in use of his time by the introduction of a machine. But if there are more machines for the same number of people, more can generally be produced. Thus, production per head increases by introduction of machines - especially automatic machines.

Machines are, in reality, non-living workers, since they have been produced by workers. They can be referred to as "objectified labor", since they are the result of labor of a group of workers. It is possible, for a particular farm or factory, to increase productivity by improving the work habits of the workers and by introducing better systems of management, etc. But other factors remaining constant, when these non-living workers are introduced into the production process, the production per living worker, or productivity, increases.

Value of a Product

It is important to know how the value of products is determined, since products are widely produced and distributed in society and are exchanged for other products. To understand the rules for such exchange or trading of products, the concept of value needs to be investigated. The value of a product is the labor expended in producing it. This should be obvious, since all vehicles, trains, aircraft, agricultural products and whole cities are created by human effort. They do not come into being by some kind of magic! Only naturally occurring resources are not produced by human effort – these include all naturally occurring vegetation and non-human life forms. Obviously, if more effort is expended in creating a product, it would be valued more by a community, as long as the effort expended is necessary for creation of the product. If the product is not produced by one person and is, thus, a socially produced item, then the labor involved in its production has to be socially necessary labor. It is easy to see this if we are thinking of products like fruits collected from trees growing naturally in the wild. But, if things are produced with the help of tools and machinery installed in buildings, or on land, this fact does not remain obvious. But if we realize that the tools and machinery and buildings, used in the production process, have already been produced by using living labor, and are really "objectified labor", then it is possible to see this truth. The concept of "objectified labor" needs further explanation.

Metals are produced by mining ore of the metal by setting up a mine. The value of the ore is the labor used in its extraction. The value of the metal is the labor used in extraction of the amount of ore used in its creation and the labor used in converting the ore into metal. The value of a tool is not the ore or metal used in its construction, since the ore occurs naturally and its value, in its location in the crust of the earth, is zero. The value of the tool is the labor used in extracting the required amount of ore, that used in converting the ore into metal and that used in shaping the required amount of metal into a tool. Thus, the tool incorporates these components of labor used in its creation. The value of the tool is the total labor used in its creation – from the ore form, as it exists in the earth, to the tool form. That total labor is incorporated into the tool, or is "objectified in" it. Another way of saying this is that - a tool is the objectified form of the specific types of labor used in its creation.

…. **"In the early stages of society, the exchangeable value of these commodities, or the rule which determines how much of one shall be given in exchange for another, depends solely on the comparative quantity of labour expended on each.**

…. **'The real price of every thing,' says Adam Smith, 'what every thing really costs to the man who wants to acquire it, is the toil and trouble of acquiring it. What every thing is really worth to the man who has acquired it, and who wants to dispose of it, or exchange it for something else, is the toil and trouble which it can save to himself, and which it can impose upon other people.' 'Labour was the first price— the original purchase-money that was paid for all things.' Again, 'in that early and rude state of society, which precedes both the accumulation of stock and the appropriation of land, the proportion between the quantities of labour necessary for acquiring different objects, seems to be the only circumstance which can afford any rule for exchanging them for one another' "**….

David Ricardo (1817), *The Principles of Political Economy and Taxation.* Urbana, Illinois: Project Gutenberg. Retrieved October 16, 2019, from www.gutenberg.org/ebooks/33310.

When a worker uses a tool to create a new product, the worker's labor is incorporated into the product and also the proportional amount of the labor objectified in the tool is used in the new product. The amount of objectified labor used in the creation of the product would depend on how many products can be created in the lifetime of the tool. Thus, the value of such a product is the living labor and the proportionate amount of objectified labor incorporated into the product. Machines are just more sophisticated tools. Also, buildings serve the same purpose as machines, but their life span is relatively longer. This process of production would be discussed later in detail, in Chapter-7.

More valuable products require more labor to produce them and the labor can be a mixture of living and objectified, i.e., non-living labor. Skilled labor is more valuable than unskilled labor, since human effort, or labor, is required to become a skilled worker, like a machine operator, an engineer, a doctor, or a scientist. Workers of the same category may have different levels of skill in doing their jobs and, thus, the value of their labor may differ somewhat, but we should be thinking of ideal workers with uniform skills in each category, so that we can analyze the basic elements of the whole production process, without getting confused by details that do not affect the fundamentals.

The value of anything in the world is zero if no labor has been expended in producing it, i.e., if it occurs naturally, like virgin land, sunshine, or air. Now a days, we are used to thinking of the value of a product in terms of its price, but this is really a misconception, although the price of a product tends to reflect its value in the long term, in most cases. We would look at these details and the concept of price in more detail, later on in this chapter.

The Origin of Ownership and Inequality

We have free use of the resources of our planet. Since we have not created the sun, our planet or its specific resources that we need and use, like sunshine of the sun, or the planet's land, air, its plant and animal life, or its mineral resources, there is no economic value that could be assigned to these resources unless labor is expended in modifying them. For example, virgin land has no economic value.

…… **"And thus without supposing any private dominion, and property in Adam, over all the world, exclusive of all other men, which can no way be proved, nor any one's property be made out of it; but supposing the world given, as it was, to the children of men in common, we see how labor could make men distinct titles to several parcels of it, for their private uses; wherein there could be no doubt of right, no room for quarrel.**

…… **Nor is it so strange, as perhaps before consideration it may appear, that the property of labor should be able to over-balance the community of land. For it is labor indeed that puts the difference of value on everything; and let anyone consider what the difference is between an acre of land planted with tobacco or sugar, sown with wheat or barley, and an acre of the same land lying in common, without any husbandry upon it, and he will find, that the improvement of labor makes the far greater part of the value. I think it will be but a very modest computation to say, that of the products of the earth useful to the life of man, nine-tenths are the effects of labor; nay, if we will rightly estimate things as they come to our use, and cast up the several expenses about them, what in them is purely owing to nature, and what to labor, we shall find, that in most of them ninety-nine hundredths are wholly to put on the account of labor."**

John Locke (1690), *Second Treatise of Government*. Project Gutenberg. Retrieved October 16, 2019, from www.gutenberg.org/ebooks/7370.

However, if a piece of land is prepared for cultivation, by clearing it of naturally growing vegetation, it is leveled for easy irrigation, a well may be drilled on it to make water available for crops to be planted on it and buildings may be constructed on it for storage of grain, then the value of the land is increased in proportion to the labor and effort spent in these processes. Similarly, naturally occurring petroleum or oil has no value and it is available free for someone who can find it. However, as labor is expended in finding it, in digging wells and preparing the wells for production and for making preparations for storage of the oil produced, value is added to each barrel of oil that is produced as a result of these efforts and the value of each barrel of oil is proportional to the labor spent in producing it, including the proportional share of the labor, incorporated in the equipment, that is used for this purpose. This applies to not only the oil that is produced but to all minerals extracted from the earth's crust and all kinds of agricultural production on the surface of the planet.

"The first person who, having enclosed a plot of land, took it into his head to say this is mine and found people simple enough to believe him, was the true founder of civil society. What crimes, wars, murders, what miseries and horrors would the

human race have been spared, had someone pulled up the stakes or filled in the ditch and cried out to his fellow men: 'do not listen to this impostor. You are lost if you forget that the fruits of the earth belong to all and the earth to no one!' "

Jean Jacques Rousseau, *A Discourse Upon the Origin and the Foundation of Inequality Among Mankind:* Project Gutenberg. Retrieved October 15, 2019, from www.gutenberg.org/ebooks/11136.

However, we have come to establish property rights in the natural resources of the planet and also in the land used for any purpose. This applies to states that claim certain territory, and are constantly competing with each other; and also, to individuals, within states, who compete with each other. However, even now, in proportional terms, the price of land does represent the value of the land and improvements on it. It remains to be determined as to why and how we have come to assign property rights to these natural resources and have thus assigned value to such resources which logically should be held in common and should have no value or price. The reasons for this are very simple - we, like other animals have occupied by force natural resources that we deem necessary for our existence, including land above everything else, and we have maintained these claims by force throughout our evolutionary history. Our species has grown in numbers and has come to dominate all other species on this planet. Thus, we claim that the earth belongs to us, and to no other species. Further, each of our states claims certain territory as "its own" by right and is willing to defend "its territory" by military force, if necessary. The present territory of these states has, in all cases, been acquired as a result of a history of violence and genocidal warfare with tribes, nations and the other, past or present-day, states, except those who were the first inhabitants of a territory.

...... "Before representative signs of wealth had been invented, it could hardly have consisted of anything but lands and livestock, the only real goods men can possess. Now when inheritances had grown in number and size to the point of covering the entire landscape and of all bordering on one another, some could no longer be enlarged except at the expense of others; and the supernumeraries, whom weakness or indolence had prevented from acquiring an inheritance in their turn, became poor without having lost anything, because while everything changed around them, they alone had not changed at all. Thus they were forced to receive or steal their subsistence from the hands of the rich. And from that there began to arise, according to the diverse characters of the rich and the poor, domination and servitude, or violence and thefts. For their part, the wealthy had no sooner known the pleasure of domination, than before long they disdained all others, and using their old slaves to subdue new ones, they thought of nothing but the subjugation and enslavement of their neighbors, like those ravenous wolves which, on having once tasted human flesh, reject all other food and desire to devour only men" ......

...... "I have tried to set forth the origin and progress of inequality, the establishment and abuse of political societies, to the extent that these things can be deduced from the nature of man by the light of reason alone, and independently of the sacred dogmas that give to sovereign authority the sanction

of divine right. It follows from this presentation that, since inequality is practically non-existent in the state of nature, it derives its force and growth from the development of our faculties and the progress of the human mind, and eventually becomes stable and legitimate through the establishment of property and laws. Moreover, it follows that moral inequality, authorized by positive right alone, is contrary to natural right whenever it is not combined in the same proportion with physical inequality: a distinction that is sufficient to determine what one should think in this regard about the sort of inequality that reigns among all civilized people, for it is obviously contrary to the law of nature, however it may be defined, for a child to command an old man, for an imbecile to lead a wise man, and for a handful of people to gorge themselves on superfluities while the starving multitude lacks necessities."

Jean Jacques Rousseau, *A Discourse Upon the Origin and the Foundation of Inequality Among Mankind*: Project Gutenberg. Retrieved October 15, 2019, from www.gutenberg.org/ebooks/11136.

What applies to states also applies to individuals within these states. Each state has a large population for the available land, so the land has been forcibly occupied by different families and individuals and continues to be so occupied in many cases. In the United States, for example, land was acquired by colonizers by riding on horses - a family was given ownership of land that a man could encircle by riding a horse, thus creating a "homestead". This was possible because the land had been forcibly acquired from the native inhabitants by the colonizing powers, by mass murder of native inhabitants and eviction of those still remaining on it. The creation of other colonies in North and South America and Australia followed similar patterns of acquisition of "land ownership". Some such "owners" of land later found it expedient to exchange part or whole of "their" land for other resources and products of real value. So that a market in land has emerged and a title to such a piece of land is available for a price. This title is still supported by society and the state by force, since most states continue to be ruled by members of the propertied, ruling classes, who are willing to maintain such "ownership", guaranteed by its "laws", by the power of the state. What applies to land, also applies to all other natural resources. However, a market in sunshine, air and waters of the oceans has not developed only because it is impossible to occupy all these natural resources by force and deny others the use of these resources.

"Will setting one's foot on a piece of common land be sufficient to claim it at once as one's own? Will having the force for a moment to drive off other men be sufficient to deny them the right ever to return? How can a man or a people seize a vast amount of territory and deprive the entire human race of it except by a punishable usurpation, since this seizure deprives all other men of the shelter and sustenance that nature gives them in common?"

Jean Jacques Rousseau (1920), *The Social Contract & Discourses*. Urbana, Illinois: Project Gutenberg. Retrieved October 14, 2019, from www.gutenberg.org/ebooks/46333.

Our desire for possession of natural resources comes from our natural dependence on the land for survival, like it is for all other species. Animals like chimpanzees, wolves, or skunks, mark "their" territory" and fight over it although they do not have the intelligence of human

48

beings to know why. It is something that their evolutionary experience has taught them. It is part of their genetic heritage. Land means survival for them. Similarly, land means our livelihood, we seek food, shelter and safety from it in different ways and it's a matter of survival, so it is natural that we would get attached to it, like other living beings. Tribes especially develop a strong attachment to "their" land, which they have come to possess in the struggle for their survival. The same applies to individuals.

...... **"To understand political power, right, and derive it from its original, we must consider what state all men are naturally in, that is a state of perfect freedom to order their actions, and dispose of their possessions and persons, as they think fit, within the bounds of the law of nature, without asking leave, or depending upon the will of any other man. A state also of equality, wherein all the power and jurisdiction is reciprocal, no one having more than another; there being nothing more evident, than that the creatures of the same species and rank, promiscuously born to all the same advantages of nature, and the use of the same faculties, should also be equal one amongst another without subordination or subjection;"............**

...... **"Before the appropriation of land, he who gathered as much of the wild fruit, killed, caught, or tamed, as many of the beasts as he could; he that so employed his pains about any of the spontaneous products of nature, as any way to alter them from the state which nature put them in, by placing any of his labor on them, did thereby acquire a property in them; but if they perished, in his possession, without their due use; if the fruits rotted, or the venison putrefied, before he could spend it, he offended against the common law of nature, and was liable to be punished; he invaded his neighbor's share, for he had no right, further than his use called for any of them, and they might serve to afford him conveniences of life"......**

...... **"The same measures governed the possession of land too; whatsoever he tilled and reaped, laid up and made use of, before it spoiled, that was his peculiar right; whatsoever he enclosed, and could feed, and make use of, the cattle and product was also his. But if either the grass of his enclosure rotted on the ground, or the fruit of his planting perished without gathering, and laying up, this part of the earth, notwithstanding his enclosure, was still to be looked on as a waste, and might be the possession of any other"...........**

John Locke (1690), *Second Treatise of Government*. Project Gutenberg. Retrieved October 16, 2019, from www.gutenberg.org/ebooks/7370.

We use many resources, naturally occurring on the planet, all our lives and we get them free. We walk on land, breath air, sit in the sunshine and swim in the waters of rivers, lakes and oceans. We cannot survive without drinking water. All these resources, which are essential to our lives and well-being, occur naturally on our planet and they cost us nothing.

In most societies land is bought and sold for a price, even if it is naturally occurring virgin land. There is no real justification for this property declaration, but it is a real process. To claim that we own land, that some piece of land is the property of some individual or state is to make an irrational claim. We, as human beings and all our ancestors were born, as an elementary form of life, on this planet. We return to the soil of this planet as dust when we come to the end of our individual lives. We do not own this planet. We have not created it, nor have we created its land, oceans, rivers, lakes, mountains, forests and animals. We can have no claim to these natural resources as individuals or states. *It is the planet which, in reality, owns us.*

Fair Basis of Exchange

If the value of a product is the labor expended in producing it, then, it is obvious that products should be exchanged on the basis of their value, if the exchange is to be fair and just. If products are exchanged for other products, obviously, the bargaining is based on the two parties' understanding of how much effort, or labor, has been used in the production of the two products being exchanged, say rice for sugar. This kind of exchange of goods is referred to as barter trade and is not flexible enough for some purposes, e.g., one side may want to dispose-off its surplus rice but may not want to obtain sugar in its place at the current time. Then, some way has to be found to record the half of the exchange transaction, i.e., the sale of rice by one party, without purchase by it of some product. In practice, thus, money has been used for exchange of products. In the earlier stages of our development, food grain was used as money, i.e., sale and purchase of products was done in terms of buckets of food-grain, say wheat, as is done even now in some areas of the world. Thus, a person could sell several buckets of rice for a certain number of buckets of wheat, and, then, buy an amount of sugar that he needs, in terms of buckets of sugar, paying for this purchase of sugar in terms of a certain number of buckets of wheat. Thus, the sale of rice and purchase of sugar may involve three parties, if the exchange occurs between producers only, or when a shop-keeper facilitates the two transactions. The shop-keeper would, thus, maintain, a certain amount of wheat, as money, and would pay the sellers of rice and sugar in terms of wheat and receive wheat from the buyers of sugar.

Obviously, barter trade and the use of food-grains as a medium of exchange was not a convenient method, since the food grains are heavy and easily get spoiled. Thus, the medium of exchange ultimately shifted to gold. Gold is a scarce commodity and has always been considered valuable because of its ornamental use. It does not get spoiled, also. Thus, gold was used as a medium of exchange, or money, for a long time. In modern times, printed money, and money issued by different states, has come into use and trade is no longer conducted in terms of gold, though reserves of gold are still maintained by states and individuals since the value of a state's currency can fluctuate wildly.

We can see that the exchange of products would tend to occur on the basis of value of each product involved, if barter trade is the mode of exchange, but the process would involve bargaining and, thus, the exact value of the products cannot be said to be the basis. When gold or another commodity is involved as money, the process would become more inaccurate, even if no external forces are involved in the transactions. In reality, other forces were always involved when exchange of commodities took place on a large scale. It depended on the social system and its power relations. We shall look at those details in later chapters. The use of paper currency has made the process even more inaccurate, because the process of creation of money is controlled by governments of different states, who can create money at will. This throws the concept of exchange on the basis of value virtually "out of the window". Thus, at present and in recent times, local as well as inter-state trade has never occurred on the basis of value of products, but on the basis of their prices which are determined by other forces. Thus, modern trade has become inherently unfair and abusive.

Value and Price

As we have described before, at present, trade does not take place on the basis of value of products. Buyers buy products on the basis of the usefulness that they perceive, or are made to perceive. Suppliers of products tend to control the supply of their products so that their prices can be kept high. They resort to hoarding of products, if necessary, to create an artificial scarcity so that the products can be sold at higher prices. Advertising is used to induce buyers to buy more of some products or to buy new products. Thus, deception and use of psychological methods of inducement have become a rule rather than an exception in trading. In the absence of estimation of the value of products, products are sold on the basis of perceived value. The perception of value is influenced by several factors. First, the value of a product is influenced by the quality of labor input. The quality of labor affects the ultimate desirability of the product - its look and feel as perceived by the buyers, its ability to meet the exact needs of the buyers, etc. The perception of price is also influenced by the product's beauty and finish, as felt by the buyers, which may not be related to its utility. *This perceived value, which in general is a false value, is called the price.* However, prices are not necessarily false indicators of the value of products. *Prices fluctuate, sometimes wildly, but their mean, generally, does tend to be determined by their values.* Those products, which are scarce, do have prices assigned to them, which have no relation to their value based on their labor content. For example, gold is a rare metal. It generally occurs with copper and its cost of production, in terms of labor, is not very different from that of copper. However, the price of gold is many times that of copper. Similarly, oil, gas, anthracite-coal and uranium are scarce and the result is that their prices are many times more than their value. The prices of products, also, vary with the society and state they are produced in. In primitive societies, wages of labor are extremely low and those in advanced societies are much higher. Thus, prices of many products, even if they are not scarce in nature, vary widely from state to state. The costs of transportation also play a part in this, since the locations where products are consumed may be thousands of miles away from their production centers. Spoilage of perishable products, like vegetables and fruits, also affects prices by location. All this happens in a jungle-like atmosphere, which has come to be regarded as an economic market as discussed in the next chapter, in detail.

Corporations

Originally exchange of products, or trade, was carried out by individuals who were themselves producers of the products. Later on, this activity was taken over by merchants, who would buy the products from producers and sell them to consumers. This generally involved transport of products and travel by the merchants over short or long distances, depending on the kind of products and the location of consumers they were dealing with. Such trading operations were generally called individually owned businesses, or "sole proprietorships". Later on, two or more individuals started to jointly own and run their businesses, as "partnerships". When the level of production and trade increased further, "joint-stock companies" were created, to be owned by several individuals. In fact, such companies were set up, even, by various imperial powers, to create and manage their imperial possessions, or colonies. For example, the "East India Company" created the British imperial possession in South Asia, called "British India", and managed it for its owners. Thus, the joint stock companies became "social groups" involved in economic activities of all kinds, with *maximum profit for their investors, or share-holders, as their only motivation.* Such companies were legally protected from liabilities arising from improper management or unexpected risks. The losses in such cases, were borne by tax-payers, or the government of a state.

The "joint-stock companies" are, now, called "Corporations". Corporations are engaged in producing goods in factories, mines and farms and in distribution of those goods. They are also involved in providing services like transport and banking. Corporations are basically privately-owned unaccountable tyrannies. Investors control their operations and workers have virtually no control over their management. A corporation may be involved in any sector of an economy, or in several sectors at the same time. Corporations, whose shares are not listed on a stock-exchange, are said to be "privately-owned" If their ownership is wholly or partly split up into shares and those shares are offered in a stock-exchange for ownership by individuals, or other corporations, such corporations are said to be "publicly owned". We would refer to all the different legal forms of businesses as "Corporations", or "companies" for simplicity.

Chapter- 5

The Market Mechanism

The "State of Nature"

Like us, some other animals have developed the skills to make tools and to create products like "nests", or other living spaces, but ours is the only species which has developed the skills involved in exchange of products. To investigate the origin of the exchange of products in modern times, we have to go back to the origin of economic skills developed by mankind. In the hunter-gatherer stage, the "products" consisted of what fruit could be gathered from trees and the animals which could be hunted for food or their furs for protection from the extremes of weather. These products had value since it took effort and labor to gather fruit and to hunt animals. At this stage of human evolution, the conditions of life of human beings were not much different from those of other animals.

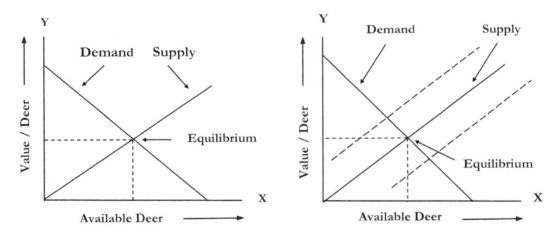

Figure 5-01 – Supply & Demand   **Figure 5-02 – Change in supply**

Political thinkers refer to the existence of mankind at this stage as existence in a *"state of nature"*, under which a condition of equality of all human beings existed. Obviously, complete freedom existed at this stage and human beings were equally free at that time. Every human being could do anything he or she wished, including robbing and killing of other human beings, stealing anything from another family, including the wives and daughters of other men. Thus the *"law of the jungle"* prevailed under these conditions, i.e., under the *state of nature*. Even now, lions, for example, fight other lions and kill them or make them run away and then take over possession of the defeated lion's females, often killing their children by their former mates. The fighting in this case is directly over females, but indirectly over habitat also. Lions also kill the babies of cheetahs whenever they find them. They want the cheetahs to leave "their", i.e., the lions', territory. This conflict is over habitat and its resources, e.g., deer inhabiting the territory. Suppose a certain number of human families lived on a certain piece of land, during the hunter-gatherer stage, i.e., in a state of nature, and had access to a certain number of fruit trees and animals also living on the same or adjoining pieces of land, then their conditions of life would be very similar to those of the lions of today. The *"state of nature"* and *equality* would exist, and the *"law of the jungle"* would prevail under those conditions.

To investigate the working of the *"state of nature"* under the *law of the jungle*, let us assume many families of deer live in two valleys, out of three, in a mountainous region. Several families of lions also live on the same land. One can imagine grass growing in the three valleys and deer living by eating this grass in two of these valleys. Whenever more rains would occur, the grass would grow in abundance and there would be plenty of grass for the deer to eat. Thus, the deer population would increase if more than normal rains happen to occur and the population would tend to decrease with below normal rains. The lions would hunt the deer - rain or no rain. Under these conditions, we can see that the lions would be competing with each other in hunting the deer and the deer would be competing with

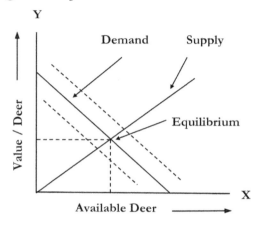

Figure 5-03 Change in demand

each other for grass, or we can say that the two valleys would compete with each other in production of deer. The deer would also be competing with each other for survival, since the death of one may mean the survival of another and vice versa, while being hunted.

Suppose the population of lions in the two valleys is the average population over a span of a few years and it is a period of average rain also over the same period of time. Then, suppose, during this period it takes a lion an average of three hours to hunt a deer, then the "value" of a deer is three hours of the lion's labor, under these conditions. Thus, the population of lions would tend to capture a certain number of deer based on their demand for deer meat under these conditions. If, due to some reasons, it takes less effort and time, on the average, for a deer to be hunted, then the "demand", or the quantity demanded, would increase and the lions would tend to catch more deer. Since, several factors would affect the demand for deer, like the desire for deer meat, the ease or difficulty of hunting other animals, the age and health of the lions in their population, the shape of the demand curve would not be a straight line, but for simplicity we can represent it so, sloping downwards, as shown in Figure 5-01. On the other hand, if it is more difficult to hunt deer, then more deer remain alive and available. If it is easy to hunt deer under the described conditions, then less deer tend to remain available and alive. Thus, the supply of deer from the two valleys tends to increase with the increase in their "value", or difficulty in hunting. Thus, the quantity supplied by the two valleys can be represented by a curve that slopes upwards, as shown in Figure 5-01. Like the demand curve, the shape of the supply curve is also not fixed, i.e., it is not a straight line. It would depend on several factors like the availability or scarcity of grass that the deer eat, productivity of female deer and the competition for resources with other species which grow on grass and other resources that the deer need. *Thus, the supply of deer and demand for them have opposite tendencies - the lions tend to kill as many as possible while the deer tend to resist being killed, as much as possible.* Other conditions being constant, these tendencies would reach an equilibrium at a certain "value" of deer, indicated as the point of intersection of the supply and demand curves in Figure 5-01, which shows how much effort or time of a lion is required, on the average, to hunt a deer.

If during the same span of a few years, mentioned before, there are plentiful rains, the deer population would grow faster and, thus, the supply curve would shift to the right as shown in Figure 5-02, indicating that more deer are available for hunting, for a certain "value" corresponding to the value during average rains. The opposite would happen if during this span of years, rains are below average. The value of deer would also fall if the same deer and lions begin to live in the third uninhabited valley. This would increase the number of valleys, or producers of deer. If more families of lions arrive in the same two valleys then the competition between the lions would increase for deer meat. Thus, the "value" of deer would increase, shifting the demand curve rightwards, as in Figure 5-03. A decrease in the lion population would have the opposite effect.

This is what happens in conditions of the "state of nature". This is also what is always happening around us - tigers and lions eating deer, deer eating grass, lions hunting and eating wildebeest, birds hunting and eating worms and insects and other animals eating each other. In addition to using non-living resources of the planet, most life forms live on other life forms, especially animals, continuously competing with each other and killing and destroying each other. This is the state of nature. Thus, the state of nature consists of a huge number of *"markets"* where the *law of the jungle*, or the *law of supply and demand*, reigns. If we look at the value of everything being hunted at a certain instant, then the set of these values at which the numerous "markets" have reached *"equilibrium"*, in what in reality is a *state of chaos*, is referred to, collectively, as the *"ecological balance"* of the planet. The important thing about these markets is that everyone is completely free to kill and destroy others, without regard to what is right or wrong. *In other words, these markets are completely free, and are devoid of any morality whatsoever.*

Over the course of our evolution we have learned that killing, raping, and robbing each other is not the best strategy for survival of a group, and that living peacefully, or at least non-violently, is a more practicable strategy, since this means security for all members of a group. So, the jungle has progressively become less free over time and what we have now, is a restricted market in which we have placed certain constraints on ourselves, as regards these and other similar activities, as far as our dealings with other human beings are concerned. But our behavior has not changed much as regards other living things on this planet. We certainly are the most vicious killers of other species at this time. Hundreds of millions of other animals and plants are killed by us daily for food, or for their furs, or just for "sport", and most of us do not feel any remorse about this.

The Market Mechanism

Several markets are affected by our changing behavior. These include markets related to our economic activities, our search for mates, i.e., our sexual activities, and what, now, we call sports and recreation activities. The entire evolutionary history of mankind, before the development of civilizations along river banks, is generally referred to as the *age of savagery*. As indicated before, this is a very long period over which mankind evolved from existence as families in the hunter-gatherer stage to extended families, to nomadic tribes to tribes engaged in agriculture. Throughout this period, men killed each other and robbed each other of whatever fruit and animals had been gathered by a family or a tribe, or families destroyed each other over possession of females. Thus, the *economic market* of products was totally free and no restrictions were imposed by the social groups about how products were acquired or produced. Similarly, there were no rules how females were acquired and possessed by men. In later stages of the development of tribes, when one tribe was killed by another tribe, the young women and children, of the destroyed tribe, were not killed but were enslaved by the conquering tribe. Thus, at this stage the *sexual market*, or the mating market, was also totally free and men could acquire mates by consent, or force, including by murdering other men, and taking over their mates. In earlier state of its development, the *labor market* was also controlled purely by the law of the jungle and was totally free. Thus, people could be captured and enslaved freely. They could then be forced to work for a minimum of food, needed for their continued existence.

The conditions of the various markets remained the same during the era of feudalism, as they were during the age of savagery. However, as the *group survival skills* developed, it became clear that murdering each other in a social group was not a permissible activity. A group cannot survive and breaks up if there is no sympathy, courtesy, affection, or loyalty to the group among its members. These are *group survival skills* that mankind has developed to create and maintain bigger and bigger social groups. Similarly, theft and robbery came to be regarded as not permissible activities. As these restrictions were adopted by societies, the markets also became restricted markets and their rules also changed. Enslavement of men and women continued till very recent times, and slaves and concubines were a norm in many societies only a few hundred years ago. Although polygamy still persists, this practice and slavery are generally looked down upon now. Even child labor has come to be regarded as an abusive activity.

In most species of animals, children learn basic survival skills by engaging in mock combat, in racing and maneuvering. These "sports" help to train the young individual for their future struggle for survival. Human beings do the same in the beginning of life of a new individual. Then the process becomes more complicated. In the age of savagery, the "games and sports" included physical competition, and competition in marksmanship with bows and arrows, and other elementary weapons, as they were developed by mankind. This activity can be called "sport" with another objective. Similarly, horse, boat and chariot races were engaged in at various stages of evolution of our social groups. Singing and dancing in war and celebration of victory, caricature of public figures, are other such activities. With time, we have given up the more violent and abusive "sports" like gladiator fights, but some degree of violence remains in the *"sports and entertainment markets"* today. Some kinds of fist fights, like boxing are still considered sports, so are free-style wrestling and "kushti" (wrestling South

Asian style). These sports and singing attract large numbers of viewers and are among numerous entertainment markets.

Thus, the *economic market,* including the *labor markets,* the *sexual markets* and *sports and entertainment markets,* have all become restricted markets. However, abusive relationships continue in all these markets and, as civilization develops, it is becoming clear that more constraints need to be imposed on them, so that societies may have more social justice and less freedom of abuse.

The sexual market, or the *market of inter-gender relations,* needs to be explored further to identify the abusive relationships that continue even at this time. The basics of how and why the two genders came into being and their traditional roles, on an evolutionary scale, have been described in Chapter-1. Also, the basic sexual attitudes of the two genders have been described there. We can see that the bodily make-ups of the two genders reflect the traditional roles they have played during the evolutionary struggle of human beings and how sexual selection has affected them. Men are, in general, physically strong because they have been responsible for the security of the family and tribe. They have also learned other survival skills needed in competition with other men, like skills of hunting, climbing, running and warfare in general. Men have learned to control their emotions and to plan and act to protect their families, even under extreme stress. Women have learned to be more emotionally intelligent in dealing with men, especially, in determining who is a reliable partner and who is not. They have also learned to be patient in face of stress. Women have traditionally spent time preparing food, having children and looking after the offspring. They have not faced physical competition like men do. Men have faced competition in other activities, like finding the best mates and as many mates as possible, where mainly physical toughness and agility are involved. Thus males, from generation to generation, got continuously physically challenged, in the early stages of evolution. It makes sense that, at those stages, the more muscularly developed males would survive the challenge. The females most likely selected the most physically tough males, thus the more physically tough the male was, the better his chances of surviving, and making sure his children survived. Similarly, the more physically tough males gathered more females than less physically tough males and passed on their genes to the next generation this way. With time, however, the tendency of the females to look for reliable partners has become an increasingly more important factor. But, still, men have continued to be physically stronger. This has given more power to the males over females.

When human beings settled down to life on river banks, engaged in agriculture, the mode of production changed drastically and slowly women also became involved in agriculture. But, still, the imbalance of, perhaps, millions of years continued and is still continuing. In Europe and North America, the arrival of the industrial age, brought a revolutionary change to society and its basic mode of production. As shortages of labor developed due to the demands of the factory system, or war, women were forced out of their families and into the production process. Patience in the face of repetitive jobs becomes a skill, or an advantage, men being much less patient. So, professions requiring patience, like large scale manufacturing, nursing and the health care industry in general, attracted more women. This has brought a major change in the balance of power between the two genders.

We are, now, in what is called the information age. The industrial or manufacturing age is behind us. And now, in socially advanced societies, production is mostly becoming a mental process. Everything is becoming more and more mental, and computerization and automation are taking over all economic activities. So that the differences of productive abilities in the workplace, between men and women, are becoming smaller and smaller, except for the basic biological differences and the disproportionate burden of reproduction that women still have to bear. We would take a deeper look at inter-gender relations in later chapters.

The Market of Goods and Services

We have discussed the concept of the market and how markets have emerged in the process of human evolution. As we indicated in Chapter-1, markets exist in all aspects of life on earth and in all aspects of human activity. Basically, when we refer to the economic market, we are referring to markets of various products and also to the labor market which is the foundation of all economic production.

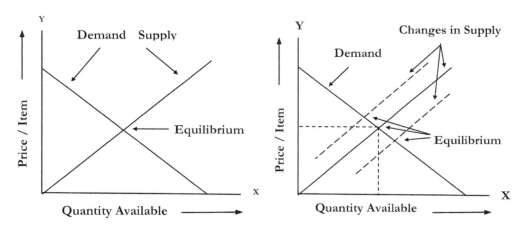

Figure 5-04: Price Determination **Figure 5-05: Change in Supply**

The only thing that is different between the markets existing in a state of nature and the free economic market is the presence and use of money in the transactions that take place in the market. One of the assumptions in the definition of a market is that there are numerous buyers and sellers, so that buyers, or sellers, cannot arbitrarily dictate the price for a product. Also, it is assumed that all buyers know what price all sellers are asking for. These assumptions are never fulfilled in a real market. The market remains a jungle, part of which remains unknown to the buyers and, even, the sellers. Thus, some sellers may be able to dictate prices to buyers, some times. Also, in a market, we do not exchange products on the basis of their value but on the basis of price in terms of money. The price increases if the number of the products in the jungle, or the market, decreases. The opposite happens if the number of products increases, which may happen as result of an increase in the number of producers, or increased production by the existing producers. If the market is completely free, then the price of a product is also determined by the demand for those products. The increase in demand does not, however, mean that the need for the products increases. It means that the consumers or the buyers of the products have more money to buy the products. Thus, increase in the number of buyers, or the increase of money in the hands of existing buyers may have the same effect. Demand refers to the ability to buy, for those buyers who have sufficient money to be able to buy products to meet their own needs.

The increase in the numbers of specific products offered for sale in a market may be due to a number of factors, the factors being unpredictable. Similarly, the increase in demand may be due to a number of unpredictable factors. For example, rain may have an effect on production of some agricultural products. Thus, demand and supply of the products is in a state of constant chaos. In addition to supply and demand, money that becomes available to buyers and suppliers is also constantly changing. Thus, the price of one kind of product is determined by a number of chaotic factors. It does tend to reflect the value of those products but fluctuates wildly around this value.

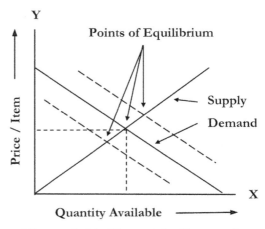

Figure 5-06: Change in Demand

Thus, the so-called law of supply and demand is the law of the jungle in which supply and demand both are in a state of chaos as befits a jungle. The price is the equilibrium where the demand and the supply curves intersect, as described before. The "equilibrium" is, however, constantly shifting. All that the sellers are doing is to grab as much money from the buyers as they can, and the buyers want as much as they can get by paying the least that they have to. So, a constant struggle goes on over prices of products, between buyers and sellers in the market, or the jungle.

Figures 5-04 to 5-06 show the working of the "Law of supply and demand". The operation of this law is very similar to the "Law of the Jungle", as described before. Demand is the inverse relationship between the price of a product or service and the quantity that consumers are willing and able to buy during a given period, all other things being held constant. When the price of a product or service rises, the quantity sold goes down. When the price of a product or service falls, the quantity sold goes up. Thus, the demand curve has a negative slope, i.e., if price is plotted on the vertical axis and the quantity on the horizontal axis, then the demand curve slopes downwards to the right.

Demand for a product depends on several factors, including the income and number of consumers, prices of other goods - even weather. Thus, the demand curve is not linear, but can be represented this way for simplicity. Supply is the relationship between the price and the quantity of a product that producers are willing to offer, at that price, during a given period. Producers tend to offer more quantity at a higher price. Thus, supply can be represented by a curve that slopes upwards on the same axes as for the demand curve. Figure 5-04 shows the equilibrium price as the point of intersection of the demand and supply curves. Figure 5-05 shows what would happen to the price if the supplies become more or less abundant, because of more or less producers entering the market or the current producers producing more or less in competition, causing a rightward or leftward shift of the supply curve, respectively. The same effect would result if the price moves up or down, respectively. Figure 5-06 shows similar effect of the drastic increase or decrease of demand, because of more or less consumers entering the market or the current consumers having more or less money to spend, causing a rightward or leftward movement of the demand curve, and the consequent effect on the equilibrium price. The same effect would result if the price moves down or up, respectively.

In earlier stages of its development of a free market, robbery, piracy and theft were also free to occur, so that supplies to the market could be freely diverted from specific suppliers to specific "consumers". In a free market, production of goods had to be completely free also, so that deception in production and trading could also be totally free. In modern times these freedoms have been curtailed although they still do occur to some extent.

The Labor Market

Along with markets of goods and services, a market in human labor also exists. In earlier stages of its development, the labor market was also controlled purely by the law of the jungle and was totally free. Thus, people could be captured and enslaved freely. They could then be forced to work without wages. Thus, human beings could be purchased like commodities. In fact, their ability to work, or labor-power over a lifetime was being purchased. They were kept alive by their masters so that their ability to work could be put to maximum use, without any wages, except provision of subsistence living. Formal slavery has been abolished in most societies. It does exist at this time, however, though mostly in hidden or semi-pure forms.

"Consequently, it appears that the capitalist buys [the workers'] labor with money, and that for money they sell him their labor. But this is merely an illusion. What they actually sell to the capitalist for money is their (maximum ability to work, or) labor-power."

**Karl Marx (1845), Wage Labor and Capital,
www.marxists.org/archive/marx/works/1845/theses/index.htm, accessed October 17, 2019.**

Thus, the labor market is no longer a free market as it used to be in the past. In recent times, in some developed states, laws have also been enacted to protect workers from dangerous chemicals and health hazards, dangerous working conditions, long hours of work, etc. Also, minimum wages and maximum working hours have been fixed by law. Thus, free labor markets have been subjected to restrictions in the more civilized states, and are no longer totally free. However, exploitation of workers continues, by payment of only part of the value of their labor. This exploitation occurs because of the Law of the Jungle as applied to the labor "market". The labor markets in the underdeveloped states continue to be virtually free and highly-free abuse of workers continues in those states, at this time.

"And exploitation, the appropriation of the unpaid labor of others, has quite as often been represented as the reward justly due to the owner of capital for his work; but never better than by a champion of slavery in the United States, a lawyer named O'Connor, at a meeting held in New York on December 19, 1859, under the slogan of 'Justice for the South.'

'Now, gentlemen,' he said amid thunderous applause, 'to that condition of bondage the Negro is assigned by Nature... He has strength, and has the power to labor; but the Nature which created the power denied to him either the intellect to govern, or willingness to work.' (Applause.) 'Both were denied to him. And that Nature which deprived him of the will to labor, gave him a master to coerce that will, and to make him a useful... servant in the clime in which he was capable of living useful for himself and for the master who governs him... I maintain that it is not injustice to leave the Negro in the condition in which Nature placed him, to give him a master to govern him ... nor is it depriving him of any of his rights to compel him to labor in return, and afford to that master just compensation for the labor and talent employed in governing him and rendering him useful to himself and to the society."

New York Daily Tribune, November 20, 1859, pp. 7-8., www.marxists.org/archive/ accessed October 2012.

Chapter-6

Intensive Capitalism

Intensive Socialization of Production

Advanced feudalism in Europe began to change in the eighteenth century. A new *mode of production* began to take hold. The factory appeared as an exciting new way of accumulating wealth for the already rich. While agricultural production required large areas of land for production of food and other raw materials, a factory could be established on a, relatively, small piece of land. This, combined with concentration of large numbers of workers and the use of newly discovered machines, made it possible to generate large amounts of capital, or profit, on the factory floor in a relatively short period of time. Since large numbers of workers were involved in such factories and groups of workers were involved in producing parts of the final products that came out of a factory gate, one person could not claim to have produced the finished product or even a part of it. The product, thus, was the product of a social group and production of this system can, thus, be described as *intensely socialized production,* since socialized production on much smaller scale had existed even in the economic system of *feudalism.* We would discuss the working of such factories, mines, or agrobusinesses, etc., or "production centers" in more detail in the next chapter, where the economics of this new system would be analyzed. Distribution of products, for ultimate consumption by other factories or individuals, was another activity that changed dramatically during this period. Distribution involved transport of products over short, or very long, distances to the final consumers, or intermediate storage centers, from which the products had to be transported further to the ultimate consumers. The final consumers could be individuals buying goods from stores, or shops, or the products could be machines needed by other factories for use in the process of production itself, in which case they were delivered to such factories.

"Every great city has one or more slums, where the working-class is crowded together. True, poverty often dwells in hidden alleys close to the palaces of the rich; but, in general, a separate territory has been assigned to it, where, removed from the sight of the happier classes, it may struggle along as it can. These slums are pretty equally arranged in all the great towns of England, the worst houses in the worst quarters of the towns; usually one- or two-storied cottages in long rows, perhaps with cellars used as dwellings, almost always irregularly built. These houses of three or four rooms and a kitchens form, throughout England, some parts of London excepted, the general dwellings of the working-class. The streets are generally unpaved, rough, dirty, filled with vegetable and animal refuse, without sewers or gutters, but supplied with foul, stagnant pools instead. Moreover, ventilation is impeded by the bad, confused method of building of the whole quarter, and since many human beings here live crowded into a small space, the atmosphere that prevails in these working-men's quarters may readily

be imagined. Further, the streets serve as drying grounds in fine weather; lines are stretched across from house to house, and hung with wet clothing."

Friedrich Engels (1845), *Condition of the Working Class in England in 1844*. Translated by Florence, Kelley, Wischnewetzky. Urbana, Illinois: Project Gutenberg. Retrieved October 15, 2019, from www.gutenberg.org/ebooks/17306.

Initially, using agricultural raw materials, and later on, other materials obtained more and more through fishing, cutting of forests, or mining, production began to move to the towns and cities. At the same time, surplus agricultural workers were also pushed towards the towns by their economic needs, as their small land holdings, and traditional lands for common use, were taken over by big landlords. Later on, mechanization of farms made them surplus and their labor unnecessary for agricultural production. This created ideal conditions for exploitation and abuse of these workers, who had no homes and who desperately needed to earn to stay alive and to feed their families.

"Since capital, the direct or indirect control of the means of subsistence and production, is the weapon with which this social warfare is carried on, it is clear that all the disadvantages of such a state must fall upon the poor. For him no man has the slightest concern. Cast into the whirlpool, he must struggle through as well as he can. If he is so happy as to find work, i.e., if the [ruling class of owners of business and industry] does him the favor to enrich itself by means of him, wages await him which scarcely suffice to keep body and soul together; if he can get no work he may steal, if he is not afraid of the police, or starve, in which case the police will take care that he does so in a quiet and inoffensive manner. During my residence in England, at least twenty or thirty persons have died of simple starvation under the most revolting circumstances, and a jury has rarely been found possessed of the courage to speak the plain truth in the matter. Let the testimony of the witnesses be never so clear and unequivocal, the [ruling class of owners of business and industry], from which the jury is selected, always finds some backdoor through which to escape the frightful verdict, death from starvation. The [ruling class of owners of business and industry] dare not speak the truth in these cases, for it would speak its own condemnation. But indirectly, far more than directly, many have died of starvation, where long-continued want of proper nourishment has called forth fatal illness, when it has produced such debility that causes which might otherwise have remained inoperative brought on severe illness and death. The English working-men call this "social murder", and accuse our whole society of perpetrating this crime perpetually. Are they wrong?"

Friedrich Engels (1845), *Condition of the Working Class in England in 1844*. Translated by Florence, Kelley, Wischnewetzky. Urbana, Illinois: Project Gutenberg. Retrieved October 15, 2019, from www.gutenberg.org/ebooks/17306.

As this new mode of production took hold, the population of the cities grew very fast and the wages of workers fell because of the large numbers of unemployed and the competition between them. The factory owners were able to take advantage of their own power, and the powerlessness of the workers, and were able to make them work for twelve hours or more a day, seven days a week, and to force their wives and children to work also. The wages that were paid to such families were not sufficient even for subsistence.

This new system of production came to be known as "Capitalism", since it resulted in exploitation, and the resulting capital formation and accumulation, at a rate never seen before in history. Thus, it can be described as a system dominated by capital and owners of capital, although this had been the case, to a lesser degree, since the beginning of civilization. The main characteristics of this system, and the new mode of production, were the concentration and socialization of production only, ownership remained in the hands of individuals, i.e., individual industrial owners, who mainly were the former feudal landlords and aristocrats. In the beginning of this change, ownership was not socialized even within the ruling class. The socialization of production, in time, spread further in society, leading to further specialization of labor and intensification of exploitation, within a society which was largely feudal at the time.

> **"When one individual inflicts bodily injury upon another such injury that death results, we call the deed manslaughter; when the assailant knew in advance that the injury would be fatal, we call his deed murder. But when society places hundreds of [destitute working people] in such a position that they inevitably meet a too early and an unnatural death, one which is quite as much a death by violence as that by the sword or bullet; when it deprives thousands of the necessaries of life, places them under conditions in which they cannot live – forces them, through the strong arm of the law, to remain in such conditions until that death ensues which is the inevitable consequence – knows that these thousands of victims must perish, and yet permits these conditions to remain, its deed is murder just as surely as the deed of the single individual; disguised, malicious murder, murder against which none can defend himself, which does not seem what it is, because no man sees the murderer, because the death of the victim seems a natural one, since the offence is more one of omission than of commission. But murder it remains. I have now to prove that society in England daily and hourly commits what the working-men's organs, with perfect correctness, characterize as social murder, that it has placed the workers under conditions in which they can neither retain health nor live long; that it undermines the vital force of these workers gradually, little by little, and so hurries them to the grave before their time. I have further to prove that society knows how injurious such conditions are to the health and the life of the workers, and yet does nothing to improve these conditions. That it knows the consequences of its deeds; that its act is, therefore, not mere manslaughter, but murder, I shall have proved, when I cite official documents, reports of Parliament and of the Government, in substantiation of my charge."**
>
> **Friedrich Engels (1845), *Condition of the Working Class in England in 1844.* Translated by Florence, Kelley, Wischnewetzky. Urbana, Illinois: Project Gutenberg. Retrieved October 15, 2019, from www.gutenberg.org/ebooks/17306.**

The new system brought seeds of social change with it. Workers, from far flung areas in a state, were being brought together on the workshop floor. They could exchange knowledge of their experiences with each other. The hold of feudal ideas, including religion, began to be broken. Also, sympathy with, and reliance on, each other grew and the industrial workers began to feel the power of being together, to face the abuse of the ruling class of industrial owners. Their political and social awareness grew and so did their class-consciousness. Families of the workers effectively became groups of slaves engaged in a struggle for survival. Both men and women had to work to survive on meager wages and the children had to work too. Thus, for these workers, all semblance of family life began to disappear.

The concentration of people in the cities increased the risks of spread of infectious diseases, the risks being further increased because of lack of sanitation in their living quarters near factories and towns. Diseases of epidemic proportions, like the Great Plagues, began to break out. The life expectancy of the population became shorter and shorter. Life expectancy in England, for example, had gone down to about 35 years in the second half of the nineteenth century. About half of the number of children were dying before reaching the age of 13 years. The working people had become virtual slaves. The ruling class did not care if workers died, and a large-scale social murder of the population was in progress. All that the ruling class wanted was an ample supply of workers to man its factories. More workers were available than the manufacturing factories could employ and the extra supply of laborers from the countryside, was "welcome", as it was lowering wages to below-subsistence level. So, the deaths of some of the workers did not matter, from the point of view of profit-making and capital formation. Thus, the wages paid to workers, at that time, can accurately be described as *starvation wages* and the workers themselves as *wage-slaves*.

"The [class of industrial owners], historically, has played a revolutionary part. The [class], whenever it has got the upper hand, has put an end to all feudal ties that bound man to his "natural superiors", and has left remaining no other nexus between man and man than naked self-interest, than callous "cash payment". It has drowned the most heavenly ecstasies of religious fervor, of chivalrous enthusiasm, of philistine sentimentalism, in the icy water of egotistical calculation. It has resolved personal worth into [price], and in place of the numberless indefeasible chartered freedoms, has set up the single, unconscionable freedom -- free trade. In one word, for exploitation, veiled by religious and political illusions, it has substituted naked, shameless, direct, brutal exploitation. The [class of industrial owners] has stripped of its halo every occupation hitherto honored and looked up to with reverent awe. It has converted the physician, the lawyer, the priest, the poet, the man of science, into its paid wage-laborers. The [class of industrial owners] has torn away from the family its sentimental veil and has reduced the family relation to a mere money relation."

Karl Marx & Friedrich Engels (1848), *The Communist Manifesto.* **Urbana, Illinois: Project Gutenberg. Retrieved October 15, 2019, from www.gutenberg.org/ebooks/61.**

The system of intensive capitalism took about three hundred years to become the dominant economic system of Europe, North America and Japan. It reached its zenith at the end of the nineteenth century. During this period the societies of the two continents were completely transformed. The extreme overflow of labor and population began to decrease as some of the working people emigrated and the rest were killed off by the system at an increasing pace. This *intensive death rate* reached its peak when, ultimately, intensive capitalism became the dominant system in society. The population in the countryside had decreased steadily and large urban centers had developed. Women had to join the labor force in increasing numbers.

> **"But with the development of industry the [class of industrial workers] not only increases in number; it becomes concentrated in great masses, its strength grows, and it feels that strength more. The various interests and conditions of life within the ranks of the [industrial workers] are more and more equalized, in proportion as machinery obliterates all distinctions of labor, and nearly everywhere reduces wages to the same low level. The growing competition among the [industrial owners], and the resulting commercial crises, make the wages of the workers ever more fluctuating. The unceasing improvement of machinery, ever more developing, make their livelihood more and more precarious; the collision between individual workmen and [industrial owners] take more and more the character of collisions between two classes. Thereupon the workers begin to form [trade unions] against the [industrial owners]; they club together in order to keep up the rate of wages; they found permanent associations in order to make provision beforehand for these occasional revolts. Here and there the contest breaks out into riots."**
>
> Karl Marx & Friedrich Engels (1848), *The Communist Manifesto*. Urbana, Illinois: Project Gutenberg. Retrieved October 15, 2019, from www.gutenberg.org/ebooks/61.

Growth of population slowed down due to the high death rate and the economic system began to go through increasingly violent periods of boom and bust. Even then, the surplus of labor did not become much less so, even in times of boom and high employment. European powers continued to go to war with each other for control of their imperial possessions and colonies they had set up on other continents. The conflicts between the European powers became acute in the beginning of the twentieth century and huge global conflicts developed over possession of resources of the colonies. In the course of this period of about two centuries, several violent upheavals occurred. The *French Revolution, in 1789,* brought a new class of industrial owners to power in France and this process spread to rest of Europe in the following decades. In 1871, the *Paris Commune* was created as a result of a rebellion by the people of Paris against the rulers of France. These events would be described in some detail in the following sections, as they constitute important milestones in the history of capitalism.

Empires in the Era of Intensive Capitalism

Intensive capitalism, during the course of its evolution, led to an enormous increase in production and concurrently brought wages down to subsistence levels, thus tremendously increasing the already excessive profit, or tribute extracted by the industrial owners from the people. As the new mode of production spread throughout Europe, huge numbers of agricultural workers found themselves unemployed due to mechanization of agriculture, causing the demand for the increased production to fall. The owners of factories needed new markets to dump their goods in. Foreign lands were violently attacked and occupied, and their people were subjected to mass murder, enslavement and callous exploitation. Huge numbers of subjugated people were slaughtered and Europeans were moved to the new colonies. These new methods of imperial abuse continued till the middle of the twentieth century.

European Imperial Expansion

Socialization of production and the intensive exploitation of labor had produced a further impetus towards domination and abuse of other states and their populations, because what could be done on home territory could be done even "better", with no fear of criticism and opposition, to the people of foreign lands, captured and enslaved by force. Empires had existed since the early stages of human civilization and the only motivation of these empires was the loot and plunder of other peoples, by extraction of "tribute". In the eighteenth century, European states had continued the "tradition" of establishing and expanding their empires. Large areas of Asia, Africa, Australia, North America and South America had been conquered. Most of the people of these continents had been killed and new colonies were established there and people from the European states had been moved to these continents. Later on, when labor shortages developed in the colonies, the enslaved people from Africa were moved there to meet the demand for labor to build the economy of the new states. North and South America, South Africa, Australia and New Zealand were the areas where control of land and agriculture was the primary objective at the earlier stages of colonization. Also, Eastern Africa was colonized to some extent. For example, the British Empire, at its zenith, controlled about one quarter of the global human population and about a quarter of the land area of the planet. This genocidal process continued till the middle of the twentieth century.

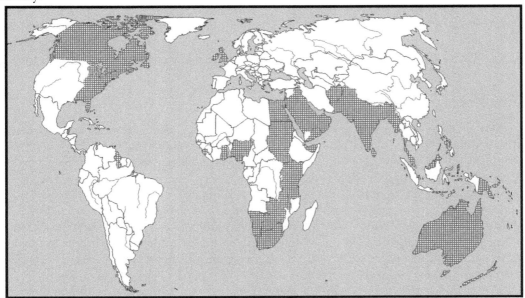

Figure 6-01: Major Territories Incorporated into the British Empire at One Time or Another – includes Bantustans (referred to as "Protectorates", etc.)

(Map based on world map at: https://www.d-maps.com/carte.php?num_car=13184&lang=en)

The acquisition of new colonies did not have much effect on the conditions of life of the people of Europe, since the movement of people from Europe to the colonies was a slow process and the movement of the people from the countryside to the towns in Europe was much faster. Thus, the concentration of the population in the towns and cities of Europe continued to keep wages down and the rate of intensive profit (*RIP*) within Europe, and the rate of capital formation, remained very high. However, with time the death rate of the population in the cities began to rise, because of the concentration of large numbers of rural families in the cities and the under-nourishment, malnutrition and the terribly cramped and unhygienic conditions under which they were forced to live, leading to break-outs of many diseases and exposure of the population to extreme weather conditions. Due to these conditions, ultimately the birth rate of the population in the cities also went down. Naturally, families of workers under all kinds of economic stress could not afford to have many children, when the children they had were dying at an increasing rate. Ultimately, as industry grew, coupled with the deaths of millions of people in the European states, the surplus labor in the market began to decrease, but this did not have any significant effect on wages till around the beginning of the twentieth century, when the death rate had increased further, as about seventy million people were killed in the two world wars and other conflicts which occurred in the period between those wars. Only then workers were able to resort to more and more strikes and demand higher wages.

As the European empires expanded, the methods of exploitation also changed and became more sophisticated. The imperial powers adopted more violent methods to destroy local industry in the captive areas and states and these subjugated peoples were pushed into more abusive economic relationships with the imperial states. This process was aided by the improvements in communications and transport. At the end of the nineteenth century, only a few European states, Japan and the United States remained independent states and the rest of the world was being abused one way or another by these states. Mankind had never before seen such pain, suffocation and darkness.

The French Revolution

Social change is a very slow evolutionary process and, in general, social changes do not take place suddenly. The ruling class may feel the need for some changes in social relationships as desired by the people. In such cases, a change may occur gradually and peacefully. But if the need for a change is acutely felt by the people but the ruling class resists such a change, the frustrations of the people may accumulate over a long period of time and the people may rise in a rebellion ultimately. If the required changes are not expected to fundamentally alter the relationship of the ruling class and the people, then the changes may occur smoothly, leaving the ruling class in its position of power. But if the changes are of a fundamental nature, especially if they impact the economic relationships of the two classes, then the ruling class may be completely overthrown as a result of the rebellion. Such a dramatic change is referred to as a social and political revolution. Thus, a revolution may be looked at as "bottled-up" evolution resulting in a major social explosion once social pressures have become overwhelming, resulting in a complete overthrow of the existing ruling class and its replacement by another.

In 1789, a powerful social explosion rocked France. In the next few years, the French revolution removed the aristocratic class from absolute power. It was the first such event in Europe. Instead of the aristocrats, technocrats like lawyers, accountants, engineers, administrators and owners of industry came to power. This class began to sweep the old ruling class aside and became the new ruling class of France. Feudal aristocrats continued to hold some power in French society, under the control of capitalists, but lost it almost completely over the next century. The same happened to the Church, which lost its huge land-holdings and its position to the new class. France was the first state in Europe to reach this stage, i.e., a stage wherein the new mode of production had become the dominant mode of production and where the new industrial ruling class had become the dominant class. Other states of Europe followed France in this development. The monarchy in France was changed into a constitutional monarchy, the king was killed and ultimately Napoleon sized power and brought war and more violence to the rest of Europe.

The revolution had brought dramatic changes in the whole social system. The French revolution is considered a milestone in the development of the economic and social system of the world. It is considered to be the beginning of the dominance of the new system of *intensive capitalism*, in Europe, which brought great developments of industry and commerce to France, along with impoverishment and misery for its people.

The Paris Commune

Many social explosions of different intensity occurred in Europe during the nineteenth century. In 1871, a violent social explosion rocked Paris, after the siege of the city by German forces. In the first quarter of that year, during fighting between the forces of Germany and those of the French regime, a power vacuum had developed. The French regime had withdrawn to Versailles, but the German forces only occupied a small territory close to Paris because of the higher priority matters related to German unification. People of Paris, including the members of the National Guard, rebelled on March 18, 1871, and organized themselves into a new National Guard and declared war on the French regime in Versailles.

On March 26 the Paris "Commune" was elected. Members of the Commune were chosen by universal suffrage. Several organizations, or political parties, elected their representatives to rule the city and the city parliament, composed of these representatives, began to function in a democratic way. Majority of the members were, naturally, working men, or acknowledged representatives of the working people. On the same day all foreigners elected to the Commune were confirmed in office, based on the idea that "the flag of the Commune is the flag of the World Republic"! The Commune was both an executive and a legislative body.[1]

The Commune abolished conscription, the standing army and the police. All educational institutions were made accessible to the people for free education without any religious interference. Judges were to be elected by the people, like the other members of the Commune. Arrangements were made for rents being paid by working people and all pawnshops were closed down. It was decided that the salary received by any employee of the Commune was to be the same as that of any other worker. Separation of State and religion was decreed, and all religious property was declared national property. The guillotine was brought out by the National Guard, and publicly burnt, amid great popular rejoicing. Factories were ordered to be restarted by their employees, as co-operative societies. Night work of bakers was abolished.[2]

In the middle of May, 1871, the army of the French regime, which had retreated to the West of the city under pressure, returned and violently re-imposed the rule of the capitalists, and their fellow feudalists of the time, on the city. When the city had been recaptured by the capitalists, its streets were lined with tens of thousands of dead bodies of its citizens, but it left a lasting impression on those who were working for social democracy at that time. More than thirty thousand of inhabitants of Paris had been killed. About thirty-eight thousand workers were taken prisoner.[3]

The Paris Commune lasted about two months. In its brief life, it had demonstrated what it means to have a *"government of the people by the people"*, *as Marx put it.* This was a form of government that could not be removed, on the basis of popular vote, by any party belonging to the ruling capitalist and feudal classes. It could, thus, be referred to as a kind of "dictatorship of the people".[4]

Marx had renamed his group as a 'communist" group and used this term when the "Communist Manifesto" was issued in 1848. He studied the events that Paris went through in 1871 and reconfirmed that the "state", which had been set up in Paris then, i.e., the Paris Commune, was the ideal of his group of "communists", who had been formerly known as "Social Democrats".

The Great Genocide

The process of European expansion, and colonization of the newly discovered continents of Australia and North and South America was accompanied by immense levels of violence and bloodshed. As colonies were established in North and South America, the native tribes were systematically exterminated. Lands were cleared in stages and European settlers and colonizers were moved into those areas under the protection of the military. At every stage the native people resisted the advance of colonization but they were primitive in the level of their economic and military evolution, all being in the tribal state of development with virtually no scientific skills. Although it was possible to reach an accommodation with the natives, it was never seriously tried and the resort to force was considered the only "solution" to all problems. Some temporary tactical concessions were made as a matter of strategy but virtually no agreements arrived at were respected and the process of slaughter of all individuals, including women and children, was moved forward region by region. In the United States, with a constitution that talked of the equality of all men as created by their Creator, laws were passed by some states which offered rewards for killing of native men, women and children! The second amendment of its constitution guaranteed the "right to bear arms" for its citizens, so that they could carry out this "heroic" task efficiently!

Some natives also died because of the new diseases brought along by the Europeans, and this is always stressed by all those who write history from a European point of view, but this was, obviously, not the main cause of the deaths of whole tribes. Some native tribes did survive, in North, Central and South America and these were those who had the good fortune to be dwelling in the jungles, deserts and mountains of the Americas. Obviously, ultimate contact with the Europeans did not kill these tribes. They have survived and grown in population. They, and people of mixed native and European race, now form about one third of the population of the forested and mountainous states of Central and South America. In North America, hardly a quarter million natives, or people of mixed race, have survived.

The genocidal activities of the Europeans, also, brought death and destruction to the tribes of South Africa and Australia and even to others who happened to be living in a climate not considered desirable by the Europeans. It is difficult to estimate the total number of deaths during this genocide, but one can easily see, that if the natives had been allowed to live, they would have probably numbered more than a quarter of mankind at this time. The *Great Genocide* ultimately led to the expansion of "Europe" into almost four and a half continents - Europe, North and South America, Australia and almost one-third of Asia. This "Greater Europe" is, now, euphemistically referred to as the *"West"*. Never before, in the history of mankind, a genocide of this magnitude had ever happened, and it is to be hoped, would never happen again. This tragedy would always remain a burden on the conscience of mankind.

The Imperial System of Exploitation

As empires developed and virtually covered the whole world, four techniques were used by modern imperial power for abuse of the people of the world. These are briefly described in the following sub-sections.

Colonization

This was the first choice of the imperial powers. North America, South America, Southern Africa, New Zealand and Australia were subjected to colonization. The native people were systematically killed and the land was "cleared" for settlement of people from the imperial powers. Virtually all the native people of North America were slaughtered and most of the people of South America also met the same end. Many tribes of Southern Africa were murdered along with many tribes of aborigines of Australia and New Zealand. New settlements of people from Europe were set up in these lands. The colonies grew in size slowly and became independent states in the later stages of this era. The United States, Canada, Mexico, Brazil, Chile, Argentina, and other South American states, South Africa, Australia and New Zealand, are some of the states which came into being by this process.

Initially, the mass killings of native people proceeded with great speed and the European settlers were brought to the colonies to settle there, but, later on, it became clear that this was an "economic" mistake. The new colonies were so huge in area that sufficient Europeans could not be found for this purpose. Europeans who did not have much capital could not move to the new colonies. Only those with some "resources" could. Thus, to deal with the situation created by one crime, another crime was undertaken as the "solution" - and the enslavement of African people started. Slaves were transported from Africa to the United States and Brazil, etc., to replace the human beings who had been killed. Also, some colonial powers chose not to kill all the native people. They, only, settled Europeans on lands with a desirable climate. If the natives were inhabiting fertile land in such colonies, then they were pushed off those lands, as the lands had suddenly become the Imperial Power's property. "His Majesty's property", is how the British referred to those lands, e.g., in "British East Africa".

Local people were forced to prepare habitations for the colonizers and to perform any tasks assigned to them by their masters. They were subjected to all kinds of abuse in this process. In the Congo, the Belgians used to cut off hands and arms of those who disobeyed. In Australia, the natives were shot and forced to jump to their deaths. The British shot and hanged those who resisted. Later on, when resistance to slavery and its cruelties became unmanageable, a process of movement of people from other imperial possessions to colonies was started and workers were brought, mainly from China and India, on contract, to meet the labor demand in the colonies, such as British Guyana, Fiji, etc. This process was changed, around the middle of the twentieth century, to a system of general immigration. Some imperial powers, even at this stage, tried to restrict immigration to white people of European origin, but, later, due to competition between the colonies for such immigrants, had to liberalize rules to allow for people of other origins, although the preference for European settlers remains, even today.

Bantustanization

During the establishment of empires, some imperial powers found that cost-effective exploitation of a conquered territory was not feasible because its terrain was not suitable for agricultural development, or it had a harsh climate for settlement. Even if the territory had the desired mineral resources, it could still be considered unsuitable for colonization for the same reasons. Sometimes, it was simply too costly to directly rule a tribal population, because it could offer stiff resistance to occupation due to the hilly terrain of its territory. In such cases, the imperial powers set up "tribal homelands", or "Bantustans", for such tribes on their traditional tribal territory or some other non-productive land, and let the tribal chiefs and their families rule the territory, in return for privileges to be provided to those imperial powers to use the tribal manpower, or extract the required mineral resources, as "tribute". Sometimes, as in the case of the racist regime of South Africa, it was considered desirable to deny citizenship rights to the members of a tribe. To achieve this objective, such a tribe was declared to be self-governing and "independent" – and, hence, not part of South Africa! If more than one tribe existed on a territory to be converted into a Bantustan, then the most powerful tribe was chosen for this "special treatment". This Machiavellian scheme, based on a "community of interests" between the ruling family of a tribe and its imperial power, was widely used by Britain and other imperial powers like the United States, South Africa and Australia.

South Africa, which was originally a Dutch colony, saw widespread, genocide of its native people. Whole tribes had been slaughtered by the colonizers. By the time Britain grabbed this colony from Holland, it had already set up three additional homelands in Southern Africa - Bechuanaland, Basutoland and Swaziland, which were basically used as reservoirs of labor, to be exploited by South Africa. After South Africa was declared an independent state, dominated by a racist regime, the new regime set up more than ten Bantustans, or "homelands", for the native tribes on the territory of South Africa, in order to deny citizenship rights to tribal people, while exploiting their labor. After a long struggle, when the native Africans won freedom and the racist regime was dismantled, the "homelands" were also dismantled after a short time. However, the originally British-created homelands, or Bantustans, still exist as "independent" states of Botswana, Lesotho and Swaziland (renamed "Kingdom of Eswatini", in 2018).

Similarly, several Bantustans were created by Britain on the Arabian Peninsula, in order to exploit their oil resources cost-effectively. In each case the ruling family of a tribe was placed in control. These Bantustans still exist, as "independent" states. The United States has not successfully created any Bantustans outside its territory. Those within its territory are referred to as *"Reservations"*, set up for the remnants of the native tribes, who were annihilated during the *Great Genocide*. It has, however, taken over control of the British-created Bantustans of the Middle East - Saudi Arabia, Jordan, Kuwait, Bahrain, Qatar, Oman and United Arab Emirates. The United Arab Emirates (UAE), better referred to as "United Arab Bantustans (UAB)", consists of seven Bantustans - Abu Dhabi, Dubai, Ajman, Sharjah, Fujairah, Ras al-Khaimah and Umm al-Quwain. All these Bantustans are now generally referred to as "Sheikdoms". They were called "Trucial states" before.

In South Asia, or "British-ruled India", Britain created scores of Bantustans. Most were known as "Princely States", where one family was, in general, kept in power. Some were called, "Tribal Agencies", with a British "Agent" dealing with the respective tribe. These agencies were, in general, different from the "Princely States", in the sense that, although only one tribe's population was included in them but one family was not found to be dominant in the tribe. A British "Agent" was appointed to deal with such Agencies, which have existed till recently.

"When those states which have been acquired are accustomed to live at liberty under their own laws, there are three ways of holding them. The first is to [destroy] them (thus making them ready for colonization); the second is to go and live there in person (with some of your people); the third is to allow them to live under their own laws, taking tribute [from] them, and creating there within the country a [government] composed of a few who will keep it friendly to you. Because this [government], being created by the [imperial power], knows that it cannot exist without [its] friendship and protection, and will do all it can to keep them, and a [country] used to liberty can be more easily held by means of its citizens than in any other way, if you wish to preserve it."

Niccolò Machiavelli (1640), *The Prince*. Translated by Luigi Ricci, Urbana, Illinois: Project Gutenberg. Retrieved June 23, 2020, from www.gutenberg.org/ebooks/57037.

Kuwait was created as a separate Bantustan by Britain on land carved out of the Turkish possession of Iraq, which was part of the Ottoman empire. After the first world war, Kuwait was declared a separate "protectorate" of Britain, as all the Arab territories of the Ottoman empire were divided up, mainly, between Britain and France. When oil was discovered on its territory in late 1930s, Britain imposed an "agreement" on the Sabah family/tribe, under which it was to receive a royalty of less than five cents per barrel of oil pumped out of

Kuwaiti oil fields. The family was required to invest half of the money, so received, in Britain – to cement its relationship with its imperial power. By the mid-1960s Kuwait's "Sovereign Fund" had thus reached an amount of more than ten billion dollars! In early 1960s, when General Karim Kasim of Iraq threatened to reunite Kuwait with Iraq, Britain promptly sent its military forces to defend its Bantustan – as Machiavelli would have expected. In 1991, when Saddam Hussein launched an attack on Kuwait and established Iraqi control over it, the US forces came to restore the Sabah family back into power. The conquest of Iraq also followed. The histories and status of all the Bantustans in the Persian Gulf are quite similar and the results are also the same.

United States, currently, the chief imperial power of the world, has joined Britain, to claim its "rightful" share of the "tribute" from the Bantustans in the Persian Gulf. At this stage, it has more control over those Bantustans, especially Saudi Arabia, as compared to Britain. United States had also created Bantustans headed by certain families, in Guatemala and Nicaragua. However, those were eventually dismantled by their people and have become "normal" independent states of Central America. They are, however, subjected to another form of exploitation, as we will discuss in the following sections.

Panamization

This technique of subjugation of peoples was developed and perfected by the United States in response to the constraints that the people of the United States placed on its ruling class. It is a variation of the Bantustanization scheme, the objective being to make it appear that the created, controlled and subservient imperial possession, or pseudo-state, is an independent entity. This was done because the people of the US had a very negative view of colonization and colonial powers, since the US was originally a set of British colonies, which became independent after a violent struggle against their ruling imperial power. When intensive capitalism developed in the US, it also *naturally* developed imperial ambitions and, militarily, it came into conflict with other imperial powers over expansion of its territory and acquisition of colonies. Territory under the control of the United States was expanded by military force, threat of use of force and financial inducements to competing imperial powers to relinquish their control over the territories it desired. The competing imperial powers included Britain, France, Russia and Spain. France, Russia and Britain responded to financial inducements, but Spain resisted militarily. Mexican territory was acquired by the United States by the use of military force and, similarly, after a war with Spain, its former colonies, namely Port Rico, Cuba and the Philippines, etc., were taken over by the United States.

Panama was created around 1900 by the United States because of its need to build a canal between the Atlantic Ocean, or the Caribbean, and the Pacific Ocean, to facilitate trade between its Eastern and Western coasts and with other states of the world. At that time Panama was part of the territory of Columbia. The United States approached the government of Columbia to allow it to build a canal to link the two oceans. Colombian rulers had their own demands in this regard. A Treaty between the two states was proposed. It would have permitted the United States to lease a six-mile wide strip of land, on either side of the proposed canal, in perpetuity by paying an annual fee. The treaty moved through the United States Senate, but the Colombian Senate held out for more money. The United States decided that Columbia was asking for more than it wanted to pay for this project, and it decided to use force. In accordance with his policy of using the "big stick" to achieve his expansionist designs, President Teddy Roosevelt sent a US gunboat to the shores of Colombia and, at the same time, conspired with a group of Colombians to create a new "state" of Panama. Also, military means were used to prevent access to the area by Columbian military forces. Thus, the required territory was taken from Columbia by force. The individuals who were found to form a puppet "government" for the territory, then, declared independence of a new "state" of Panama. After the new "revolutionary government" had been installed in Panama, the United States had its puppets sign an "agreement", transferring all rights to the territory that was to be used to build the Panama Canal, to the United States in perpetuity. The government of the "revolutionary puppets" also agreed to "allow" the US government's military and police forces to be deployed or stationed in Panama, if and when required. In return for signing this agreement, the members of the puppet government were paid about ten million dollars. The "government" of Panama was ultimately allowed to set up a police force to control its territory. It was not allowed to set up a regular military. The police force, later renamed the "National Guard" for some time, effectively remained under the control of the military of the United States. Thus, a system of control and exploitation was set up in this territory using military force. The US "embassy" in Panama actually managed the country to make sure there was no threat to the control of the Panama Canal. Managed elections were ultimately held in which the residents of the territory "elected" a government. The canal was built and most of the benefits of the canal went to the rulers of the United States and their corporations. However, the US did allow some financial gain of the canal to go to the local "revolutionary" ruling class that it had created, thus creating a permanent interest for the class in this US project. Some of these benefits did end up going to the people of the territory also. The United States referred to Panama as an "independent" state, not as an imperial possession, although that is what it really was and is. It should be noted that the US did not put a tribe into power in Panama. In fact, no tribe existed there at that time. Thus, Panama never became a Bantustan.

This technique of "Panamization" is somewhat different from "Bantustanization". It is more sophisticated in terms of creation of a "community of interests" between the local ruling class and the ruling class of the US and can be less abusive since it leaves the possibility open for the local people to claim, and try to obtain, real independent status for themselves. The presence of the pseudo-representative government also gives the people a feeling of participation in the government of the "state".

The United States had control of Panama Canal in perpetuity but tensions naturally emerged between the imperial power and its possession. President Carter abolished the treaty in 1977 and agreed to transfer the control of the Panama Canal to Panama by the end of the twentieth century. During President Reagan's administration, a local military dictator, General Noriega, tried to be assertive regarding control over the canal. As a result, Panama was invaded by US forces and Noriega was overthrown. He died recently in a US prison. The canal has remained, effectively, under American control. The common interests between the ruling classes of Panama and the US have grown with time. Cross investments by the ruling classes and share-holding arrangements in their companies are used as instruments for creation of the "community of interests", but the local ruling class is kept subservient to the US ruling class. The same applies to the local police/military, which is not allowed to have the weaponry or strength to challenge the military control of the US.

This technique of "Panamization" was also used by the United States in Cuba, Porto Rico, Japan, South Korea, Taiwan and the Philippines and all these states were "Panamized" by it. Only Cuba and the Philippines have managed to break out of this imperial stranglehold. Taiwan being a part of China, the United States is likely to lose its control completely to China itself. At present a process of "de-Panamization" is in progress. The Panamization of Japan happened at the end of the Second World War. At the end of the Second World War, Korea was divided between the Soviet Union and the United States, the latitude of the 23rd parallel was decided to be the dividing line, before the occupation. The Soviet Union moved its troops into the Northern half of Korea, after it had forced the Japanese occupation forces in the province of Manchuria in China, to surrender. It organized the military of North Korea and gave it all the arms it needed and withdrew its forces out of Korea. The United States kept some forces in South Korea as it Panamized the state. In 1950, the North attacked the South to take it over and to drive the American forces out, but failed in this attempt. The 23rd parallel, thus, remains the line of control, or a de-facto border, to this day. Porto Rico has also remained an American possession and has been given the status of a de-facto constituent province, or "state", of the United States.

In 1991, Iraq was invaded by the United States but was not Panamized at that time. When the Shia Arabs in the South of Iraq rebelled and the Saddam regime crushed the rebellion, the US encouraged this foolish action by the rulers of Iraq, so that their unpopularity would grow. After huge numbers of people were killed, or were starved to death by "sanctions" imposed by the US and Iraq's rulers became sufficiently unpopular with its people, it was considered ripe for Panamization. In 2003, it was invaded again and placed under full occupation. Saddam Hussein's dictatorial regime was very unpopular in the Shia segment of Iraqi society. The process of Panamization was, thus, initiated with creation of a new Shia army and civil administration overseen by a huge "US Embassy". The original army of Iraq, consisting mostly of Sunni Muslims, was disbanded. Shia opponents of the Saddam regime, especially those with economic and financial interests in the West, were recruited as officials for the new administration. The people of Iraq have continued to resist the occupation of their state. The Sunni officers of the disbanded army joined a new terrorist organization, named ISIS/ISIL, and started fighting the new Iraqi regime with devastating consequences for Iraq. The United States wants to Panamize Iraq to extract most of the income from its vast oil reserves, for the benefit of the ruling class of the US. However, it seems that the ultimate objective of the US policy in the Middle East is to Panamize Libya and all the Arab states east of it. This would include whole of the Arabian Peninsula. Iran is also to be Panamized after its planned conquest. The imperial dream is that the Panamized Middle East would have its Head Quarters in the Jewish colony in Palestine and would consist of all oil-rich territories of the Middle East. Currently, the regional Head Quarters of the US military are in Qatar, while Israel, effectively, remains an extension of the United States – to be used for any action, that the US finds politically difficult to perform directly, to move this project forward.

Captive-Market Subjugation

Whenever colonization was not considered feasible, the technique most commonly used by the imperial powers, for the exploitation of subjugated peoples was to set up a captive market in the captured territories, and to exploit the local agricultural and mineral resources. For example, Britain set up such a cloth market in "British India", i.e., South Asia. Britain manufactured cloth, and the people of India were forced to buy it at a high price from British manufacturers. India, of course, was not allowed to import cloth from any other source. Also, the British destroyed the existing textile industry in India, so that Indians had no choice but to wear the clothes made out of British-manufactured cloth, or to use coarse cloth made by village artisans. Similarly, the products of India were exported to Britain at the lowest prices, effectively determined by the imperial power. Later on, some mining and manufacturing industries were also set up in South Asia and the same market mechanism, and terms of trade, imposed by Britain in its favor, were used for economic exploitation of the local population.

At a later stage of the British imperial rule of South Asia, as wages of workers in Britain went up because the huge surplus of labor had decreased there, workers were hired in India to work in the textile factories in Britain and the usual starvation wages were paid to them. This further increased the rate of intensive profit (RIP) of the owners of textile factories in Britain. In this way, by means of this controlled market and controlled trade between the imperial power and its subjugated people, huge amount of capital was transferred to Britain from India. When the British arrived in India, the standard of living of the people of South Asia (the British referred to this as "British India") was not very different from the people of Britain and the local population was perhaps 60 million people. Over the period of British occupation, lasting about two hundred years, the standard of living of the population of South Asia fell steadily. This happened because of the economic exploitation and a concurrent increase in the South Asian population. The opposite happened to the standard of living of the population of Britain, although most of the capital accumulated, went to the ruling class of Britain. Agriculture was the main industry in South Asia at that time. Now, the population has reached about one and a half billion people, of which three-fourth are living below the poverty line. One fourth of the population is living in "extreme poverty", with income of less than $1/day per person.

Emergence of the United States

History of the United States is closely linked with the history of the British Empire. The US was, initially, a set of British colonies which became independent during the Great Genocide. At that time, the blood of the natives of North and South America, Australia, New Zealand and Africa, especially South Africa, was being shed freely and without any mercy. The United States emerged in this bloodbath of the local population as an independent state. It had huge resources of naturally occurring fertile land and started, mainly, as a producer of agricultural products.

Colonization of North America

The process of colonization of the Americas started with the discovery, in 1492, of the "New World" by Columbus. At that time, the native population of North and South America consisted of hundreds of tribes speaking many different languages. The native population of North America alone, perhaps, numbered more than 30 million at that time. Spain was the first to establish colonies in Central and South America. Then, Portugal "stumbled" into South America, as its "explorers" lost their way while trying to reach India, and established what is now called Brazil. Spanish colonies were also established in what is now the South-Western United States. Agricultural land in North America was plentiful, and had the ideal climate for agricultural production. Colonization basically dispossessed the natives of the land they had, and on their graves the states of Canada, United States and Mexico came into being. The states of South America, to a great extent, had a similar history, although all the local inhabitants, fortunately, could not be slaughtered. Natives survived in mountainous and heavily forested or desert areas which Europeans did not want. Most of the fertile lands were captured by Europeans, and slaves were brought in from Africa to develop those lands, since the natives could not be easily controlled or enslaved and were, therefore, subjected to large-scale massacres.

In North America, Holland, France, Spain and Russia initially established their colonies. Britain established its colonies on the Eastern Shores of the continent by means of "Joint Stock Companies" (or Corporations) in which the British ruling class invested heavily. Each of these companies was meant to colonize a specific area, by exporting the unwanted poor of Britain, and to "earn" a high amount of profit for its owners. The Virginia Company was the first one to be created, to set up the Jamestown colony in 1607, in what later came to be known as the "state" of Virginia. The Plymouth Company and other such companies followed - on the lines of East India Company, the East Africa Company, New Zealand Company, etc., which were established for other parts of the world. The Eastern shores of North America were colonized first by such companies. 90% of the native population of the colonies "disappeared" during the first 60 years of this process.

When shortage of labor was felt, because human sweat and blood was necessary for the planting, cultivation, and harvesting of cash crops, in addition to slaves "indentured servants" were brought from Europe. These "servants", or virtual slaves were offered free transport to North America in return for five or so years over which they had to provide wage-less labor to pay off their debts. They were provided food and lodging only, just like slaves, during this period of "indenture". Because of rampant abuse, most of these indentured servants could not live to complete the terms of their contracts. Female servants were especially subjected to harassment and abuse. Another similar scheme was to import those English citizens who were imprisoned for debt. The idea was to reduce the prison population in Britain and use the labor of such prisoners in the colonies. The "state" of Georgia was specially created, in1733, by using such prisoners.

As the colonies grew, the cost of occupation also increased. To reduce the cost, local "governments" were set up - "elected" and controlled by local property owners. In each colony, the legislature consisted of two houses. The lower house consisted of local property owners elected, or selected, by the local ruling class. The members of the upper house were nominated by the British government, like the House of Lords in Britain. The governor was appointed by the British king and represented the colonial authority of Britain. Laws were passed to transfer the cost of occupation to the colonies and increase the RIP for the British "investors". These laws included taxes on goods imported into the colonies, bans on setting up certain industries, controls on currency and creation of money, taxes on legal documents ("Stamp taxes"), etc. The governing property owners of the colonies resented these laws as they reduced their own capital accumulation. This dissatisfaction grew with time, especially after the "seven-year war" between the British and the French, and the resulting Treaty of Paris, in which the French lost their colony of Quebec and its adjoining areas to Britain and a large territory, around New Orleans, to Spain.

United States as a Nascent Imperial State

Resentment against the British king and the British ruling class grew with time and the popular slogan of "No taxation without representation" became a rallying cry for a rebellion against the British. As tensions increased, the rulers of the thirteen colonies met to condemn British actions and to plan strategy. These grand meetings were called "Continental Congresses" The thirteen colonies set up a "continental army" during the second continental congress, in 1775, and appointed George Washington as the commander of this army.

The colonies declared independence in Philadelphia during 1776, accusing the British king of being a despot, and of trying to establish an "absolute tyranny". Fighting between the two armies followed. It continued for several years. France joined the rebellious colonies in an alliance to defeat Britain as a matter of revenge. It, even, sent its troops to support the rebellion, once it became clear that the colonies were in a position to prevail. The Continental Congress continued to control the war effort against the British as a de facto central government, although it was not given the power to levy taxes but continued to depend on the states to make contributions for this purpose. However, it did have the authority to print money and to engage in diplomatic activities. It also had the authority to resolve conflicts between colonies. Nine out of thirteen colonies had to approve a law for it to be enacted. The creation and maintenance of the army and the war effort was its main responsibility. A rudimentary constitution was framed later on, in November, 1777, embodying these "Articles of Confederation", which were formally ratified in 1781. The central government, at this stage, had only a unicameral legislature, with no judiciary or executive. The legislature was elected. However, at this stage most powers of government rested with the colonies. Finally, the thirteen colonies became independent "states", when the Treaty of Peace was signed in Paris, in 1783, between the Continental Congress representing the colonies, and the British authorities.

Economic crises had developed as the new state called the "United States" came into existence. The economic problems faced by the Congress deeply touched the lives of most Americans in the 1780s. The war had disrupted much of the American economy. On the high seas the British navy had a great superiority and destroyed most American ships, crippling the flow of trade. On land, where both armies regularly stole from and robbed the local farms in order to find food, farmers suffered tremendously.

When the fighting came to an end in 1783, the economy was in a bad shape. Exports to Britain were restricted. Further, British laws prohibited trade with Britain's remaining sugar colonies in the Caribbean. Thus, two major sources of colonial-era commerce were eliminated. A flood of cheap British manufactured imports that sold cheaper than comparable

American-made goods made the post-war economic slump worse. Finally, the high level of debt taken on by the states to fund the war effort added to the economic crisis by helping to fuel rapid inflation. The people of the United States began to rebel against the ruling class. Some of these rebellions led to armed conflicts and were suppressed by military force. In an effort to deal with the economic problems, the Congress and the thirteen states began to adopt conflicting policies. While some of the economic problems were brought under control by these measures, new ones began to arise and the need for a strong central government began to be felt, acutely.

The states, also, adopted a variety of state constitutions based on their perceived needs and ideas about representative government. While some states kept the basic structure of the colonial administration, others like Pennsylvania, radically modified their constitutions. South Carolina, basically, retained the colonial administrative structure. The property qualifications for voting were retained and even higher qualifications for holding office were adopted, making 90 percent of, even, white men ineligible to vote. Pennsylvania abolished the upper house of the parliament, corresponding to the "House of Lords" in Britain, as its members were wealthy men and aristocrats. It also abolished all property qualifications for voting and for holding office as a representative of the population of the state. However, its constitution limited the voting rights to white male tax-payers only.

A convention of delegates from all states was finally called in the city of Philadelphia to decide the form of the new central government. The Convention was presided over by George Washington, who had resigned earlier as the commander of the Confederate army, and its proceedings were kept secret. The basic conflict that developed at this stage was between the large and small states, as the small states wanted to retain the "one state, one vote" principle, while larger states wished their population to be considered for their participation in the central government. Ultimately, a federal system of government was adopted with a bi-cameral legislature. The lower house, called the House of Representatives, was to be elected by the population of the federation, each state having a number of seats corresponding to its population. The upper house, called the Senate, was to have equal representation for each state. Two senators were to be elected for this purpose, by the House of Representatives of a state.

The federal government was to have three independent branches with checks and balances provided in the constitution, to prevent the emergence of a tyranny. The Executive branch was to be headed by the President, who was to be elected by an "electoral college" consisting of delegates from all states, corresponding to the population of each state. The winner of the majority of votes in a state was to be assigned all the delegates of the state and the loser would receive no delegate votes. This provision was to ensure a clear winner of the presidential races. The Legislative branch was to consist of the Congress with its House of Representatives and the Senate. The Judicial branch was to consist of the Supreme Court and its subordinate courts, for resolution of disputes between the other two branches of government, the citizens of the federation and other economic and political organizations and associations. Property qualifications for voter eligibility, and for holding office, were abolished and all white, tax-paying men were to be eligible to vote. Also, effectively, slave owners were to be given additional voting rights based on the number of slaves they owned.

For each batch of five slaves, each slave owner was to be awarded three additional votes. The total number of votes, calculated this way, were to determine the size of a state's representation in the federal House of Representatives. Thus, all slaves and women were ineligible to vote. Also, poor people, who were not taxed, or were not paying taxes, were held ineligible for this purpose. The native people were to be totally annihilated and, thus, no voting arrangements were needed for them!

When the constitution had been approved by the constitutional convention, the process of ratification by the states was started. It took about two years for the constitution to be ratified by 11 states. Thus, by that time, more than the required 9 votes had been obtained. The remaining two states took longer to ratify the constitution. The federal administration came into being in 1789 and George Washington became the first President of the United States. Thus, the thirteen "states" surrendered their sovereignty to the federation, thenceforth called the "United States", and the federation became a sovereign state. The former "states", in effect, became constituent provinces of the new state. They, however, continue to have some degree of autonomy in some areas and are called states of the Union to this day.

The US constitution was designed to ensure that the US would become a "nascent" imperial power in which the dominance of the ruling class of plantation and industrial owners would be ensured and real democracy would be limited, i.e., the rule of the people by the people and for the people would be denied and kept under the control of the ruling class. It seems that framers of the constitution were inspired by the Roman Empire and the new state was expected to expand, accordingly, towards the West coast. The framers were very concerned about the people laying claims to "their" property and the properties of their fellow members in the ruling class. Their aims were not surprising, because most of them were slave owners or slave traders and had huge landholdings and plantations where slaves were forced to work. James Madison wanted to "protect the minority of the opulent against the majority". Alexander Hamilton referred to the ruling class and the people as the "few" and the "many". He wanted "a distinct, permanent share in the government" for the "few". George Washington was the richest plantation owner and slave owner of the group. In addition to his leadership in other areas, this helped him in becoming the first President of the new state [5]

The United States had emerged as a new constitutional republic at this point, but the issue of individual rights had been left undefined. Similarly, the issues of slavery and voting rights of female citizens had been left unresolved. At the time of independence, the population of the United States was a little more than three million, of which the majority was of British origin. About 20 percent were slaves and a much smaller minority of natives also existed within its borders. Individual rights were defined by the first ten amendments to the constitution and these ten amendments are now referred to as the "Bill of Rights". The Bill of Rights defines rights of individuals regarding freedom of speech and expression, freedom of conscience and belief, freedom of assembly, protections against unfair search and seizure of property; and ensures fair trial of individuals accused of crimes. However, it deals with the "right to life" in a narrow sense of this phrase.

Trade relations of the United States changed as a result of independence. After initial setbacks, trade expanded, local industries were given protection from foreign competition, the debts of the states were taken over by the federal government and financial institutions like the Bank of North America and various departments of the federal government were created. As the economic and military power of the United States increased, it looked westwards to expand its territory. This was resisted by the native population and European colonial powers. Huge territory of Louisiana was purchased from France for 15 million Dollars and the control of the United States was extended dramatically with time. This brought the state into conflict with Spain, Britain and Russia. After wars with the native tribes, resulting in their massive slaughters and genocide and wars with Mexico and Britain, the boundaries of the state reached the West coast of North America. As more and more territory was acquired, the flow of immigrants also increased and large numbers of colonizers moved to the West, laying "claims" to agricultural lands. In 1812, the United States invaded Canada with the hope of conquering all of North America. This invasion failed and, in fact, caused a counter-invasion by the British, in which they were even able to attack Washington D.C. and burn the White House. The previous US-Canada border was then restored. A two-party system emerged with the two parties named Federalist and Democratic-Republican parties. With passage of time, the Federalists disappeared and their agenda was partly taken over by the Democratic - Republican Party, which also split up into what now-a-days are called the Republican and Democratic parties.

During the first half of the nineteenth century, property requirements for voting were gradually abolished, state by state, and economic status disappeared as a foundation for citizenship. By 1840 more than 90 percent of adult white men possessed the right to vote. Not only that, voters could now cast their votes for more offices. Previously, governors and presidential electors had usually been selected by state legislatures as part of a strategy that limited the threat of direct democratic control over the highest political offices. The growing participation by the people during the first decades of the 19th century changed this and increasingly all offices were chosen by direct vote. The United States, thus, became the world leader in allowing popular participation in elections, as the suffrage was expanded to include more and more of white men. The right to vote, however, remained restricted for white women and all African Americans.

Colonization of North America had been started when most of Europe, especially Britain, was going through a transition from advanced feudalism to intensive capitalism. The system of colonization led to the annihilation of the natives and their replacement by Europeans and later, by African slaves. Those who moved to the new colonies were already familiar with factory life, but by using Joint-stock companies, Britain had also set up the system of Controlled Market Subjugation in the colonies on the Eastern shores of North America. Plantations were set up in these colonies and agricultural production was encouraged. Most industries were not allowed to be set up. Instead, the colonies were forced to import finished products from Britain. This maximized the RIP for the members of the British ruling class, who were the share-holders of the companies which had set up those colonies and who owned all the industries in Britain itself.

As the United States achieved independence, the system of controlled-market subjugation also came to an end. At this stage, the US was basically a producer of agricultural products, especially tobacco and cotton. Thus, industries, based on cotton and tobacco, began to be set up. Cotton ginning and weaving were introduced into the South and cotton textile mills were set up in the North. Thus, the Slave-owning and Slave-free states began to become inter-dependent. Different techniques were adopted by the ruling class in the US, in an effort to maximize RIP. These included the introduction of distributed production activities, in which some of the work was done in homes and semi-finished products were brought to the factories (the centers of intensive capitalism) for final finished product creation. Women were employed in cotton ginning and weaving factories, since they could be paid less, thus maximizing RIP for factory owners. However, none of these techniques could compete with the use of slavery and availability of slave labor over the lifetime of a slave. This naturally created a conflict within the ruling class, when, in some states, slavery was legal and in others it was illegal. Behind all the "moral" arguments lay the rivalry between slave owners, who could achieve very high RIP, and those to whom this "privilege" was denied. As more territory in the West of the North American continent was colonized, and more states were formed, this issue appeared as competition between slave-owning and slavery-free states. On several occasions it almost led to war. In the meantime, slaves continued to rebel against their masters and all those rebellions were violently suppressed.

As industrialization grew, so did the need for transportation of raw materials and financing of new "ventures". Thus, banks were set up and their branches increased in numbers. Also, canal and road networks were expanded accordingly. The railroads were built by contractors, who were lavishly paid by the government, in terms of land. Thus, tens of millions of acres of fertile land were given to the railroad companies as payment. The construction of railroads led to increased mining of coal. Mining also increased in general because of the lower cost of transportation by rail. Coal and iron began to be mined extensively. By late nineteenth century railways covered the whole territory of the United States, from coast to coast. The state-owned banks were abolished and hundreds of "private" banks, owned by the ruling class, were set up. Most of this "infra-structure" was created by the state. A new business entity was created, by law. It was and is called the "Corporation", to protect the owners of financial and industrial institutions from risks and extending privileges to them that were not available to the people in general. These are now referred to as "Limited Liability Companies", or simply "Corporations". This process of industrialization and evolution into full-fledged intensive capitalism took about a hundred years. During this period, almost half the population of Ireland immigrated to the United State. Very large numbers of Germans and other North Europeans also immigrated. In 1819, the territory of Florida was purchased from Spain. In 1845, Texas was annexed. It had previously become a "Lone Star Republic" on its own. The formal annexation of Texas led to war with Mexico. As a result, additional territories between Texas and California were conquered by the US and a sum of 15 million Dollars was paid for some additional territory near the border. The Oregon territory was acquired and the British claims were excluded from it. Gold was discovered in California in the middle of the nineteenth century. It led to a rush of immigrants towards the West coast. Mining of copper, silver, zinc and other minerals also increased. This period also was a period of innovation. Many new machines and technologies were developed. A horse-drawn reaper (*harvester combine*) and a riding plough (*tractor*) were developed, Morse invented his Morse code and the electric telegraph. Sewing machine and the milling machine (a machine-tool) were invented. The electric bulb, the elevator, the telephone, cable cars and the subway system came into being. The *Steamboat* was developed and the process of rubber vulcanization was invented, leading to large scale manufacture of tires.

During the nineteenth century, the forced removals of native tribes and their mass killings continued. Only those native tribes who lived on desert or hilly lands, which the colonizers did not want, were spared this mass slaughter. Even these tribes, like the Navajos of the Grand Canyon, were greatly reduced in numbers both because of military action and spread of diseases, for which it took some time for immunity to develop. "Treaties" were signed, as a tactical matter, with these tribes, guaranteeing them some of their lands, but these treaties were then routinely violated. The tribes that still survived, though in miniscule size and very few in numbers, were pushed into barren and useless lands and declared to be "free" within these Bantustans (referred to as "*Reservations*"), where they have remained as an insignificant source of labor for the regular American economy. The total population of these Bantustans is, now, about a quarter million, including people of mixed race.

As the territory of the United States was expanded and new states were created, the competition between slave-owning and slavery-free states continued to grow. Now, the tensions existed not only because the slave-owners in the Southern states had the special privilege of buying and keeping slaves and using slave labor to enrich themselves, but the number of slave-owning states and the number of slave owners threatened to increase, thus increasing the "inequality" between the plantation and factory owners of the North and the South. These tensions ultimately exploded into civil war in 1860, when several states of the South seceded and raised a "confederate" army, to defend "their right" to enslave and abuse people. Abraham Lincoln had just become the President. As the violence started in South Carolina, he issued a proclamation "freeing" the slaves in the seceding Southern states - over which his government had no control. This was a tactical matter, since this proclamation was not applied to other slave-owning states which had not seceded. The Civil War lasted about four years and the seceding states were reunited by force with the rest of the United States. The constitution was amended to end slavery. This, however, did not dramatically change the lives of African Americans. Though "freed", they were forced, for economic reasons, to work at the same plantations where they used to work before. Their lives did not improve much. In fact, the lives of most became worse than before as they were subjected to discriminating and humiliating restrictions of all kinds, including restrictions on what buses they could travel on and what part of a bus they could sit in, what public wash-room they could use and what schools their children could go to. From slaves, they, effectively, became "wage slaves" and most of them remain so even now, being subjected to discrimination, though illegal, of all kinds. The end of the Civil War, did give the African Americans the freedom of movement to some extent. Many ended up moving to other "free" states of the North, thus becoming wage-slaves of the North. Similarly, their right to vote did become a reality. This did increase their political power with time. The Civil Rights movement of the 1960s, ultimately, brought the most humiliating discriminatory practices to an end.

With the end of the Civil War, the United States was able to focus on development of its agriculture, mining, transportation and industries. Thus, the stage of intensive capitalism began to make fast progress. Trade between the United States and other states of the world grew very fast and the economy of the United States grew at a high rate, because of this peaceful period, while the rest of the world, especially Europe, was facing increasing violence and wars. This period was also the time of a great increase in corruption and is referred to as the "Gilded Age". At this time, the "*establishment*", consisting of plantation and factory owners, politicians and military men, dominated the political and economic system of the country. During the late 1860s, under an implicit threat of war, Russia ceded the territory of Alaska to the United States, for 7.2 million Dollars (1.9 Cents/acre!). With this acquisition, the continental United States reached the present extent of its territory.

As intensive capitalism made progress, it was increasingly resisted by workers and formation of labor unions accelerated. In the beginning, these unions, their office holders and members had to face a lot of suppression and violence by the ruling authorities, but the labor movement kept up its march, demanding better treatment for workers. In October 1884, a convention held by the *Federation of Organized Trades and Labor Unions* unanimously set May 1, 1886, as the date by which the eight-hour work day would become standard. On Saturday, May 1, thousands of workers went on strike and rallies were held throughout the United States, with the cry, "*Eight-hour day with no cut in pay*." Strike breakers, protected by police, were brought in by the authorities to suppress the movement and violent incidents followed. Some workers were labeled as anarchists and four were, ultimately, hanged to death. Now, virtually the whole world (except the United States) celebrates the First of May as Labor Day, in their memory.

Two issues still remained to be resolved at the end of the nineteenth century - the issue of women's right to vote and the right to vote for the people who were denied this right because they were too poor to pay taxes. The struggle for women's right to vote continued throughout the nineteenth century. Some states granted this right to women, while others did not do so for a long time. Ultimately, the nineteenth amendment to the constitution resolved this issue in 1920. The right to vote was extended to all citizens 18 years old, or older, in 1971. Those who were disfranchised because of their inability to pay taxes were also given this right at the same time.

United States as a Major Imperial Power

As the nineteenth century came to an end, the United States had greatly expanded its territory. Its economy had grown to be bigger than that of any European state. Its trade with European, South American and Asian states had grown dramatically. All this became possible because it had inherited a European population with business, industrial and agricultural skills, a territory that had huge agricultural and mineral resources; and a political set up that ensured it internal and external peace and stability to foster innovation and to grow into a modern state. It developed the same industries as in Europe, but quickly they became much more technologically advanced than European industries, including an armament industry.

As the United States developed further imperial ambitions, the Hawaiian Islands were annexed by the United State by the end of the 19th century. The United States had initiated a

"Big Stick" policy towards Latin America; the "Big Stick" was the American navy. The Spanish-American war was started by the United States in 1898 in the Philippines. The United States invaded Cuba at around the same time. The war lasted about ten weeks. A peace treaty was signed between the two states in Paris. As a result, the United States received the Spanish colony of the Philippines and the islands of Puerto Rico and Guam. Spain was paid a sum of 20 million Dollars as compensation for its losses. Cuba was declared free, but it had to surrender the territory of the Guantanamo Bay for establishment of a naval base by the United States. Also, it had to agree not to enter into any agreements with any European power "that may endanger its independence" and the United States was allowed to land its troops into Cuba, if it considered it necessary. Thus, Cuba had become a de facto imperial possession of the United States, like Hawaii, Guam, or Puerto Rico. During the following period of expansion, of about one hundred years, the United States invaded the Latin American states more than thirty times, as part of the "Big Stick" policy.

The need for a canal to link the Atlantic and the Pacific oceans was being acutely felt, as the American navy was used in both the oceans. Also, business activities required this connection. It was decided to build the Panama Canal for this purpose. To obtain the rights to the territory, the United States had to negotiate with Colombia, but Columbian authorities desired more than the US government was willing to pay. Thus, the US grabbed the territory from Columbia by military force and created the new "state" of Panama, using some local puppet "revolutionaries", as described before. The Colombians were powerless to stop this project. In 1914, at the cost of about three hundred and fifty million dollars, the Panama Canal was opened for business.

Social Democrats and others inspired by the slogan of socialism began to organize themselves, around the beginning of the twentieth century. Ultimately the Socialist Party of the United States was established and it initially made great progress towards becoming a part of the politics of the country, till it faced repression after the Russian Revolution, along with its splinter group, the Communist Party of the United States.

Summary and Conclusions

As we look at the history of the United States, we can see that it inherited a lot from its creator - Britain. At the time of independence, majority of its population was of British origin. It consisted of thirteen colonies/states, which were administered by British legal arrangements. Its provinces, or states, had an elected parliament, consisting of a lower house with members of the local ruling class, elected by the local owners of plantations and factories. The upper house of the parliament consisted of members appointed by the British ruling class, on the lines of the members of the House of Lords in Britain. In addition, each state had a governor appointed by the British king. There was a small British military contingent stationed on the Eastern shores of North America, but it was under the control of the British king, British parliament and the local governor and had no political role of its own. Thus, like in Britain, the British colonies had a tradition of civilian supremacy. This tradition was inherited by the United States, where even during the uprising against the colonial power, the army, led by George Washington, was under the civilian control of the Confederacy. The population of the United States was mainly European, the native population was being annihilated as the Great Genocide continued. The African slaves did exist, but they had no political role in the pseudo-representative government of the United States that existed at that time. The European population mainly consisted of people with some capital, who could pay for their move to the new continent. The debtor slaves and "indentured servants" were only a tiny minority. At that time, Europe, especially Britain, had evolved to the advanced Feudal stage and intensive capitalism had become dominant in its economy. Thus, the original British colonizers were quite knowledgeable and politically conscious individuals, who were aware of the new technologies and new industries being developed in Britain and the rest of Europe. As the population of the United States increased and large numbers of the people from Ireland, Germany and other European states immigrated, a population with many agricultural, industrial and scientific skills came together. Many individuals had immigrated to the colonies because of religious persecution. As immigrants of different European nationalities and religious beliefs came together, quite a high level of tolerance for national and religious differences developed because of the fluid nature of the colonization process and the fact that the economic condition of all immigrants was dramatically improving, because of the free acquisition of land and the ready availability of other economic opportunities. Thus, we can see that the United States inherited the traditions of representative government, civilian supremacy and religious tolerance from English, Irish and other European nations. These traditions and qualities had already made the small British state into a global imperial power. Tolerance for people of different European national origins developed because of coming together of the immigrants from all over Europe.

Many of the framers of the constitution of the United States, including George Washington and Jefferson, were slave owners. However, they had been exposed to the thinking of great European political thinkers and, by the very conditions of the rebellion against Britain, were compelled to think about rights of individual states in the global community of states, and the rights of human beings in general. Thus, the constitution they developed, in time, demonstrated their far-sightedness. Jefferson had written the Declaration of Independence, which declared that all men were created equal by their Creator. George Washington had chaired the constitutional convention. The constitution had laid down that the slave owning states were entitled to three votes for every five slaves they owned, while slaves and women were not entitled to vote. Thus, effectively, a black man was declared to be equal to three-fifth of a white man. Of course, women were not created equal and did not count, since white men represented them and their interests. The only way to reconcile these statements is to say that "only white men are human"! This is the kind of mentality the framers of the constitution had at the time when the United States came into being. Still, they were able to look beyond their prejudices and frame a constitution that ensured that the United States would have relative internal peace and would be able to resolve its internal contradictions with time. That is exactly what happened. The framers of the constitution were owners of large estates. They wanted a pseudo-democratic system of government, but did not want the people to encroach upon "their" wealth, including the lands they had come into possession by the use of force. They also intended the new state to be an imperial power. The design of the constitution does reflect these objectives and has served its purposes well, but the growth of real democracy may, ultimately, lead to a fairer distribution of wealth within the United States. The imperial character of the US is likely to persist much longer. During the second half of the twentieth century, the social inequality decreased substantially. It has, however, increased dramatically after 1980. The issue of slavery could not be resolved peacefully and the United States had to go through a civil war. However, the issues of civil rights for former slaves, extension of the right to vote to women and to all adults, were all resolved without large-scale violence. Many of the provisions of the constitution had been left vague so that the Supreme Court could make its contributions to the laws of the state as its society evolved. It has done that in an admirable way, while preserving its independence and increasing its stature in the eyes of the people. Citizens and institutions of the United States have consequently developed a high level of respect and esteem for it. In recent times, however, the court has been filled with more conservative judges, who have begun to work according to the wished of the "founding fathers", who drafted the constitution of the US. To these members of the judiciary, the wishes of the slave-owner founders have become more important than the current needs of society and the wishes of the citizens at this time. One example of this change has been the "Citizens United" decision, which declares corporations as persons!

As the United States developed after independence, its territory increased dramatically. Its original agricultural economy developed into an industrial economy and rail, road and canal transport systems were developed. The search for minerals indicated that the state had been very lucky in occupying lands that had large deposits of valuable mineral resources. This greatly increased the economic power of this state. As the territory and the economy of the United State expanded, peaceful and stable social and economic conditions were created, especially on its Eastern territory. These were characterized by high levels of tolerance for free scientific thinking, free from national, religious and class prejudices. Under these conditions, the potential economic rewards for discovery and development of new technologies also increased dramatically. All this encouraged and promoted innovation and many new technologies were developed, during the nineteenth and early twentieth century. These included the large-scale production of the automobile, which greatly changed the size and layout of American cities and dramatically affected the productivity of industries. The United States became the world leader in many industries and its economy became the largest economy in the world.

Zenith of Intensive Capitalism

As the process of socialization of production progressed, manufacturing industry spread all over Europe and, to some extent, to colonies and to other subjugated areas of the world. The system of intensive capitalism reached its zenith in Europe, around the beginning of the twentieth century. In 1901, the European imperial powers, Japan and the United States ruled the rest of the world, except the Spanish colonies and Brazil, which had become "independent" colonies. Figure 6-02 shows the imperial domination of the world at the close of the nineteenth century. All people of the world were being subjected to intense exploitation. Followers of Marx, the Communists, became more vocal agitators for revolution. It was a time of intense debates and political violence. Talk of revolution and social explosions was in the air. Germany was the focus of these speculations. No one, had expected that the Tsarist Empire would explode like no state had exploded before in history, but against all expectations of the "social democrats" and revolutionaries, Germany did not go through a revolution.

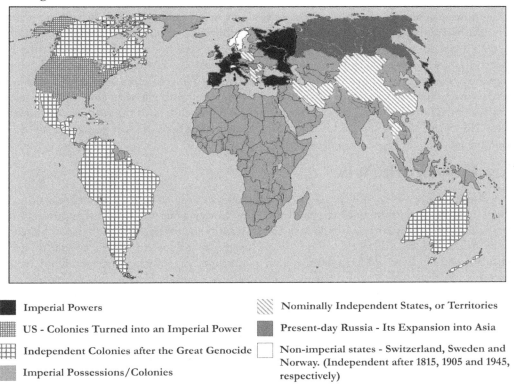

■ Imperial Powers

▓ US - Colonies Turned into an Imperial Power

▦ Independent Colonies after the Great Genocide

▨ Imperial Possessions/Colonies

▨ Nominally Independent States, or Territories

▨ Present-day Russia - Its Expansion into Asia

☐ Non-imperial states - Switzerland, Sweden and Norway. (Independent after 1815, 1905 and 1945, respectively)

Figure 6-02: Imperialism during and After the Great Genocide

(Map based on world map at: https://www.d-maps.com/carte.php?num_car=13184&lang=en)

At the beginning of the twentieth century, as a result of the great genocide, many of the native nations and tribes in the territories occupied by imperial powers, had been annihilated almost completely. Colonies had been established on their native lands, especially in North and South America, Australia, New Zealand and Southern Africa. The system, characterized by blind pursuit of capital accumulation at the cost of immense human suffering, as described before in this chapter, was referred to by Karl Marx as "Capitalism". We refer to it as "*intensive capitalism*" to distinguish it from the next stage of evolution of the social system in Europe, etc. The full "maturity" of *intensive capitalism* in European imperial powers had, naturally, driven the ruling classes of those powers to seek and exploit human beings outside their states and to grab more natural resources there. The reasons for this tendency are explained in the next chapter. All the four techniques, described before in this chapter, were being employed extensively and the ruling classes of those imperial powers were accumulating wealth by these means and living luxurious lives characterized by conspicuous consumption, while the people of even those imperial states continued to suffer. This, certainly, was the darkest time in the history of human civilization.

There was, naturally, continuous resistance to imperial occupation everywhere, but even under these conditions, when exploitation of the subjugated peoples had reached its extreme, when all the accumulated capital was being transferred to the home territories of the imperial powers, there was unrest in those home territories also. Workers were protesting and organizing everywhere, in the "home" territories, against the economic and social exploitation that they were being subjected to, despite all the riches being acquired by the imperial states from abroad. Those riches were only benefiting the ruling classes of the respective imperial states. They were also being used in expanding the imperial hold of their respective states - accompanied by very high levels of violence.

The First World War

A number of secret organizations had sprung up all over Europe. These organizations had picked up arms against the institutions of the state, especially in the Russian Empire. These organizations of angry and desperate individuals were being referred to as *"terrorists"*, a term coined by the Russian ruling class. They were engaged in violent armed attacks against state institutions and individuals representing such institutions - like the police, the military, courts, taxation authorities, etc. Political parties, seeking a revolutionary change, had developed in all European and other imperial states, referring to themselves as Social Democrats, or Socialists. The social-democratic followers of Marx decided to call themselves *"Communists"*, referring to the *Paris Commune* as their ideal, to distinguish themselves from the other Social Democrats. The hopes of revolutionaries in Europe, at the beginning of the twentieth century, were focused more and more on Germany, where the social democratic movement was the strongest and was being led by the Social Democratic Party of Germany, (SDP) - a political party which exists even today with its name unchanged. The policies it proposed then were far different from those it has adopted now. Its program and policies were being vigorously debated, at the time.

The ruling classes of all imperial powers were also expanding their military machines, not only due to the need to suppress the subjugated peoples in Europe and the rest of the world, but because these powerful states, as happens to a group of bank robbers once they have accomplished their job, were also afraid of each other. They were trying to secure what was in their possession, but were also thinking, and planning, to grab what was in the possession of their fellow imperial powers. Thus, especially in Europe, armaments were being manufactured, and acquired, at increasing rates. Armies and navies were being expanded. Also, all imperial powers were busy negotiating and signing public and secret "security" agreements and pacts - to defend their far-flung empires and to attack neighboring powers. Europe was the center of this struggle among the imperial powers. These imperial states could be divided into two groups - those who already had large imperial possessions in Europe and abroad, and those who did not - or those who felt they did not have a "fair" share and "deserved" more.

The states on the periphery of Europe were in the first category, because they had better access to the sea, had built large naval forces and, as a result, had occupied larger chunks of territory of the world outside Europe. The states in the center of Europe fell in the second category. The Austria-Hungarian Empire had subjugated people in Europe only. It had occupied not only the traditional territory of the Austrian and Hungarian nations, but also that traditionally inhabited, among others, by the Polish, Czech, Slovak and Ukrainian nations. Germany had expanded to occupy Polish and other national territories in Europe and parts of Africa.

As we have described above, Europe in 1914 had become a powder keg, ready to explode at any time. Other non-European imperial powers, like the United States and Japan, were also fully armed and ready for a conflict with the other imperial states. The event that initiated war was the assassination of the heir to the throne of the Austria-Hungarian Empire, in Serbia, by an organization known as the Black Hand. Austria-Hungary declared war on Serbia when Serbia refused its demands related to the assassination of the heir to its throne. Russia started to mobilize its army in support of Serbia. Germany viewed the Russian actions as acts of war against Austria-Hungary, and declared war on Russia. France, then, declared war against Germany and on Austria-Hungary following the German declaration. Germany swiftly invaded Belgium with the objective of conquering all territory up to Paris, causing Britain to declare war against it. With Britain's entry into the war, her colonies also were made to enter the conflict. These included Australia, Canada, India, New Zealand and the Union of South Africa. Japan also entered the war soon after this. Italy, although allied to both Germany and Austria-Hungary, sided with Britain and its allies against her two former allies.

Thus, this global conflict started with the German invasion of Belgium and France. At one stage, in December, 1914, the Germans offered to evacuate Belgium in exchange for Belgium Congo (Now named the Democratic Republic of Congo), but Britain rejected this offer. Austria-Hungary and Russia entered the conflict soon after. After the German march on Paris was stopped, the western front settled into a static battle of attrition with a trench line that changed little until 1917. Hundreds of thousands died in the trenches to gain only a few feet of movement of the frontline. Automatic weapons and poison gas were extensively used in this fight. As a result, casualties were very high on all sides. In the East, the Russian army could not make much headway against Austria-Hungary and Germany. The Ottoman Empire had joined the war in 1914, followed by Italy and Romania. Thus, additional fronts had been opened and the war had become a truly global war between imperial powers.

In less than three years, all European Imperial Powers had been weakened by the war. Germany was encircled but the front lines were not moving anywhere. The United States had, by now, grown into the biggest economic power in the world. Its trade relations were focused on Europe and South America. It was dominating South and Central America and had imposed the Monroe Doctrine in these regions, which meant that it had the exclusive "right" to exploit these regions. In May 1916, sensing the weakening of the European imperial powers and adverse effects on its trade relations that a continuing conflict in Europe would have, President Wilson proposed a "universal association of states" to be called the League of Nations. Superficially, the purpose of this League was to ensure "collective security" and to prevent wars in the future and to provide a forum to resolve conflicts as they arose in the future. This was part of his "missionary diplomacy" in the context of the Monroe Doctrine. It meant that the United States had the "moral responsibility" to deny recognition to any Latin

American government that was viewed as hostile to American interests. Wilson wanted to extend this concept to the whole world. Mexicans, specially, were not pleased by this new proposal. The United States had conquered one-third of Mexican territory in the nineteenth century and had sent its troops into Mexico shortly before, but was now presenting this missionary interpretation of the Monroe Doctrine as a guarantee of territorial integrity of all states and as an example of "international cooperation"! European powers did not give a response to this proposal initially, but Wilson felt that they would have to accept his proposal by the time the war ended, since they would then be financially beholden to the US.

Clearly, the United States was seeing the conflict as an opportunity to dominate Europe. As the war progressed, the Tsar of Russia was overthrown in February, 1917 and major changes began to occur there. Those seeking a revolution had split up into two groups - they were referred to as the Reds ("Bolsheviks") and Whites ("Mensheviks"). The Reds were joined by the Social Revolutionaries, who were focusing on land reform. Committees (or "Soviets") had been set up all over the Empire, mainly in the towns. Now, the process was extended to the whole territory of the state. A relatively small group headed by Trotsky also joined the Reds. Russia left the war in 1917, after the Reds seized power in October of the year, and creation of the Soviet Union was taken in hand. Leadership of the Reds was in the hands of Lenin. Trotsky was assigned the task of setting up the *"Red Army"* since the Tsar's Russian army had been disbanded.

United States had declared a policy of neutrality at the beginning of the war, but, noticing the possible consequences of the global conflict, including the danger posed by the revolutionary changes in Moscow, and with hopes of imposing a treaty of "collective security", it entered the war in 1917.

The Russian Revolution brought a new kind of state into existence. A state like the Soviet Union had never existed in human history. It eventually brought dramatic changes to the global imperial system. But, initially, this event did not have much effect on the progress of the First World War. The war continued and the new regime in Moscow withdrew the Russian forces from the conflict in Europe and declared its intensions to work for peace. It declared that it had no territorial designs on any state and did not want any other state to have such intentions towards it. A decree of peace was issued a day after the Revolution, appealing for a *"democratic peace"*. But these declarations did not have any effect on its enemies - especially Germany. The German high command agreed to a cease-fire and negotiations for a peace treaty, at Brest-Litovsk. Trotsky represented the Soviet Union. The Soviet Union was in a very weak position. Its old army had disappeared. The new Red Army was in the process of being created. Germany was the most powerful state in Europe and it was not possible for the Soviet Union to fight or win a war against it. The German generals demanded that the Soviet Union agree to the annexation of the entire Baltic area, the Western part of Belorussia and the entire Ukraine and pay a huge "indemnity" The Soviet Union ultimately agreed to such terms and the *Treaty of Brest-Litovsk* was signed in 1918. The conflict between the Reds and the Whites, however, continued to grow. The Germans began to support the Whites with the hope of completely destroying the Soviet Union.

In the beginning of 1918, President Wilson presented the war objectives of the United States. The objectives were summarized as fourteen points, the achievement of eight of which was a must. These eight included open diplomacy, freedom of the seas, general disarmament, the removal of trade barriers, impartial settlement of colonial claims among imperial powers, the freedom of Belgium from German occupation, the evacuation of Russian territory and the establishment of a "Universal Association of Nations". German armies were driven back towards German territory and their surrender occurred in late 1918.

By the war's end, four major imperial powers - the German, Russian, Austro-Hungarian and Ottoman Empires - had been defeated and the last two empires had disappeared. A new revolutionary state, called the Soviet Union emerged from the Russian Empire, while the map of Central Europe was completely redrawn into numerous smaller national states. A total of about seventy million armed men were involved in this global conflict. War raged in Europe, Africa, Asia - around Japan, and on the seas - especially in the Atlantic. When it ended, it left death and destruction everywhere. More than *fifteen million* human beings were dead. A huge amount of capital, in the hands of the European imperial powers, accumulated by the exploitation of human beings everywhere, had gone up in smoke. Intensive capitalism had led to intensive warfare, intensive blood-shed and intensive destruction - a dubious achievement of "capitalism" at its zenith. This was the beginning of the end of intensive capitalism in Europe - though this was not clear at that time.

After the fighting was halted through a cease-fire, a peace conference was convened in Paris to work out a settlement of war claims. The United States proposed the "League of Nations". The purpose of this organization, ostensibly, was the prevention of future conflicts but the motivation behind it was also the extension of the power of the United States into Europe. When the negotiations ended, Germany alone was blamed for the war and the punitive Treaty of Versailles was signed. According to this Treaty, all German colonies were taken away and were distributed among the victors. The German army and navy were to be restricted in size. German merchant fleet was taken over by Britain and all German foreign assets were seized.

In addition to other economic penalties, Germany was to pay a large amount in cash or kind, immediately. Additional reparations were to be determined later. Several territorial adjustments were also made to German borders, *dividing the German nation* and creating large German communities in Czechoslovakia and Poland. East Prussia was separated from the main German territory by provision of a "corridor" to Poland for access to the Baltic Sea. Also, coal-rich territory bordering France was handed over to France with a provision for a referendum. In total, Germany lost about one-eight of its pre-war territory in Europe. France was afraid of the size of the German nation and its proportionately large economy, which, in the future, was likely to lead to dominance of Europe by Germany. Its preference was to break up Germany into several parts - as had been the case before Bismarck's regime unified most of the German nation. The other imperial powers did not accept that, because such a division would have left France as the dominant power on mainland Europe - a situation that Britain, among others, did not want to arise either. The territories, acquired because of the collapse of the Ottoman Empire during the war, were treated somewhat like the former German territories. The European territories surrendered by the Ottoman Empire were allowed to become independent and several small European national states came into being. Arab territories wrested from the Ottoman Empire were taken over by Britain and France. At this point, the French possessions in Arabia included Morocco, Algeria, Tunisia and Syria. The rest of Arabia, including the Arabian Peninsula, was handed over to Britain. A Commission was set up to supervise the implementation of these provisions of the Treaty.

Karl Marx and His Radical Ideas

We have to carefully consider the radical ideas that Karl Marx put forward during the second half of the nineteenth century, because *these ideas gave rise to the Russian revolution during the First World War and had far-reaching consequences for the whole world in the next century*. Karl Marx was born in 1818 in Germany. His parents had converted to Christianity from Judaism. He studied mostly at the University of Berlin. There, he obtained his doctorate in Philosophy in 1841. His Doctoral Thesis was on the Philosophy of Democritus and Epicurus. His exposure to Greek philosophy had an effect on his scientific thinking and the kind of terminology he later developed in his writings. After graduating, Marx naturally looked for work, but his radical ideas made an academic career impossible. When he could not find work in Berlin, he started a newspaper named *Rheinische Zeitung*. His views about the government of Germany and its ruling class generated deep hostility in the government and his newspaper was banned. Finding himself with no hope of finding a job in Germany, Marx left Germany for France with his wife. There he met Frederick Engels, who was the son of a German textile-mill owner. Engels had acquired an intimate knowledge of the economic and social system of that time and the condition of the working class, while working in his father's textile mill in Manchester. The two men became life-long friends. They saw the developments of those days in France. A war was on between Germany and France and had led to the creation of a workers' "state" in Paris for a short time. That event, which later became known as the "Paris Commune", had an intense effect on their thinking. From that time on, Engels undertook to support Marx so that he could devote himself to his writings about the social system that had developed in Europe at that time.

After the Paris Commune had been crushed with large-scale bloodshed of the working people of Paris, Marx had to move to Belgium. Marx and Engels wrote the "Communist Manifesto" there. Marx was not tolerated in Belgium either and had to move to Britain. Britain tolerated this "crazy man" for the rest of his life. He continued to write and spend most of his time in the British Library in London, where the space he used to occupy is marked now for visitors. His family life was very tragic. His wife had inherited some property and this helped the family in its subsistence living for some time. During that time Marx's sole income was derived from articles he wrote for the then socialist New York Tribune. He was the European correspondent of that newspaper for about ten years in the mid-nineteenth century.

Engels, working again in the Manchester factory, faithfully contributed to Marx's support. Marx and his wife had seven children, but four of their children died before becoming adults, due to poverty and sickness. Two of his daughters committed suicide after marriage. Despite hardships of his life, Marx remained steadfast in his determination to complete his work. He maintained about forty notebooks at his home and when he would finish writing about some topic, he would hand over his papers to his wife. Only his wife could understand his terribly illegible writing. She used to transcribe his writings into one or another of the notebooks. Sometimes, Marx would take time off from his writing to take part in a meeting of workers. In 1864, he joined the International Workingmen's Association. He continued his work till the end of his life. No revolution took place in his lifetime. About three decades after his death, suddenly it was realized that his ideas were the ideas whose time had come, as the Tsarist Empire exploded in a revolution led by his disciples!

Marx devoted most of his time to an analysis of the social system as it existed in Europe in the middle of the nineteenth century. He named the system 'Capitalism" to indicate that it was driven by an insatiable greed to accumulate maximum amount of capital, without any moral consideration whatsoever. We refer to this stage of human social development as *intensive capitalism*. It was a continuation of what we would call "*advanced feudalism"*. The insatiable greed of the ruling classes was causing immense pain and suffering to working people and their families. Marx authored several books and pamphlets. However, his findings were published, mainly, in a three-volume book entitled "*Capital*". Here are some of the important ideas that he and Engels put forward in their writings:

"The [intensive capitalist class] has through its exploitation of the world market given a cosmopolitan character to production and consumption in every country. To the great chagrin of [reactionaries] it has drawn from under the feet of industry the national ground on which it stood. All old-established national industries have been destroyed or are daily being destroyed. They are dislodged by new industries, whose introduction become a life and death question for all civilized nations, by industries that no longer work up indigenous raw material, but raw material drawn from the remotest zones; industries whose products are consumed, not only at home, but every quarter of the globe. In place of the old wants, satisfied by productions of the country, we find new wants, requiring for their satisfaction the products of distant lands and climes. In place of the old local and national seclusion and self-sufficiency, we have intercourse in every direction, universal inter-dependence of nations. And as in material, so also in intellectual production. The intellectual creations of individual nations become common property. National one-sidedness and narrow-mindedness become more and more impossible, and from the numerous national and local literatures, there arises a world literature." [6]

This is how Marx observed the beginnings of globalization of economic activities of intensive capitalism, both in terms of production of material products and intellectual creations - like writings, music, and much later, software and processes for development of medicines and technology in general.

"The mode of production of material life conditions the general process of social, political and intellectual life. It is not the consciousness of men that determines their existence, but their social existence that determines their consciousness." [7]

Human beings are social beings. We are born in social institutions - a family first of all. Our existence is social and we are connected to other individuals by social relationships, including those related to economic activities. Thus, we develop consciousness in our social existence. Since the fundamental mode of production determines the superstructure of social, political and economic processes, our consciousness is to a great extent determined by it.

"The history of all hitherto existing society is the history of class struggles." [8]

This seems to be a simple statement about human history, but it encapsulates a very long period of human evolution. We have dealt with the evolution of mankind since the emergence of homo-sapiens and up to development of *intensive capitalism* in Europe, in previous chapters. This period was spread over a hundred to two hundred thousand years. Focusing on the last six or so thousand years, we see this phenomenon in extended families, nomadic tribes, tribes settled on land and engaged in agriculture, nations, national states, multi-national states and empires. In all these stages of evolution we see that the leadership of a social group consisted of, relatively, a few individuals who used and abused the others and there was always a struggle between the rulers and the ruled, the rulers basically living a parasitic life on the work and efforts of the rest of the social group. Also, conflicts between tribes, nations and even multi-national states have been led by the ruling classes of the social groups involved - these were conflicts, in reality, between the respective ruling classes. In making this statement, Marx combined all the previous stages of social development into one stage, which he referred to as "feudalism".

"Our epoch, the epoch of the [intensive capitalists], possesses, however, this distinctive feature: it has simplified the class antagonisms: Society as a whole is more and more splitting up into two great hostile camps, into two great classes directly facing each other: [intensive capitalists] and [industrial workers]." [9]

In the middle of the nineteenth century, *intensive capitalism* was emerging in Europe with the destruction of the old order - *advanced feudalism*. Factories and mines were being set up everywhere in Europe and agricultural production was being mechanized. When the mode of production characterized by *intensive capitalism* makes progress in a feudal society, both cities and the rural areas of a state are dramatically transformed. Corporations owning one or more factories get established and continue to extend their activities. The same happens to corporations in the mining, transportation, banking and other "service" sectors of the economy. The result is that the demand for labor continues to grow in the population centers, i.e., cities and towns. The opposite effect is experienced by the rural areas. Because of the occupation of small landholdings by big landlords and growing mechanization of agriculture, the peasants are thrown out from the farms on which they may have worked for generations. These agricultural workers then end up moving to the cities in search for work. Some are able to find work. Others die along with their families because they cannot find enough to eat, or they become sick and do not receive medical attention. Contagious diseases spread rapidly in the crowded and un-hygienic conditions in which these workers and their families are forced to live. As industrial production and mechanization of agriculture advances, society splits up into two hostile camps or two great classes facing each other - the ruling class of Industrial owners that owns the mechanized farms and other manufacturing and service industries and "businesses", and the oppressed class of destitute industrial workers, or *"proletarians"*.

"The essential condition for the existence, and for the sway of the [intensive capitalist] class, is the formation and augmentation of capital; the condition for capital is wage-labor. Wage-labor rests exclusively on competition between the laborers. The advance of industry, whose involuntary promoter is the [intensive capitalist class], replaces the isolation of the laborers, due to competition, by their revolutionary combination, due to association. The development of modern industry, therefore, cuts from under its feet the very foundation on which the [intensive capitalist class] produces and appropriates products. What the [intensive capitalist class], therefore, produces, above all, is its own gravediggers. Its fall and the victory of the [industrial working class] are equally inevitable." [10]

As agriculture and other industries develop, workers are brought together in relatively small factories and are made to work for starvation wages, since there is an over-supply of labor and workers are not able to bargain for higher wages under those market conditions. They, and their family members including children, are forced to work long hours on wages that are not sufficient even to feed them. Under these conditions, workers, despite their lack of knowledge, are able to see the cruel workings of the economic system and the vicious behavior of their exploiters - the owners of business and industry. Thus, the system of intensive capitalism itself produces a class of abused workers who are ready to bury the system itself, if they can, just like grave-diggers. Marx, at his time, thought that this class of "grave-diggers" would become so large and powerful that it would crush and destroy intensive capitalism. It would seize economic and political power and would embark on establishment of a new, humane economic system. Such a revolution did take place but not where Marx had expected. It occurred in the Tsarist Empire and not in Germany, or Western Europe. As it turned out, the political and military system of control in Central and Western Europe was so strong that it did not give way under the pressure of the working class. The ruling class let loose large-scale violence by its military and police forces and mass murders were conducted of workers and those who led their labor and political organizations. This intense violence ended up burying the "grave-diggers" themselves. Even when large-scale military suppression was not in progress, tens of millions of workers and their families were dying of starvation, disease and economic abuse. Others remained confused by the mass media of the ruling classes and died quietly without any protest. This *Intensive Death Rate* and the accompanying growth of industry ultimately brought about a shortage of labor. Workers were, then, able to form labor unions and demand wages higher than those of wage-slaves.

"Does it require deep intuition to comprehend that man's ideas, views, and conception, in one word, man's consciousness changes with every change in the conditions of his material existence, in his social relations and in his social life?"

"What else does the history of ideas prove, than that intellectual production changes its character in proportion as material production is changed? The ruling ideas of each age have ever been the ideas of its ruling class." [11]

Based on the idea that the "ruling ideas of each age have been the ideas of its ruling class", Antonio Gramsci developed the concept of "cultural hegemony" of the ruling class. He argued that consent by the people to the rule of the dominant class is achieved by the spread of collections of ideas, consisting of assumptions, beliefs, values and the resulting world views. These ideas are promoted, enforced and reinforced by means of social institutions including educational institutions, media, families, religious institutions, political groups and institutions, and the laws of a state, etc. Thus, the people are socialized into the norms, values, and beliefs of the ruling class. Since the ruling class owns and controls these institutions that maintain social order, it achieves its rule over the minds of virtually all individuals in a society.

Cultural hegemony is strongly displayed when those ruled by the dominant class come to believe that the economic and social conditions of their society are natural and inevitable, rather than created by people with vested interests in particular social, economic, and political objectives. This results in the belief that the economic system, the political system, and a class stratified society are legitimate, and so is the rule of the dominant class.

Gramsci developed this concept of cultural hegemony in an effort to explain why a violent upheaval and a revolution, that Marx seemed to have predicted, had not occurred. Marx had recognized the important role that ideology played in reproducing the economic system and the social structures that supported it, but Gramsci believed that Marx had not sufficiently stressed the power of ideology to reproduce the social structures through institutions like religion, education and news and entertainment media. He felt that intellectuals in a capitalist society, because of their privileged status, were not detached observers of social life, but actually functioned as agents of the ruling class, teaching and encouraging people to follow the norms and rules established by it. Gramsci also looked at the role of "common sense" in promoting cultural hegemony. For example, the idea that one can succeed financially if one just tries hard enough, is a form of common sense that serves to justify the system. For, if one believes that all it takes to succeed is hard work and dedication, then it follows that the system of capitalism and the social structures that are organized around it are just and valid. It also follows that those who have succeeded economically have earned their wealth in a just and fair manner and that those who have not, have earned their poverty. This form of "common sense" promotes the belief that success and social mobility are strictly the responsibility of the individual only. It hides the real processes leading to class, racial, and gender inequalities that are built into the capitalist system.

The function of cultural hegemony is to "manufacture" consent of the people to be ruled by the ruling class, under what the ruling class considers its "just and legitimate laws meant for benefit of all". Thereby, the people are also expected to consent to, and accept, their own exploitation as "normal" This is distinct from dictatorial rule by force, because it allows those in power to achieve rule non-violently, by using ideology and culture. Cultural hegemony is frequently referred to as the "soft power" of a state.

"Between capitalist and communist society lies the period of the revolutionary transformation of the one into the other. There corresponds to this also a political transition period in which the state can be nothing else than the revolutionary dictatorship of the proletariat (i.e., the working people)." [12]

Marx believed that when the existing intensive capitalism was overthrown by the people, the people would set up a new economic and political system in which the working class would be the ruling class. He had defined communism as a social system in which human beings would live like one family, i.e., a system that would be based on the principle - *from each according to his ability to each according to his needs*. This is the principle on which the human family functions. Parents bring up their children. They work to the extent that they can and try to provide for all the needs of their children. When they grow old and the children grow up to be able to work then they work according to their ability and support their parents. There is no accounting of who had worked how much and who deserved more or less. All members of an ideal family make an effort to provide for the needs of everyone in the family. The basis of this family relationship is the natural love for each other. When the existing social system of intensive capitalism would be overthrown, Marx thought, the working class would not be able to immediately set up the new system on new rules. This would be done gradually. In the meantime, he thought, during the transition phase, a revolutionary dictatorship of the proletariat would exist. He did not elaborate this concept clearly, in enough detail, to avoid confusion. His thinking was that under the system of intensive capitalism, in which elections do take place and multiple political parties exist, the class of industrialists and business owners always stays in power. Thus, effectively, under intensive capitalism a *de facto* dictatorship of this class exists. His thinking seems to be that the working class would set up a system to achieve a similar result, i.e., the working class would stay in power despite elections and functioning of multiple political parties and, thus, a dictatorship of the working class ("the proletariat") would exist. When the transition period, which may last a long time, is over, and the new economic and political system has been set up, leading to removal of all major class differences, only the working class would exist. Thus, at that stage, the new communist society would emerge and there would be no need for a "dictatorship" of this class over another.

The Paris Commune had been such a dictatorship although the representatives of the people were elected to the city council. Marx had called it the "government of the people, by the people". Looking at all the changes that the Commune had brought about, Engels had remarked that the *Paris Commune was the Dictatorship of the Proletariat*. After the Russian revolution, the Communist Party of the Soviet Union, led by Vladimir Lenin, set up a military-style dictatorial system in the Soviet Union and incorrectly claimed that it was based on the concept of the Dictatorship of the Proletariat, as proposed by Marx. This led to intense disagreements and polemical debates among the various communist parties of the world.

"In place of the old [capitalist] society, with its classes and class antagonisms, we shall have an association, in which the free development of each is the condition for the free development of all." [13]

Marx thought that, after the revolution, the working class would wrest, by degrees, all capital from the class of capitalists. It would centralize the means of production in the hands of the state. In the beginning this would have to be done by despotic inroads on the rights of property and by means of measures which would facilitate the transition to a classless society. This transition stage was named "socialism". These initial measures included transfer of all land ownership to the state, a highly progressive income tax, abolition of all rights of inheritance, exclusive bank ownership by the state, centralization of all means of communication and transport in the hands of the state, extension of state-ownership of industries, free education for all children in public schools. (The exact description of these measures, as outlined in the Communist Manifesto, is given and discussed in Chapter-10, with reference to China's social system at this time.) A free society was thus expected to emerge in which each individual would be able to develop to his full potential. How these changes were actually brought about in the Soviet Union, China and other states which went through revolutionary changes, and what effect these changes had on the imperial powers and their societies, would be described later.

> **"National differences and antagonisms between people are daily more and more vanishing, owing to the development of the class of industrial and business owners, to freedom of commerce, to the world market, to uniformity in the mode of production and in the conditions of life corresponding thereto. The supremacy of the [industrial workers] will cause them to vanish still faster. United action, of the leading civilized countries at least, is one of the first conditions for the emancipation of [industrial workers]. In proportion as the exploitation of one individual by another is put to an end, the exploitation of one nation by another will also be put to an end. In proportion as the antagonism between classes within the nation vanishes, the hostility of one nation to another will come to an end."**
> [14]

This follows from the concept of a class-less society. If differences and tensions between individuals are due to exploitation of one by another, these tensions would disappear when the exploitation is brought to an end. This would also happen to the hostility between nations when the class differences within nations are ended, resulting in the end of the hostility between the former ruling classes of nations.

> **"The highest form of the state, [i.e., the state based on a ["representative government"]], which under our modern conditions of society is more and more becoming an inevitable necessity, and is the form of the state in which alone the last decisive struggle between the proletariat and the [capitalist class] can be fought out. The [state based on "representative government"] officially knows nothing anymore of property distinctions. In it wealth exercises its power indirectly, but all the more surely. On the one hand, in the form of the direct corruption of officials, of which America provides the classical example; on the other hand, in the form of an alliance between government and the stock exchange, which become the easier to achieve the more the public debt increases and the more the [corporations] concentrate in their hands not only transport but also production itself, using the stock exchange as their center."** [15]

This statement by Engels is in accordance with Marx's thinking. Thus, according to Marx and Engels, the final struggle for supremacy between the class of industrial and business owners and the class of working people, would have to be fought in the modern state structure, which includes a pseudo-representative government like that of the present-day United States, that includes multiple political parties and elections, etc., while the dictatorship of the industrial and business class prevails. Since Engels wrote these words near the end of the nineteenth century, his reference to America, i.e., the US, seems to be about what is called the "gilded age" in US history.

The Russian Revolution

As indicated before, at the beginning of the twentieth century Europe was in turmoil, boiling with unrest and discontent. Legal and underground revolutionary parties had emerged everywhere. The people of Europe were rebelling and demanding better living conditions since their lives had been torn apart by the system of intensive capitalism. Secret armed groups had formed and had started violent attacks on state institutions. This was true of whole of Europe - more so in the Tsarist Empire, which had violently suppressed all dissent for decades. Rebellions occurred there with increasing frequency and were suppressed violently.

A big uprising occurred in 1905. Workers staged strikes and protest marches everywhere. These protests were suppressed with unusual violence. At this time two underground political organizations existed in the Empire. Members of the underground Social Democratic Party of the Tsarist Empire were, mainly, leading the struggle of the working class in the urban areas and considered themselves to be the representatives of the industrial workers. The Social Revolutionaries had organized the peasantry. The peasants, in contrast to all the rest of Europe, were still in a revolutionary mood due to their intense suppression by the Tsarist regime, and were supportive of Social Revolutionaries and the Social Democratic Party. During this uprising, the Social Democrats began to split up into two groups. The Mensheviks (i.e., "those who were a minority", or "Whites") held that no revolution from feudalism to socialism was possible in the existing economic conditions in the Tsarist Empire, unless the revolution coincided with a European socialist revolution. The Bolsheviks (i.e., "those in majority", or "Reds") believed that a socialist revolution was possible despite the backwardness of Russian society. Although all political organizations were banned by the Tsarist regime, the process of organizing the workers was continued. The workers of each trade came together and chose delegates, who united to form a council of delegates, or a Soviet. Thus, a comprehensive organization of workers was created, which included all kinds of wage earners.

In February 1917, the Tsarist regime was overthrown and replaced by a provisional government of most political parties. This provisional government, headed by Kerensky, was to set up a constituent assembly to draft a constitution for the new republic. The provisional government was not able to do this and was hopelessly deadlocked because of sharp disagreements between various political parties. The main conflict was between the two wings of the Social Democratic Party of the Tsarist Empire. The contrast became more acute with time. The Whites considered it to be their task to take part in a provisional coalition government until the duly constituted constituent Assembly had formed a definite government. The Reds endeavored, even before the meeting of the constituent assembly, to overthrow this provisional government, and replace it by a government of their party. An additional ground of opposition came with the question of peace. The Whites wanted immediate peace as much as the Reds. Both wanted it on the basis of *no annexations or indemnities*. The Whites had initially been in the majority in the joint meetings of the social democrats and they wanted a general peace, and all belligerents to adopt this position. However, as long as this was not achieved, they wanted the Russian army to be kept in a state of armed readiness. The Reds, on the other hand, demanded immediate peace at any price, and were ready, if necessary, to conclude a separate peace, and they sought to enforce their views on the men of the army. They were supported by the war weariness of soldiers of the army, and the people in general, and were helped by the apparent inactivity of the dysfunctional provisional government.

The *"Russian Revolution"* started in late October, 1917, beginning with the rebellion of Russian naval personnel in the city of Petrograd (now re-named Saint Petersburg). The provisional government was removed and replaced with a government of the Reds, who had renamed themselves the Communist Party of the Soviet Union. These radical changes were based on the supposition that the Russian Revolution would be the starting point of a general European revolution, and that the bold initiative of Russia would summon the industrial workers of whole of Europe to rise. The European revolution was, thus, supposed to defend the Russian revolution and completely liberate the Russian people also. The Soviets of 1905 were local organizations confined to single towns. The process of formation of Soviets was extended to the whole territory of the state. In the economic structure of Russia, the Soviets could only attain the position of authority in 1917 by not confining themselves to the industrial workers of the towns as in 1905. This time the soldiers and peasants were also organized. With the disbanding of the army the soldiers had lost their power due to their military status. The committees, or Soviets, of 1917 were not only more numerous, but closely knit together. Single Soviets were affiliated to, and elected, a higher regional Soviet, which in its turn was to elect a Soviet for a larger region or district. Thus, these several layers of regional Soviets ended with electing a central committee covering the whole territory of the state. The Reds, who, together with the Social Revolutionaries, obtained a majority in the workers organizations after the social explosion of late 1917, proceeded to transform these Soviets into elements of a political system of government. The Soviet Union was to be the *"organized dictatorship of the proletariat"*, the only means, as Lenin expressed it, whereby the *most painless transition to Socialism* was made possible. This was to be done by denying political rights to all those who were not represented in the Soviets. Immediately after the takeover of the government by the Reds, the new regime was confirmed by the second Congress of Soviets. However, this process was opposed by a strong minority, which left the Congress in protest.

In 1918, a constitution of the Soviet Union was drafted. It laid down that not all the inhabitants of the Soviet Union, but only specified categories had the right to elect representatives, or "deputies", to the Soviets. It was decided that all those may vote *"who procure their sustenance by useful or productive work"*. Those who were excluded from the franchise included those who employed wage laborers for profit. A home worker or an electrician with an apprentice, who may have lived and felt quite like an industrial worker, had no vote. Even more workers in business and industry became disfranchised by the definition which aimed at depriving private traders and middle men of the vote. The worker who lost his work, and endeavored to get a living by opening a small shop, or selling newspapers, lost his vote. Another clause excluded from the franchise *everyone who had unearned income,* for example, dividends on capital, profits of a business, rent of property, or interest on a bank account.

Quite a number of workers, especially in the small towns, owned a little house, and, to survive, rented part of their houses. They also lost their right to vote. How big the unearned income had to be to cause the loss of the vote was not stated, leading to abuses on a large scale in the Soviets, since the Soviets themselves were authorized to conduct their own elections.

The constitution also provided that "In the interest of the working class as a whole the Soviet Union may withdraw rights from any persons or groups who misuse them to the detriment of the Socialist Revolution." This effectively declared the whole opposition to be outlaws, for every Government, even a revolutionary one, discovers that those in opposition *misuse their rights*. Yet even this was not sufficient to ensure the *painless transition to Socialism*. After the Reds got rid of the opposition of the Whites and the right wing of the Social Revolutionaries within the Soviets, a great fight broke out between them and the left Social Revolutionaries, with whom they had formed the government. Most of them were also driven out of the Soviets. Thus, the whole process of setting up of the new Soviets and representation in them was rigged to ensure that only Lenin's Reds were really represented in them. This was a huge blunder, since the Reds were a small minority in the working class and these actions generated opposition to the government from every other group.

The question remains, as to why Lenin encouraged such unnecessary and clearly undemocratic actions. During his struggle against the Tsarist regime in Moscow, he certainly had suffered much. His brother had joined a group that was carrying out terrorist attacks on the functionaries of the Tsarist regime. He was captured by the army of the Tsar and was hanged to death. Lenin was also arrested several times. Once, when he was in prison, his mother went to see him. The prison guard told her that if Lenin did not reform himself, he would also be hanged like his brother. One can easily imagine her anguish and suffering at that time. *It seems these times had an indelible impact on Lenin's mind also.* After the revolution when he was leading his party's government, many times there were disagreements and intense debates over many issues of political organization, the setting up of new government institutions and the program of the government. These included: the setting up of the new Soviets; the speed of government take-over of capitalist-controlled industry; take-over of royal property and large mansions owned by the ruling class and; collectivization of agriculture. Lenin's approach was basically democratic - the central committee functioned on the basis of one person one vote and decisions had to be endorsed by the majority. Lenin was among those who lost a vote many times and accepted it. But he consistently pleaded for extreme measures and was not willing to make compromises with those who had a minority support in the Central Committee. It seems this intolerant attitude was because of his personal experiences with the ruling class headed by the Tsar - especially *the hanging of his brother*. The result was that a one-party dictatorship was created. The "one party" consisted of only the Red faction of the Social Democrats. The Whites, the Social Revolutionaries and the "capitalists", even one-man businesses, were deprived of the right to vote. The communist Party of the Soviet Union was made supreme in every facet of state life and was declared the *Vanguard of the working class* and no other party was to be allowed to operate in the state.

It was decided that all local social and economic institutions were to be managed by the Soviets. Large farms were to be taken over by the state and their land was to be distributed to agricultural workers on those farms. All factories, businesses and educational institutions were to be taken over by the state and were to be managed by the local Soviets. The highest Soviet, i.e., Central Committee of the Soviet Union, became the legislature. It elected a small executive committee, called the Politburo, to run the government. It was headed by the General Secretary of the Communist Party. Thus, in practice, the General Secretary and his politburo ruled the country and its decisions were conveyed to the lower committees, or Soviets, for implementation.

At the time of the Revolution, more than 80% of the population of the Soviet Union was engaged in agriculture. The Soviet society was, basically, an advanced feudal society at that time. Industrialization had not yet advanced to the level it had in the rest of Europe. There were landlords who had huge landholdings and there were many small land owners called Kulaks. Share-croppers and serfs actually worked on the lands owned by the landlords. The Reds decided to collectivize agriculture under these circumstances. This was, however, based on a flawed understanding of agricultural production and the traditional rural population engaged in it. The idea behind collectivization was that when small landholdings were combined and the families, who worked on the lands separately, began working together on the combined land, this would result in savings in terms of effort required for cultivation. The farm animals could be taken care of together and could be shared for ploughing the fields. Thus, land tilling, seeding and watering could be done together. All this supposed to save labor and result in bigger crops. Thus, it was expected that, due to the *economic advantages of the new system* of land use, agriculture would show higher growth. It was also expected to create camaraderie amongst the former peasants. *However, what was not realized was that the peasants were primitive people, as is the case in any advanced feudal society, and they felt attached to their pieces of land, which had been passed to them by previous generations. They did not want to share their lands, or the work involved in cultivation, with anyone else.* Besides, working together meant that the privacy of families would be violated. Due to these reasons, collectivization was resisted by the peasants, especially by the Kulaks. Initially collectivization was voluntary, though under psychological pressure. Later on, during Stalin's period, it was forced on unwilling peasants and caused resentment amongst them and led to a drop in production. Instead of such collectivization, if land trading organizations had been set up to buy land from those who wanted to sell and selling it to those whose farms would become economical for mechanized farming, the results would have been dramatically different. Alternatively, land could have been taken over by the state by decree, without displacing the families on it. Each family could lease-back the land, it was occupying, for a long enough period of time, say, 30 years. After expiry of such leases, transfer of such land could be accomplished by continuation of the leasing process, to members of the same family or another family, by sale/transfer of lease-hold rights. Such management of agricultural land would not have been painful for the rural population and could have been designed to increase the size of holdings for mechanized cultivation. The lease rates could be designed to provide the state with tax revenue as required. This tax revenue, then, could be used to set up agriculture research stations and institutes and produce agricultural machinery and chemicals. But this was not done, causing immense suffering to the rural population.

Later on, the economic system was further refined. A central *State Planning Committee*, also called the *"Gosplan"*, was set up to formulate long term economic plans, determine movement of raw materials to the farms and factories from other farms and factories and to *determine the prices of their products*. Thus, this one-party political system required all central authorities to convey their decisions to numerous Soviets spread over the length and breadth of the Soviet Union and these local Soviets were to carry out the orders regarding raw material acquisition and pricing of their products to the local collective farms and factories. Thus, huge bureaucracies were set up to deal with various military and economic activities. The Gosplan, the military and the secret service, called the Committee for State Security, were made subordinate to the politburo. Thus, in effect, the military and civilian population of the country was to function like a set of military units under the command of the Central Committee of the Communist Party. All economic life was reorganized on these lines.

As the Soviet Union was set up, Karl Kautsky of the German Social Democratic Party emerged as one of its strongest critics. In his book entitled "Dictatorship of the Proletariat", he summarized his views on the subject as follows:[16]

"That by an object lesson of this kind in the more highly-developed nations, the pace of social development may be accelerated, was already recognized by Marx in the preface to the first edition of Capital:

'One nation can and should learn from others. And even when a society has got upon the right track for the discovery of the natural laws of its movement - it can neither clear by bold leaps, nor remove by legal enactments the obstacles offered by the successive phases of its normal development. But it can shorten and lessen the birth-pangs.'

In spite of their numerous calls on Marx, our Bolshevist friends seem to have quite forgotten this passage, for the dictatorship of the proletariat, which they preach and practice, is nothing but a grandiose attempt to clear by bold leaps or remove by legal enactments the obstacles offered by the successive phases of normal development. They think that it is the least painful method for the delivery of Socialism, for 'shortening and lessening its birth-pangs'. But if we are to continue in metaphor, then their practice reminds us more of a pregnant woman, who performs the most foolish exercises in order to shorten the period of gestation, which makes her impatient, and thereby causes a premature birth. The result of such proceedings is, as a rule, a child incapable of life."

History has shown that the Soviet Union was not a child incapable of life, but it did not have a very long life either. Seventy years, or so, is a very short life for a state. Intolerance led to civil war. Forced collectivization of farms led to inefficient agriculture. The total suppression of the economic market, ultimately, led to stagnation. The total suppression of dissent and the creation of a one-party state, on its basis, led to the dictatorship of one political party and, ultimately, to dictatorship of one individual. It led to disaster when Stalin, a mentally sick monster, became the dictator.

Civil War and the Imperial Invasion of the Soviet Union

The way the Soviet state had been established, and the level of intolerance displayed by Lenin's Communist party, resulted in completely alienating the Whites, the Social Revolutionaries, the officer corps of the old Russian Army, the capitalists, house owners of all categories and skilled workers who worked independently. Reds had seized complete control of all state institutions, despite the fact that, initially, they were a minority in the Soviets. Those who were deprived of any share in power, especially the officer corps of the Russian army, organized themselves into an armed force and an armed conflict started. Thus, revolutionaries of Russia started fighting each other and a civil war was started, in which the well-trained soldiers and officers of the former Tsarist Russian army were cooperating with the Whites. The Soviet Union was rightly considered a great threat by all imperial powers, including the US, and these imperial powers sought to put an end to it. The US, immediately, cut off the financial help it was previously extending to the Tsarist regime for keeping an Eastern front active against Germany. It also extended the maritime blocked of Germany to the Soviet Union. The civil war had provided the imperial powers with an ideal opportunity to completely destroy the Soviet Union.

At the end of 1917, the Bolsheviks effectively controlled Petrograd, Moscow and the territory between these two cities. With the fall of the Tsar, several parts of the Russian empire took the opportunity to declare their   independence. Finland did so in 1918 and a civil war also started there. The Whites in Finland were helped by the Germans, who naturally thought of Finland as their new imperial possession. The Finnish Whites were, thus, able to push back the Finnish-Russian border till Petrograd was almost in the range of their artillery. Within Russia itself, those who opposed the Reds looked to the western powers for help. For their own benefit, the western powers wanted to completely wipe out the Red menace but, publicly, they stressed the intention to "help" Russia so that an Eastern front would be created for the German army to split it once again, thus relieving the problems being experienced by the imperial powers on the Western front.

As soon as the cease-fire was reached among the main belligerents in Central and Western Europe, virtually all the imperial powers, including the United States, Japan and many European states, invaded the Soviet Union. The imperial invasion came from all sides. In the Northern territories, the British and French forces set up puppet governments led by Social Revolutionaries. Also, a combined American, British and French invasion was launched from the North-East of Finland. The Germans attacked on a wide front from Ukraine to Finland, from the territory they had already occupied after the treaty of Brest-Litovsk. Kerensky's forces had some successes due to the soldiers known as the Czech Legion. They ultimately captured so much territory that they became a serious problem for Trotsky and his task of defeating the Whites was, thus, made a great deal more difficult. The United States, and other imperial powers, besides Japan, were busy imposing and expanding their control on Chinese territory and Siberia, at that time. Britain, France and Portugal were also involved in these efforts. The Social Revolutionaries and monarchists had set up their own regimes in Eastern Russia and Siberia creating many difficulties for the Reds.

Figure 6-03: American Troops in Vladivostok.–Japanese Troops Watching

Unknown
(https://commons.wikimedia.org/wiki/File:American_troops_in_Vladivostok_1918_HD-SN-99-02013.JPEG), „American troops in Vladivostok 1918 HD-SN-99-02013", marked as public

The United States sent its troops first into Vladivostok from China, where it was in control of its puppet Chiang regime. The US troops occupied the territory around Vladivostok and were joined by troops from Australia, India and Canada. The American aim was to create a separate state in Siberia and to wrest this territory from the Soviet Union. Tsarist general, Kolchak, after several maneuvers, created his *"Republic of Siberia"* with US support and conquered a large part of the region by 1919. Japan invaded Eastern Siberia and occupied the Kamchatka peninsula and other Soviet territory close to its empire in the East. From the Black sea, British troops invaded towards the East and went towards their main objective - the oil producing region of Baku. Also, from the Black sea, British troops attacked in a North-East direction into Russia. The French attacked northwards from the Black sea, into what is now Ukraine. A full-fledged invasion of the Soviet Union from all sides was thus underway. In about a year of fighting, most of the Soviet territory had been captured by the invading armies. *Only Moscow and, roughly, an area with a radius of about 250 miles around it remained in the hands of the Red Army. Perhaps ninety percent of Soviet territory was in the hands of the Whites and the imperial powers at that time.* The Whites, however, had a terrible reputation regarding their treatment of the people under their control. Most of the population of the state had been engaged in agriculture. People had been promised land-ownership. The Whites, however, wanted to restore Russia of the "good old days". This naturally aroused hostility among the peasants under their control. As the imperial powers were forced to withdraw from Russia, the Whites were unable to resist the Red Army, which drove Kolchak and his rapidly disintegrating forces back to Siberia where he surrendered. His execution put an end to his *"Republic of Siberia"*. White forces in the south of Russia were evacuated from the Crimea in late 1920. The Soviet Union was, thus, able to recover from a disastrous situation. By the end of 1921, most of the Soviet territory had been recaptured by the Red Army, organized and led by Trotsky. Some resistance, however, continued in Central Asia and the far Eastern part of Siberia.

The Soviet Union accepted the independence of Finland. Some territory belonging to the Ukraine and Belarus remained under the control of Poland and Germany. Also, three Soviet Republics - Latvia, Estonia and Lithuania could not be re-occupied by the Red Army. The whole Royal family was executed by the Reds. The committee for state security, or Cheka (later renamed the KGB) was entrusted the job of tracking down opponents of the regime.

After succeeding against imperial powers and their forces on Russian territory itself, Trotsky faced another military threat from Poland, which had become independent in 1918. Sensing the weakness of the Soviet Union because of its civil war and the imperial invasion, it invaded the Ukraine in 1920. However, the Polish army was not able to defeat Trotsky's Red Army, but was pushed back instead. Allies of Poland in Western Europe, then, intervened and the Soviet Union signed the Treaty of Riga with Poland, with a loss of

Figure 6-04: Territory under Soviet Control after Invasion by Imperial Powers

Hoodinski
https://commons.wikimedia.org/wiki/File:Russian_civil
_war_in_the_west.svg), "Russian civil war in the west",
Changes by Tayyib A Tayyib,
https://creativecommons.org/licenses/by/3.0/legalcode

some territory to Poland. This Treaty, finally, brought the Russian Civil War to an end. About a million soldiers and several millions of civilians had died in this conflict. *Thus, the First World War really ended in 1921.* Japan remained in control of Korea and a large part of China.

To deal with the devastation caused by the civil war and the imperial invasion, Lenin issued the *New Economic Policy*. Under this policy small businesses were allowed to be set up, the grain requisitioning system was ended and farmers were allowed to sell their produce on the open market. As a result, agricultural production grew substantially. A mixed economy developed in most consumer goods sectors and the condition of the people improved in general. There was, however, criticism of this policy from the left wing of the communist party. The *New Economic Policy* was ended by Stalin, in 1926. Instead, Stalin, with his tendencies towards paranoid delusion, carried out forcible collectivization of agriculture, resulting in large scale misery and starvation for agricultural workers.

Political Changes in Asia

The Qing Dynasty, the last of the ruling Chinese dynasties, had collapsed in 1911 and China had fallen under the control of many warlords. The Communist Party of China (CPC) was set up in 1921.The period, starting from the First Opium War in 1839, up to establishment of the People's Republic of China in 1949, is referred to, by the Chinese, as the "*century of humiliation*". During this century, Western imperial powers and Japan attacked China and created their zones of occupation on its territory. Britain waged two "opium wars" with China to force it to accept opium, being produced and transported from India, to be distributed to its population. After 1911, the Nationalist Party of China, Kuomintang (KMT), led by Sun Yat-sen, tried to re-unify China by fighting against warlords. Sun Yat-sen made efforts to obtain help from imperial powers, but did not succeed. He, also, approached the Soviet leadership and the Soviet Union directed the CPC to join forces with the Kuomintang. Thus, the *First United Front* was formed by the two political parties. By 1924, however, Chiang Kai-Shek had become the dictatorial chief of the KMT. As his power increased, he became more and more intolerant of the CPC. In early 1927 the rivalry between the two parties led to a split.

The CPC and its allies in the KMT had decided to move the seat of the KMT government to Wuhan, where communist influence was strong. Chiang, on the other hand, wanted to eliminate the communist wing of the KMT. Fighting between the two parties began in Shanghai. The KMT was purged of leftists by the arrest and execution of thousands of CPC members. This is referred to as the *Shanghai Massacre*. Starting with this massacre, the Chinese Civil War was fought between forces loyal to the KMT and forces of the Communist Party of China (CPC). The war began in 1927 and ended around 1950. The conflict eventually resulted in two de facto states, the Republic of China (ROC) in Taiwan and the People's Republic of China (PRC) in mainland China, both claiming to be the legitimate governments of China. After the Shanghai massacre, the KMT had resumed the campaign against warlords and captured Beijing. At this stage a large part of Eastern China came under the control of the KMT with its headquarters in Nanjing. The activists of the CPC had to retreat to the countryside, where they established control over several areas. These areas were referred to as "liberated areas" as the CPC had set up its administrations there. KMT army continued to violently suppress them. This marked the beginning of the ten year's struggle, known as the

"Ten Year's Civil War". It lasted until Chiang Kai-shek was forced to form the *Second United Front* against the invading Japanese.

Around 1930, the KMT again decided to completely destroy all pockets of Communist activity in a series of campaigns. These campaigns failed and Chiang's armies were badly mauled when they tried to penetrate into the heart of communist controlled "liberated areas", where they were easily encircled and forced to withdraw. In 1931, Japanese forces occupied Manchuria, a territory larger than the whole territory of Japan and set up a separate "state" under a puppet regime. Soon after this, Chiang launched another campaign to encircle the "liberated areas" to cut off their supplies. Forces of the People's Liberation Army (PLA), of the CPC, managed to escape through the gaps in the encirclement. The massive military retreat of Communist forces lasted about a year and covered many hundred miles. It became known as *the Long March*. While retreating, the PLA confiscated property and weapons from local warlords and landlords, while recruiting peasants and the poor. This increased the support of the CPC among the people. However, a large number of soldiers of the PLA died during this march before reaching Shaanxi province. The long march also made Mao the undisputed leader of the Communist Party of China.

In 1937, Japan launched a full-scale invasion of China and its well-equipped troops conquered most of the territories controlled by the KMT. Tens of thousands of Chinese civilians were massacred when Nanjing was conquered. More than twenty thousand women were raped and murdered. Many civilians were decapitated as a "sport", in Genghis Khan style. Others were bayoneted to death, set fire to, or simply buried alive. After this defeat, the KMT again reached out for help to the CPC. Both Chinese parties suspended fighting with each other and formed the *Second United Front* to focus their energies against the Japanese. This brought Japan into conflict with the imperial ambitions of the United States also, which was supporting and controlling Chiang's KMT. It was clear that Japan, also, wanted to set up a large empire in East and South-East Asia. Korea had already become a Japanese possession in 1910. Japan had named this concept of a Japanese empire as the *"Greater East Asia Co-Prosperity Sphere"!* The driving force behind Japan's imperial moves, in addition to creation of a large area with a population subjected to *Controlled Market Subjugation*, was the "need" for raw materials, especially oil for its industries. The United States responded with an embargo on Japan, dealing with strategic materials, *including oil.* Throughout its military operations in China and South-East Asia, Japan created a system of sex-slaves, or "comfort women" for its soldiers to abuse local young women.

In later stages of the conflict between CPC, the KMT, Japan and the US, the KMT was fully supported by the US, while the CPC and its PLA were supported by the Soviet Union. Ultimately, the Japanese were defeated in China when the Soviet forces attacked Manchuria and Northern Korea and the PLA advanced to take over all Chinese territory in Northern China. Simultaneously, fighting was going on between the KMT and the CPC.

Political Changes in Europe

As mentioned before, the signing of the Treaty of Riga with Poland had brought the "Russian Civil War" to a complete end. The Soviet Union thenceforth focused on its political reorganization and reconstruction, although the loss of the former Soviet Republics of

Finland, Latvia, Estonia, Lithuania and the loss of Belarusian and Ukrainian territory was acutely felt.

Germany also felt the loss of its territory to France, Czechoslovakia, Austria and Poland. The Treaty of Versailles had placed the full blame for the First World War on Germany. This was an absurd proposition, which was made an article of the Treaty, since all the European imperial powers had behaved the same way as and when they got the opportunity. The invasion of the Soviet Union was clearly and openly a shared project of all the imperial powers. How could Germany be singled out for blame? The Treaty had created a divided German nation and nationalism was bound to reunite it.

On top of the war losses, Germany was singled out for huge reparations. It was required to pay most of its foreign assets as an immediate payment of its reparations. Its whole merchant fleet and its imperial possessions, or "colonies", were also grabbed by Britain. Finally, when the invasion of the Soviet Union was over and the other signatories to the Treaty were free to focus on Germany again, a hugely absurd figure for reparations was established. The purpose, of course, was to extract the maximum possible amount of capital from the German economy. The amount was so absurdly high that Germany could not pay it except a small amount and found ways to delay or avoid paying the rest. The main imperial power, pushing for maximum loot and plunder of the German ruling class, was the French ruling class, which, having failed to get Germany divided into several parts found this to be a less desirable but acceptable outcome.

Germany found the prohibition against reunification of Czechoslovakia especially insulting since the United States had previously given long lectures about "national self-determination" to the Europeans. Germany continued to negotiate and maneuver for better terms. It joined the League of Nations and succeeded in getting the military commission, supervising German disarmament, abolished. It also succeeded in getting all the territory in French possession returned to it. It then initiated a secret program of rearmament. This program was completely un-verifiable, except by the usual means of intelligence agencies.

Throughout the post-Versailles negotiations, France tried to reach some agreement with Britain, so that Britain would come to its rescue if and when it was attacked by Germany. The rulers of Britain consistently refused, because they did not want the French to become the dominant power on mainland Europe. That would have raised the possibility of France grabbing all the hoarded capital in Europe and grabbing all the territories "belonging" to European imperial powers. The memories of wars between the two imperial powers were still fresh in the minds of the ruling classes of both these imperial powers. The French rulers, of course, remembered, with satisfaction, the success of the United States in freeing itself from British control. The Statue of Liberty must have acted as it was meant to be - a clear symbol of that British defeat and French jubilation.

In 1931, a conflict started in Spain after elections were held after a long time. One group of parties started a process of land reform. It also tried to reduce the power and privileges of the Church and wanted to introduce regional autonomy, especially in the Catalan region. The other group of parties was opposed to this program. The conflict ultimately grew to such a level that the military got involved in it and a Civil War erupted. The dictatorship in Portugal supported the right-wing Nationalists. They were also supported by Hitler's Germany and Mussolini's Italy. Troops from Germany and Italy were dispatched and weapons were also supplied by the two Fascist regimes. The Left-wing Republicans were supported by international volunteers from many countries, including the United States. Writers, Ernest Hemingway and George Orwell were among them. Later on, Stalin provided arms to the Republican side. After a long and bloody struggle, the civil war came to an end in 1939. About half a million people died in this war, including about a hundred thousand prisoners who were murdered by Franco, the new dictator. Also, about 35 thousand died in his concentration camps. Their mass graves have been found, re-opened and examined by forensic experts.

.... All the human culture, all the results of art, science and technology that we see before us today, are almost exclusively the creative product of the Aryan.

....Human culture and civilization on this continent are inseparably bound up with the presence of the Aryan. If he dies out or declines, the dark veils of an age without culture will again descend on this globe.

... "With satanic joy in his face, the black-haired Jewish youth lurks in wait for the unsuspecting girl whom he defiles with his blood, thus stealing her from her people. With every means he tries to destroy the racial foundations of the people he has set out to subjugate."

....This contamination of our blood, blindly ignored by hundreds of thousands of our people, is carried on systematically by the Jew today. Systematically these black parasites of the nation defile our inexperienced young blond girls and thereby destroy something which can no longer be replaced in this world.

.... A state which in this age of racial poisoning dedicates itself to the care of its best racial elements must some day become the lord of the earth.

Adolf Hitler (1925), *Mein Kampf.* Reprinted by the permission of Houghton Mifflin Harcourt, Copyright ©1971,New York. p. 290, 383, 325, 562, 688.

Hitler became Chancellor of Germany in January, 1933. Within eighteen months of his taking office, by assassination and purges of his opponents, he became the head of a dictatorial, military regime. He started a large-scale program of expansion of German military, while talking peace at the same time, and negotiating with other imperial powers to wriggle out of the penalties of the Treaty of Versailles. He succeeded in doing this and then proceeded to reunify the German nation.

In "Mein Kampf", Hitler outlined his political philosophy. He thought that all human culture, every product of art, science and technical skill, was almost exclusively the product of German creative power. He felt that the German nation's "Aryan" superiority was being threatened particularly by the Jewish "race" that, he argued, was lazy and had contributed little to world civilization. He was worried about Jewish boys "seducing" young German blondes and thus undermining the superiority of the German race! He was disgusted by prostitution, and human trafficking in general and this, he thought, was the work of mainly the Jews! Hitler also claimed that Jews, who were only about 1% of the population of Germany, were slowly taking over the country. They were doing this by controlling the largest political party in Germany, the *German Social Democrat Party*, many of the leading companies and several of the country's newspapers. Hitler believed that the Jews were involved with Communists in a joint conspiracy to take over the world. Like Henry Ford, Hitler claimed that most Communists were Jews. Hitler felt that the Russian revolution was a Jewish conspiracy to dominate the world, since the Jews had created this ideology (perhaps he had Marx and Trotsky in mind) and the combination of Jews and Marxists had set up the Soviet Union. According to him, it was the intention of the Jews and Marxists to take over the whole world! He declared that, to defeat this conspiracy, he intended to occupy Russian land that would provide protection and "living space" for the German people and would also destroy the Jewish and Marxist attempt to control the world. A close study of "Mein Kampf" indicates clearly that Hitler was mentally sick. He was paranoid and obsessed with race. He had developed the paranoid delusion about the "superior" "Aryan" Germans and was willing to resort to violence, of a level only a paranoid person is capable of, to protect this delusional "reality".

.... The relation of the Jews to prostitution and, even more, to the white-slave traffic, could be studied in Vienna,.... When thus for the first time I recognized the Jew as the cold-hearted, shameless, and calculating director of this revolting vice traffic in the scum of the big city, a cold shudder ran down my back.

..... By the categorical rejection of the personality and hence of the nation and its racial content, it destroys the elementary foundations of all human culture which is dependent on just these factors. This is the true inner kernel of the Marxist philosophy in so far as this figment of a criminal brain can be designated as a 'philosophy'. With the shattering of the personality and the race, the essential obstacle is removed to the domination of the inferior being - and this is the Jew.

... Thus there arises a pure movement entirely of manual workers under Jewish leadership, apparently aiming to improve the situation of the worker, but in truth planning the enslavement and with it the destruction of all non-Jewish peoples.

..... If, with the help of his Marxist creed, the Jew is victorious over the other peoples of the world, his crown will be the funeral wreath of humanity and this planet will, as it did [millions] of years ago, move through ether devoid of men. Eternal Nature inexorably avenges the infringement of her commands. Hence today I believe that I am acting in accordance with the will of the Almighty Creator: by defending myself against the Jew, *I am fighting for the work of the Lord.*

Adolf Hitler (1925), *Mein Kampf*. Reprinted by the permission of Houghton Mifflin Harcourt, Copyright ©1971, New York. p. 59, 65, 320, 320

Hitler, through his studies, also developed a perverted view of what the Law of Evolution meant, as proposed by Darwin as a theory at that time. According to his understanding, the Theory of Evolution indicated that only the strong had survived in competition between individuals in the evolution of mankind, and the weak had always perished as a result. This, he thought, was what Nature intended - that the weak should perish. He, also, could not distinguish between what made an individual physically strong and what made a nation into a strong and powerful nation. Thus, he felt that if he killed all the weak people in Germany and elsewhere - the Jews, homosexuals and Gypsies - he would be helping the evolution of a superior nation into further superiority - and would, thus, be *doing the work of the Lord!*

Nobody paid attention to Hitler's insane ideas at the time, and his shrieking and shouting felt like music to the abused, insulted German nation facing extreme economic deprivation. His popularity kept on increasing with time, till he was appointed Chancellor of Germany, by the German ruling class. The United States and its European Allies had imposed reparations, punishments and "sanctions" on the German people and had created a population willing to listen to anyone who talked about honor and well-being - no matter how irrational his ideas and plans. These imperial powers had actually created the ideal conditions for the rise of a Hitler in Germany! They were blinded by their greed and lust for more loot and plunder and could not see the consequences of their own actions - but that is how imperial powers behave!

Italy, under Mussolini, felt the urgent "need" and opportunity to expand its African Empire, which, at that time, consisted of only Libya. This was obviously not sufficient to meet the appetite of Italy's ruling class, which hungered for more. Mussolini invaded Abyssinia (Now called "Ethiopia") in 1935 and initiated its conquest, completing this imperial project in about a year. He flirted with the idea of an alliance with Britain and France, but, facing hostility because of his Ethiopian adventure, he fell out with these imperial powers, although his conquest was later "legitimized" by these imperial powers, through the League of Nations.

The Great Depression
The Beginning

Before and during the First World War, the economy of the US had grown fast due to trade with Europe, which was its biggest trading partner. South America also continued to be a trading partner and its relations with the US continued to be dictated by the Monroe Doctrine. Trade and financial relations with Europe were given a much greater impetus during war-time, as the demand for its goods, especially military equipment, increased dramatically and European debt to the US kept on increasing. The greatly increased exports had enriched the ruling class of the US, which had invested heavily in real estate and the stock market, in speculative ventures. The US economy had, thus, become heavily dependent on exports and its productive capacity had increased far beyond domestic demand. This was also the time of the gold standard and all inter-state transactions were settled by using gold as a currency. Before and during the war, the US accumulated considerable amount of gold reserves, but the reserves of the European imperial power continued to go down, forcing those states to borrow from the US.

At the end of the First World War and the following imperial invasion of the Soviet Union, whole of Europe and the Soviet Union lay in ruins. Buildings, roads, bridges, railways, factories and farms had been destroyed everywhere. More than 15 million people had died in Europe alone, excluding the Soviet Union. Millions more had died in the Soviet Union due to the civil war and the Imperial Invasion. Also, the disorganized division of large agricultural farms, the resulting famine and the harsh food-grain acquisition policies of the Soviet government, during the imperial invasion of the Soviet Union, resulted in very large number of deaths due to violence and starvation in the civilian population. The Soviet Union was not a highly industrialized state at this time, but whatever industry it had, was destroyed or badly damaged. Conditions in the East were similar. Japan had not faced large-scale attacks on its soil, but it had been constantly been at war with the Chinese, Koreans and the Soviet Union and the cost, of this warfare and occupation of foreign lands, was enormous for itself and for the people of the countries that were victims of its imperial ambitions.

Collapse of industry everywhere gave rise to huge levels of unemployment and poverty. Families, who had lost male members in the war, were struggling to survive in the new harsh environment of the time. Thus, the purchasing power of the people everywhere had gone down. There was a global drop in demand of goods and services, due to the end of military conflicts and decrease in income of the people due to unemployment. This situation directly affected all the imperial powers and their economies. Germany and Britain were the worst hit European states. These two states were, also financially, the most dependent on the US. In Germany, unemployment rose sharply and Britain was less severely affected, but its industrial and export sectors remained seriously depressed. The collapsing economies of European states had a direct effect on the trade with the US. As the exports of the US dropped, unemployment began to increase in the US also, as industrial output began to drop and factories began to close down. The banks in the US also faced a crisis, as company after company in Europe collapsed causing loan defaults.

In October 1929 the New York stock market crashed, with a sharp drop in the value of its listed stocks. As the recession deepened, factories shut down and banks failed due to "bank runs". This caused the consumer confidence to crash, as many people lost their life savings. Farm income fell by about half and unemployment began to grow fast. Many banks were consequently forced into bankruptcy. The failure of about half the number of banks in the US, combined with a widespread loss of confidence in the economy, led to much-reduced levels of spending by households, thus further reducing demand for goods and services. This caused a reduction in investment in manufacturing and consequently a further drop in production of goods and services. Thus, a downward economic spiral had been set in motion at high speed. The result was drastically falling output and rising unemployment in the US, followed by similar conditions in Europe, South America and Asia. It was quite well-known that *intensive capitalism* was prone to cycles of boom and bust. Thus, a recession was to be expected when demand of goods, military and civilian, decreased after the war. But this was no ordinary recession. *Also, the imperial powers failed to manage it, although it was possible for the United States to use its gold reserves in such a way that, at least, the European powers could initiate a recovery of their economies, but this was not done.* Due to the extreme nature of this recession, affecting finance, industry and trade, and the extreme human suffering it caused, it is referred to as the *Great Depression*.

Many other countries had been affected by the depression by 1931. Almost all states tried to protect their domestic production by imposing tariffs, raising existing ones, and setting quotas on foreign imports. The effect of these restrictive measures was to greatly reduce the volume of inter-state trade. As state after state took measures against the import of foreign goods, *this accelerated the recessionary trends in its trading partners.* In Germany, unemployment had reached the same level as that in the United States by 1932. By this time, the total value of world trade had fallen by more than half.

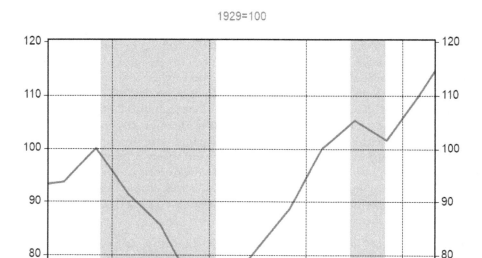

Real Gross Domestic Product

1929=100

Source: Historical Statistics of the United States/Haver Analytics

Figure 6-05: GDP Change During the Great Depression

Printed with permission of Haver Analytics. London | New York | Tokyo | Singapore

60 E 42nd St, New York, NY 10165

In Europe, the Great Depression strengthened extremist forces. In Germany, economic distress directly contributed to Hitler's rise to power in 1933. The Great Depression lasted until the beginning of the Second World War in 1939. It was the longest and most severe depression ever experienced by intensive capitalism, till that time. It is generally said to have begun with a catastrophic collapse of stock-market prices on the New York Stock Exchange in October 1929, although the US economy had gone into a severe recession much before that time. During the next three years stock prices in the United States continued to fall, until by late 1932 they had dropped to only one-fifth of their value in 1929. Besides ruining fortunes of thousands of individuals, this unexpected and sharp decline in the value of assets greatly strained banks and other financial institutions, particularly those holding stocks in their portfolios.

By 1932, US manufacturing output had fallen to about half of its 1929 level, and unemployment had risen to about 25 percent of the work force. The Great Recession in Europe after the First World War, became the Great Depression in the United States but quickly turned into a worldwide economic downturn because the size of the economy of the United States, at that time, was almost half of the economy of the whole world, and the United States had developed close economic relationships with economies of European states. The United States had emerged from the First World War as the major creditor and financier of postwar European economies which had been devastated by the war. European states, also, had huge debts. Germany and other states, defeated in the war, also had a special problem - the need to pay war reparations, in which the United States had been helping these states. So that when the US Economy crashed into a Depression, the European economics also moved from a severe recession into the Great Depression.

The Great Depression exposed the lack of knowledge and understanding of the business cycle, or the "boom" and "bust" phenomenon of intensive capitalism, and the resulting inability of governments to manage it. Ben Bernanke has written and spoken extensively about the causes of this Depression. [17]

Prior to the Great Depression, governments traditionally took little or no action in times of an economic downturn. It was believed that an "invisible hand" and market forces would somehow magically solve this problem automatically. But, this time, the "market forces" wreaked havoc on the people and the ruling classes of Europe and North America. As years passed by without a recovery, they had to listen to new ideas from economists of that time. John Maynard Keynes of Britain was one of them. Keynes studied the economic processes of the time and developed new ideas as to how to manage recessions of the future. The resulting debates eventually inspired some fundamental changes in the way recessions were to be handled. Government actions, both in terms of monetary and fiscal policy, came to assume a principal role in ensuring some degree of economic stability in states with intensive capitalism as their social and economic system.

The New Deal

In 1932, Franklin Roosevelt proposed a series of government actions to deal with the depression and won the election. His program was named the "New Deal", with the slogan that "The only thing we have to fear is fear itself". The New Deal proposed new social and economic measures, which favored the people and were intensely opposed by those who wanted the pure intensive capitalism to continue. The Communist Party of the United States had separated itself from the Socialist Party at that time and was demanding major reforms in favor of the working class. Since the banking and credit system of the state was paralyzed at that time, the new administration quickly moved to allow only solvent banks to continue operations and others were closed. A policy of moderate money creation was adopted to start an upward movement in commodity prices. Strict regulations were imposed on the stock market and the *Federal Deposit Insurance Corporation (FDIC)* was created to insure savings-bank deposits. A social service infrastructure program was started to provide employment to young men and women.

To provide relief to farmers, a program was introduced to raise crop prices by providing surplus crop insurance and by paying farmers a subsidy to compensate for voluntary cutbacks in production. As a result, farm output dropped and income of farmers increased. During those years, a severe drought also hit the Great Plains states, significantly reducing farm production in what came to be known as the "Dust Bowl". Crops were destroyed, agricultural machinery was ruined along with farm animals. About a million people, referred to as "Okies," left Texas and other bordering states. Most headed farther west to the land of myth and promise, California. The migrants were not only farmers, but also other professionals, whose lives were connected with farming. John Steinbeck has described their lives in his novel, *"The Grapes of Wrath"*

The memories of the failure to destroy the Soviet Union by the imperial invasion were still haunting the ruling class of the US. The Soviet Union was continuing to make progress in industry and military preparedness. Also, the Communist and Socialist parties of the United States were agitating for social security, unemployment insurance and health insurance for the working class. The "specter of communism" was haunting both Europe and North America. To co-opt these political parties, labor reforms were introduced during the New Deal. At the same time, the persecution of the Communist Party was initiated. Organized labor made greater gains than at any previous time in American history. In 1935, the government defined unfair labor practices, gave workers the right to bargain through unions of their own choice and prohibited employers from interfering with union activities. The great progress made in labor organization brought working people a growing sense of common interests, and labor's power increased not only in industry but also in politics. This power was exercised largely within the framework of the two major parties, however, and the Democratic Party generally received more union support. The New Deal did achieve significant increases in production and prices -- but it did not bring an end to the Depression in the US.

The Second "New Deal"

The *Works Progress Administration (WPA)* as the main agency of the so-called second New Deal, was an attempt to provide work to the working class. This was an infrastructure construction program. The *Tennessee Valley Authority (TVA)* was created to generate electricity and to help in economic development of the Tennessee Valley, a region severely affected by the Great Depression. In 1937, the *Bonneville Power Administration (BPA)* was started to generate electricity from the Columbia River in the Northwest of the US. Actors, painters, musicians and writers were employed through the *Federal Theater Project*, the *Federal Art Project* and the *Federal Writers Project*. In addition, the *National Youth Administration* gave part-time employment to students, established training programs and provided aid to unemployed youth. In 1935, the *Social Security Administration* was set up. Social Security created a system of insurance for the aged, unemployed and disabled. In Europe, Hitler had taken over in Germany in 1933. The Nazis' public-works projects and the rapid expansion of munitions production ended the Depression there by 1936. In the US, President Roosevelt won the election for his second term in 1936, with the help of a broad new coalition aligned with the Democratic Party, consisting of labor, most farmers and immigrants from Europe and Africa. But despite all these recovery programs, mass unemployment and economic stagnation continued, though on a somewhat reduced scale. Unemployment rate, still, remained at about 15 percent till the outbreak of the Second World War.

Events Leading to the Second World War

In the 1940 presidential election, for the first time in US history, Franklin Roosevelt was elected to a third term as President. The US, initially did not enter the war. At the same time, tensions mounted in Asia, as Japan joined the Rome-Berlin axis. In response, the United States imposed an embargo on Japan regarding export of oil and other materials essential to manufacture of military equipment.

In 1936, Hitler ordered his army into the de-militarized Rhineland - the border area adjoining France. The Soviet Union had moved forward with its reconstruction after the terrible destruction brought to it by its Civil War and the Global Imperial Invasion. It had, by now, completely eliminated its ruling class and was threatening to expand this revolutionary process to a global level. The ruling classes of all of Europe had begun to be haunted by the ghost of a social revolution. In this atmosphere, Britain decided that Germany, with Hitler's dream of creating a "living space" for the German nation in Eastern Europe and the territory of the Soviet Union, was a net asset rather than a threat. It adopted the policy of *"appeasement"* towards Germany, considering it *"a European bulwark against Bolshevism"*. Thus, having reached an agreement with other European powers, especially Britain, Hitler invaded and occupied Austria in 1938. Czechoslovakia, at this stage had a population of five national identities - Czechs, Slovaks, Germans, Poles and Hungarians. This multi-national state had been carved out of the Austro-Hungarian Empire without any regard for self-determination for minorities, who were, naturally, unhappy with the new borders and wished to join their national states. Thus, the invasion of Czechoslovakia by Germany, in March, 1939, was welcomed by most of its population. Britain had agreed to the return of all German-majority districts of this state to Germany, but Hitler ended up occupying the whole state, dividing it into two administrative zones. Mikhail Gorbachev has referred to these events as follows:

> **"And who handed over Czechoslovakia to the Nazis? On his return from Munich, Chamberlain said that he had brought peace to the British people, but in effect everything turned out otherwise; he had brought them war. That was mainly because the British rulers had only one thought on their minds: how to turn Hitler against the East, against the Soviet Union, and how to crush communism." [18]**

Quoted from *Perestroika* by Mikhail S. Gorbachev. pp 192. Copyright© 1987 by Mikhail Gorbachev. Used by permission of HarperCollins Publishers.

Britain continued to hope that after this conquest, Hitler would look east and focus on his dream of colonizing the Soviet Union, leaving France and the rest of European territory, west of Germany, under British dominance. Appeasement seemed to be working at that time. In the Soviet Union, several changes had taken place since the end of the civil war and the imperial invasion. Lenin had died in 1925 and was succeeded by Stalin, who had climbed into this position of absolute power using his bureaucratic skills. No one had noticed that Stalin was mentally sick. Later on, from his actions, it became quite clear that he was paranoid and extremely intolerant of dissent, or even minor differences of opinion. His paranoia was amply demonstrated by his elimination of all potential domestic rivals and murder or deportation of hundreds of thousands more who opposed him only in his paranoid imagination. These included nearly ninety percent of the senior members of the Communist Party and the Red Army! This mass murder, by Stalin, of his opponents is known as *The Great Terror*. Much later, after 1989, when Mikhail Gorbachev came into power, a commission was set up to look into the cases of all those who were affected by the *Great Terror*. It was found that more than *seven hundred thousand persons* had been murdered, without any real trial. Only about twenty thousand had minor charges against them, but were executed anyway. *The Great Terror* had greatly weakened the Communist Party and the Red Army, when another global war started in Europe. When the one-party dictatorship was being set up in the Soviet Union, Lenin had been aware that such a dictatorship could become the dictatorship of one person. But, clearly, he, or anyone else, had not imagined this kind of murderous behavior by an insane despot within the Communist Party of the Soviet Union - a party that stood for social justice and human rights for all working people of the world!

Much before the policy of appeasement could decisively turn Hitler away from Western Europe to invade Eastern Europe and the Soviet Union, Stalin had succeeded in starting negotiations with Hitler's Germany. A non-aggression pact was signed between the two states on August 23, 1939. By this pact, Soviet Union hoped to recover its territory that it had lost during the invasion by imperial powers and, at least, delay a German invasion. Hitler wanted to conquer Poland and recover German territories it had lost during the First World War. He, also, wanted to free up its troops on the Eastern front, so he could complete the conquest of Western Europe. After the pact, and its secret protocol, had been signed, the Soviet Union started the process of re-incorporation of the Baltic states of Finland, Latvia, Estonia and Lithuania, into the Soviet Union. These states were, however, formally annexed later. The Ukrainian and Belarusian territory also remained to be recovered from Poland. As negotiations were progressing with the Soviet Union, Germany offered to make a deal with Britain if its African colonies, grabbed by Britain after the First World War, were returned to it, but Britain did not give a positive response to this proposal.

As would be noticed, the First World War had not really stopped after an Armistice had been signed. Germany's maritime fleet and African colonies had been grabbed by Britain. Britain and France had taken over Arab territories formerly part of the Ottoman Empire. The Soviet Union had been invaded by the United States and other European and Asian imperial powers. Japan had expanded its empire beyond Korea and had taken over Manchuria - territory larger than its state territory. It had occupied additional territories in China. Wars between Japan and China and between Japan and Mongolia were in progress. Soviet Union had recovered control of almost all territory of the Tsarist Empire. Italy had expanded its empire beyond Libya and had conquered Ethiopia. Germany had also recovered all the territories with German populations - and more. Two big empires had disappeared - Austria-Hungarian Empire and the Ottoman Empire and a third, the Tsarist Empire, had changed into its opposite. Europe, thus, had four states (instead of seven during the First World War) - Britain, France, Germany and Italy - trying to rob each other and grab all the capital hoarded at the cost of the rest of humanity. In addition, the Soviet Union was engaged in trying to prevent another land-grab by any of these imperial powers. The United States was waiting for an opportunity to dominate Europe and East Asia, and was also deeply apprehensive about the Soviet Union and what it might do to frustrate its imperial ambitions. Thus, the tensions between the imperial powers had increased, as two of these bank-robbers had died and a third had changed into a crusader - more hoarded capital was at stake. The territories occupied by the imperial powers outside Europe, had increased and, thus, more "sources" of capital were also at stake. On the other hand, fewer robbers had survived to fight over all the capital and its sources, while under threat of complete deprivation by the Soviet Union. This was clearly a recipe for a bigger and more violent conflict - which, naturally, did break out with a much higher level of violence than ever before.

During these years, the military forces and military industries of the US were greatly expanded. Also, American factories were flooded with orders from overseas for armaments and munitions. When the US entered the war directly in 1944, unemployment dropped rapidly and the *Great Depression ended. The War had solved the problem that had defied solution for almost twelve years.*

The Second World War

The war is generally considered to have begun on September 1, 1939, with the invasion of Poland by Nazi Germany and subsequent declarations of war on Germany by France and most of the countries of the British Empire. Several countries were already at war at that time. Ethiopia and Italy were engaged in the Second Italian-Abyssinian War and China and Japan in the Second Sino-Japanese War. The Soviet Union attacked Polish-occupied Soviet territory from the East, while Hitler's forces were moving East in Poland. Poland was conquered in about a month. At this point in time, all German territory had been recovered from its neighbors and the German nation had been re-united. In addition, Poland and Slovakia had been conquered by Germany. But Hitler had no intention of stopping German expansion at this point. He wanted to create a *new order in Europe*, in which the Germans, *as a superior nation*, would rule over all other inferior nations in Europe and elsewhere, and a large territory in Central Europe, i.e., in Eastern Europe and the Soviet Union, would be colonized by the German nation. This, according to Hitler, was the *duty of the German nation, because it was a superior Aryan race*!

German Expansion into Western Europe

In early 1940, German forces invaded Norway and Denmark. Norway did offer some resistance, but the German Army prevailed and, then, Germany began its assault on Western Europe by invading Netherlands, Belgium and Luxembourg. After conquering these small states, it launched a large-scale attack on France. In the middle of 1940, France signed an armistice with Germany, which provided for the German occupation of the northern half of the country and permitted the establishment of a regime in the south with the city of Vichy as its capital. This regime, referred to as *Vichy France*, was to be superficially independent but, in fact, was controlled by Germany. The obvious collapse of France convinced Mussolini that the time to implement his pact with Hitler had come. Italy declared war against France and Britain. Italy, however, could not achieve much against France.

In the second half of 1940, Hitler's regime waged a bombing campaign over Britain. The objective was to achieve air supremacy over Britain for an amphibious landing by the German ground and naval forces. The British Navy needed to defend Britain from invasion and to retain command of the ocean trading routes. This would have ensured supply of food and raw materials for Britain and would have denied the same to Germany. Germany was the dominant land-based power in continental Europe. Its navy was relatively small and its main naval weapon during the war was to be the submarine, or U-boat, with which it attacked British, American and allied shipping. When the fall of France was imminent, US president Franklin D. Roosevelt declared that the United States would "extend to the opponents of force the material resources of this nation." After France collapsed, he pursued this policy by helping the British in their struggle against Germany. Roosevelt arranged for the transfer of surplus American weapons and other war material to Britain. A law passed by the US Congress, called the "Lend-Lease Act" not only empowered the president to transfer defense materials and information to any foreign government whose defense he deemed vital to that of the United States, but also gave him a free hand in negotiating deals with the "opponents of force." More than ten billion dollars were allocated for this purpose. This was a huge sum in those days. Initially, only European states and Chiang's regime in China received help under this program. The beginning of lend-lease enabled Britain to expand its navy and defend itself better in the air against German air attacks. As a result, Germany failed to achieve air supremacy over British air space. *Hitler noticed the improved defense capabilities of Britain and increasing US help to it. At the same time, his military analysts told him that the Soviet Union was working on plans to increase the Red Army and Air Force's capabilities to exceed those of the German forces.* Hitler, thus, decided to shift his focus to Eastern Europe and the Soviet Union with the hope that once the Soviet Union had been conquered, the enormously increased resources in his possession would, then, enable him to subjugate Britain and resist the United States. Events did not turn out according to his expectations, though.

Invasion of the Soviet Union

In the Soviet Union, speaking to his generals, Stalin had mentioned Hitler's references to his planned attack on the Soviet Union in *Mein Kampf*, and had said they must always be ready to repulse a German attack. Stalin's paranoid mind became an asset for the Soviet Union at this stage. Stalin thought that Hitler expected the Red Army to be ready for war in about four years. He, therefore, decided that the Soviet Union must be ready much earlier. Development of heavy industry, especially production of military hardware, had already made considerable progress. Defense production was further accelerated. Stalin, however, wanted to delay the war as much as possible. Germany invaded the Balkan region in early 1941 and conquered it in collaboration with its allies. Around the same time, Stalin's regime and Japan signed an agreement, in which the Soviet Union pledged to respect the territorial integrity of the Japanese colony of "Manchukuo", while Japan made a similar pledge about Mongolia. For

Stalin, the purpose of this agreement was to free Japanese forces so that they could enter into conflict with the United States and other imperial powers, while the Soviet forces in Siberia could be moved to the European theater to face the expected German invasion.

Beginning in June 1941, about four million soldiers belonging to Germany and its allies invaded the USSR along an almost two-thousand-mile front. *Operation Barbarossa* was the code name for the invasion. *This has been the largest invasion in the history of warfare till now.* The ambitious operation was driven by Adolph Hitler's persistent desire to conquer the Soviet Union for colonization. He had a plan for this purpose -- *The entire population of the cities of the Soviet Union was to be starved to death, by encircling those cities and laying siege to them till their whole population died of starvation. The rural population was not to be killed, initially, so that agricultural production could continue. This was expected to create an agricultural surplus to feed the population of Germany and would, also, have allowed for replacement of the urban population in the Soviet territories, by the "superior" upper-class German population! According to Hitler's thinking, these plans were meant to totally annihilate the Soviet people – thus destroying the "Marxist, Jewish conspiracy" against his "Aryan" nation!*

Operation Barbarossa was named after Frederick Barbarossa, a medieval Roman Emperor. The military operation itself lasted from June to December 1941. The strategy, that Hitler and his generals had developed, involved *three separate army groups* assigned to conquer three separate regions of the Soviet Union. The main German attacks were conducted along historical invasion routes. *Army Group North* was to march through the Baltic States into northern Russia, and either take or destroy the city of Leningrad (now called Saint Petersburg). *Army Group Center* was to advance to Smolensk and then Moscow, marching through what is now Belarus and the Western and Central regions of Russia. *Army Group South* was to invade and conquer the heavily populated agricultural center of Ukraine and, then, to continue eastward over the southern Soviet Union to the river Volga - with the objective of conquering the oil-rich Caucasian region.

The German invasion of the Soviet Union was exciting news for the US government. Its status and implications were discussed in detail at all government levels and the objectives of the US policies were clearly stated.[18] The US had no sympathy for either Hitler's Germany or Stalin's Soviet Union and its was desired that both sides suffer maximum casualties and property damage. The US was to help the losing side initially, i.e., the Soviet Union, if it asked for help. The level of military and economic help was to be adjusted to obtain optimum results desired. For the same reasons, US operations on the Western front were to be delayed till Hitler's armies were in retreat, or were under severe pressure.

According to Hitler's original invasion plan, it was expected that Moscow would be conquered *within four months.* Having destroyed most of the Soviet Air Force in its forward bases, German forces quickly advanced deep into Soviet territory using blitzkrieg tactics. The Soviet defense forces were overwhelmed and the casualties sustained by the Red Army were enormous. Within one month, Army Group Center had managed to penetrate the defensive positions of the Red Army near Minsk and then continued its move towards Moscow.

On the Northern front, Finnish military forces occupied territory on the North of Leningrad, while German Army Group North took up positions to the south of the city. Both German and Finnish forces, thus, encircled the city. The purpose, as planned, was to maintain a blockade, preventing the defenders from receiving any food or supplies. It had been calculated that the city would reach starvation within a few weeks. The population of Leningrad was informed of the danger by the Soviet government and over a million citizens were mobilized for the construction of fortifications. Several lines of defense were built along the perimeter of the city. Hitler's priorities were clear - *Leningrad first, Donetsk Basin second, Moscow third.*

As Operation Barbarossa progressed, the *Atlantic Charter* was announced after a secret meeting, between Roosevelt and Churchill, in August, 1941, in Newfoundland. It was later agreed to by all the imperial powers allied with the US. It laid the basis for later discussions in the Bretton Woods Conference. President Woodrow Wilson had formulated "Fourteen Points" as the US aims in the aftermath of the First World War. Similarly, Roosevelt also set a range of ambitious goals for the postwar world even before the US had entered the Second World War. The United States, as the dominant imperial power, wanted to re-open and control the world economy, so as to have unhindered access to markets and raw materials of all states. The ideal goals of the war were, thus, formulated somewhat as follows:

Figure 6-06: Operation Barbarossa

Source Wikimedia Commons. Released into PD by Creator. Modified by Tayyib A Tayyib. Retrieved, October 14, 2020.

- Restoration of self-government to those deprived of it. No territorial changes to be made against the wishes of the people of those territories. Global cooperation to secure better economic and social conditions for all. Freedom from fear and want.

- Free access to raw materials. Reduction of trade restrictions. Freedom of the seas.

- Establishment of a wider and more permanent system of general security. Disarmament of "aggressive" states. Abandonment of the use of force.

United States visualized the *"new world order"*, under its domination, to be based on these principles. Britain was not willing to allow unrestricted access to its empire and its resources to the United States and other imperial powers, but was in no position to resist. The United States visualized a global controlled market and was able to extract this agreement. US representatives studied with their British counterparts the reconstitution of what had been lacking between the two world wars: a system of international payments that would allow trade to be conducted without fear of sudden currency depreciation or wild fluctuations in exchange rates - ailments that had nearly paralyzed global intensive capitalism during the Great Depression. Moreover, it was felt that without a strong European market for US goods and services the US economy would be unable to sustain its growth.

Within August 1941, German forces captured the city of Smolensk, an important stronghold on the road to Moscow. The strong Soviet defense of the Smolensk region lasted for about two months. This delayed the German advance until September, effectively disrupting the blitzkrieg. At this stage, Moscow was vulnerable, but Hitler ordered the attack to turn north and south and eliminate Russian forces at Leningrad and Kiev. Hitler felt that Germany urgently needed the food and mineral resources located in Ukraine and the oil from oil fields of Baku. The conquest of those oil fields would also have greatly cut off oil supplies to Soviet forces. Thus, the German army was ordered to first secure Kiev and the region around it and then to move towards Moscow, after Leningrad had been conquered. The attack on Kiev inflicted another significant defeat on the Red Army and the Soviet forces had to abandon Kiev. Although a big victory for Germany, the Battle of Kiev set the German blitzkrieg even further behind schedule. By this time, the German army had been weakened considerably.

The main objective of the German offensive towards Stalingrad (or "Volgograd"), on the Volga River, was to capture the oil fields of Baku. The offensive, however, stalled near Stalingrad, facing determined Soviet resistance. The German troops were not prepared for the Russian Winter and a large part of the Army Group South was trapped as Soviet forces encircled and isolated it, but Hitler refused to order a retreat. Ultimately, a large number of the total of about 250 thousand troops froze to death and the remaining units of the German Army had to surrender to Soviet forces.

Hitler did not want to accept the surrender of Leningrad. The plan was to continue the siege and shelling of the city, *starving as much of its population as possible*. The defenders of the city tried to protect the civilian population of the city and evacuated about one and a half million residents. The Baltic Fleet of the Soviet Navy continued to offer resistance to the siege of the city, but could not prevent its encirclement by the Germans and all supply routes to Leningrad and its suburbs, were cut off.

Hitler still expected to finish the war before winter by taking Moscow. In late 1941, Army Group Center launched its final offensive towards Moscow. Hitler was hopeful of crushing Soviet forces before the coming winter. Moscow was the most important target, and it was expected that the city's surrender would lead to the collapse of the Soviet Union. However, both the armies were quite exhausted at that time.

The defenders of the city were about one million, but their equipment losses had been fully replaced with new military equipment from factories that had been moved toward the Ural Mountains. This had happened due to the emphasis by the Stalin regime on relocation of heavy industry. With all the men at the front, women in Moscow dug anti-tank trenches around the city. The Soviet command began constructing a triple defense ring surrounding the city, forming the *Moscow Defense Zone*. Contrary to German expectations, the encircled Soviet forces did not surrender and the German offensive ultimately got bogged down. Melting of snow turned roads into stretches of mud. Offices of the Communist Party and other civil and military offices were evacuated from Moscow, leaving only a skeleton staff behind. Limited damage was caused by the air attacks. At this point in November 1941, Soviet Union was declared eligible for lend-lease by the US. Leningrad was, then, under siege. Civilians and soldiers were dying daily by the thousands. Army Group Center of the German forces had reached Moscow and heavy fighting was in progress. Army Group South had conquered Kiev. *Thus, the American objective, of letting Hitler kill as many Soviet civilians and soldiers and causing as much destruction as possible, had been achieved. Also, a large number of German soldiers had been killed.* There was a possibility that Soviet resistance would collapse and the *US Government's desire, to have as much German destruction as possible, would not be fulfilled.* In fact, this would have meant that Hitler's plan to conquer the Soviet Union and then to attack Britain, was moving in the right direction. This could, ultimately, result in very high casualties for the military forces of the US and its allies. About ninety percent of the German army had been deployed on the Eastern front and, by prolonging the conflict further, while ensuring a German defeat, optimal casualties and destruction on both German and Soviet sides could be achieved, without any cost to Britain or the United States. Thus, *this was considered the ideal time to declare the Soviet Union eligible for help through Lend-Lease.* American deliveries of military equipment and other supplies to the Soviet Union began shortly afterwards.[19]

At the end of 1941, Soviet troops launched a strong counteroffensive that drove German forces away from the outskirts of Moscow. Operation Barbarossa had failed. In anger, Hitler took personal charge of the German army, setting most experienced German officers against him. The Red Army had offered stiff resistance, destroying everything before withdrawing from any area. As their supplies began to run out, the German forces were ordered to retreat, but retreat was also very difficult. Russian troops started to attack the Army Groups from the flanks. Huge German losses resulted.

Artillery bombardment of Leningrad and its siege lasting two and a half years caused the greatest destruction and the largest loss of life ever in a modern city. Deaths of up to *one and a half million soldiers and civilians* occurred. Many of those who were being evacuated also died during evacuation. The siege of Leningrad was the cruelest siege in world history, *as genocide was the main objective. World has never seen terrorism on such a scale in any city.* Fortunately, the encirclement was broken by the Red Army and, against Hitler's wishes, all the inhabitants of Leningrad did not die of starvation.

The name *"United Nations"* had been coined by President Roosevelt, to describe the countries fighting against Germany and its allies, under American control. It was first used officially on Jan. 1, 1942, when 26 states joined in the *Declaration by the United Nations*, pledging to continue their joint war effort and not to make peace separately. In this Declaration, the

Allies of the US pledged adherence to the charter's principles. The Declaration stated *"Their faith in life, liberty, independence, and religious freedom, and in the preservation of human rights and justice in their own as well as in other lands, has been given form and substance as the United Nations." This was a remarkable statement by a group of imperial powers and their allies. These long-term abusers of mankind were claiming moral superiority in terms of "liberty, freedom and human rights"!*

Attacks on Eastern and Western Fronts

Starting in 1942, the British and American air forces systematically bombed industrial plants and cities all over Germany – for about three years. As a result, much of urban Germany was reduced to rubble by 1945. In North Africa, several see-saw battles were fought between British, Italian and German forces and the British had to retreat to Egypt. With the arrival of help from the US, the British forces were able to push the Germans back, leading to their surrender. In Europe, the Germans, while retreating from Stalingrad, mounted an offensive at Kursk, which resulted in *the biggest tank battle in history*. Stalin kept asking the allies to open a second front in Western France, so that the German pressure on Soviet forces may be reduced, but Britain wanted a priority to be given to opening a front on the "soft underbelly of Europe", i.e., in Italy and the Balkans, so that the Soviet Union may not dominate Eastern and South-Eastern Europe after the end of the War. The opening of the Second front was, thus, delayed to suite both British and the US, since the delay ensured maximum Soviet and German casualties and minimum casualties for the United States and its allies. In middle of 1943, Allied forces landed in Sicily. Mussolini was deposed and the Italian military surrendered, but the German troops stationed in Italy seized control of the northern part of the peninsula and continued their resistance.

The *First Quebec Conference*, held in the middle of 1943, was a highly secret military conference held at Quebec City between Churchill and Roosevelt, hosted by the Canadian prime minister. The imperial powers, allied with the US, began planning for the invasion of France and decided to continue the buildup of American forces in Britain for this purpose. It was, also, decided to intensify the bombing of Germany and the war effort against Japan. Also, a decision was made to develop the atomic bomb and a secret agreement was signed between Britain and the United States, to share nuclear technology. Tensions had developed in Palestine due to the colonization process. A joint statement was issued to "calm" the Palestinians.

The *Tehran Conference* was opened in late 1943. This was the first conference between the three heads of government - Roosevelt, Churchill, and Stalin. The purpose was to coordinate their strategy. Stalin wanted the Polish-Soviet and Polish-German borders to be finalized. He, also, continued to push for the second front against Germany. It was agreed that this front would be opened in May 1944, in coordination with the Soviet attack on Germany's eastern border. It was agreed to support Iran's government and the governments set up by the Soviet Union in Eastern European countries were also discussed. Finland was told to negotiate a peace agreement with the Soviet Union, rather than surrender unconditionally.

Even before the start of the Second World War, it had become clear that the League of Nations was not capable of effectively dealing with inter-state conflicts. The United States had not become a member of this organization. The Soviet Union had been expelled and Germany had withdrawn from it. In 1939, the State Department of the US had considered the creation of a new organization for "maintenance of world peace", and the requirements of such an organization so that it could function according to the long-term objectives of the United States. Roosevelt had, on many occasions, talked about a concept he had developed about maintaining world peace. He had hoped that all aggressive powers could be disarmed while only *"four policemen"* could remain armed to maintain world peace - i.e., United States, Britain, Soviet Union and China. France was, however, not included in this group - perhaps because it was too weak or because it was under occupation at that time. The need for an international organization to replace the League of Nations was first stated officially on Oct. 30, 1943, in the *Moscow Declaration*, issued by China, Britain, the United States, and the Soviet Union. In 1944, those four countries drafted specific proposals for a charter for the new global organization, to be known as the *United Nations Organization*, or the UNO. All the states that had ultimately adhered to the 1943 declaration and had declared war on Germany or Japan by Mar. 1, 1945, were called to the *founding conference* held in San Francisco (Apr. 25–June 26, 1945). Drafted at San Francisco, the UN charter was signed on the last day of the meeting and ratified later.

The Soviet advances continued against the forces of Army Group North and Army Group South. German forces were driven out of the Baltic States and Belorussia. Smolensk was retaken by the Soviet forces in 1944. Allied soldiers landed in France, which was taken over within two months. The US troops crossed into Germany and the Germans launched an unsuccessful counterattack in Belgium and northern France, known as the Battle of the Bulge.

Preparing to rebuild the international economic system according to the new world order, as World War II was still raging, delegates from 44 states allied with the US, gathered in the town of Bretton Woods, in the US state of New Hampshire. This conference was called the *United Nations Monetary and Financial Conference*. An agreement was negotiated and signed on July 22, 1944. *The Bretton Woods system* of monetary management included establishment of new financial institutions and specification of rules and procedures for regulating global commercial and financial relations among the world's major states. As part of this system, planners established the *International Monetary Fund (IMF)* and the *International Bank for Reconstruction and Development (IBRD)*. The IBRD is, now, part of the World Bank Group. These organizations became operational in 1945 after a sufficient number of countries had ratified the agreement. The main provision of this system was to tie every state's currency to the US Dollar, with a fixed exchange rate.

The *Second Quebec Conference* was a high-level military conference held in September, 1944, in Quebec City. Agreements were reached among Roosevelt, Churchill and the Canadian prime minister, regarding the Allied occupation zones in defeated Germany, continued US economic aid to Britain and the British naval participation in the conflict against Japan. In early 1945, the US and British air forces virtually destroyed the city of Dresden, killing tens of thousands of people, in an effort to prevent the industries of the city, along with their workers, falling into Soviet hands. The Red Army reoccupied Ukraine and Eastern Europe, including the Balkan region. Thus, the Soviet Armed forces were finally poised to attack Germany proper. Operation Barbarossa had ended in a disaster for Germany.

Operation Barbarossa was the largest military operation in human history in both manpower and casualties. Its failure was a turning point for the German army. Regions covered by the operation became the sites of some of the biggest battles, deadliest atrocities and highest casualties. The German forces captured about three million Soviet POWs. Most of them never returned home alive. They were deliberately starved to death in German camps as part of a *program to reduce the Eastern European population*. It is quite clear that this war was waged in the most inhumane fashion possible and violated all standards of civilized behavior in war.

The War in Asia

Japan had conquered "French" Indochina in 1940-41, while heavy fighting with the two Chinese armies was going on. As mentioned before, an agreement signed by the Soviet Union and Japan had, basically, freed up Japanese forces for action against the US. As a result, the Soviet Union, also, had moved its forces to the European theater. Feeling more confident as a result of the agreement with the Soviet Union, the Japanese Navy made its surprise attack on Pearl Harbor, Hawaii, on Sunday morning, December 7, 1941. The Pacific Fleet of the United States sustained heavy losses as a result. The main objective of the attack was to do long term damage to the United States, so that Japan would have enough time to establish its long-planned Southeast Asian empire and to create viable defenses against the US. This reasoning was unrealistic since Japan had no allies in Asia and it did not have the capability to wage war with the Chinese and other nations of South-East Asia and with the US at the same time, while the economy of the US alone was far bigger than Japan's and the US was already mass producing military equipment for its air, ground and naval forces and forces of its allies in Europe. Japan could not be expected to succeed in this gamble.

As planned by the Japanese rulers, the attacks on Southeast Asian states were continued. Hong Kong Malaya, Singapore, Philippines, and Burma were conquered. Tens of thousands of ethnic Chinese, perceived to be hostile to the new regime, were killed. The Japanese then seized the key oil production areas of Borneo, Central Java, Sumatra, etc., defeating the Dutch forces. They were welcomed ecstatically as liberating heroes by the oppressed Indonesian natives. The Japanese then consolidated their lines of supply by conquering key islands of the Pacific. Having, thus, almost completed the *"Greater East Asia Co-Prosperity Sphere"*, the Japanese rulers felt that they could withstand the coming US offensive. Events proved them to be wrong. During 1943 and 1944, the forces of the United States and its allies, steadily advanced towards Japan. The Philippines were reoccupied. An intense bombing campaign over Japanese mainland was undertaken by the US. This was combined with naval interdiction of Japanese supplies. The bombing campaign led to the deaths of hundreds of thousands of civilians. This, however, did not succeed in persuading the Japanese military to surrender.

The End of the War

The *Yalta Conference*, of the three leaders of the allied powers, was convened on February 4, 1945, at Yalta in the Soviet Crimea. The most important issue before them was the management of territories recovered from Nazi Germany. Other issues included reparations, division of Germany itself and the treatment of war criminals. The creation of the Trusteeship Council was also discussed along with the possibly affected Trust territories. The policy of demanding Germany's unconditional surrender was reaffirmed. Plans were made for dividing Germany into four zones of occupation (American, British, French, and Soviet) under a unified control commission in Berlin and for war crimes trials. The Soviet Union committed to declare war on Japan two to three months after the end of the war in Europe. The Charter of the United Nations was also discussed and approved. It was decided to give the right to vote to Belorussia and Ukraine, in addition to the Soviet Union. It was decided to hold a *"United Nations" conference* in the US. Further, it was agreed that the five nations, which would have permanent seats on the Security Council, should consult each other on the question of territorial trusteeship, prior to the conference. The conferees decided to ask China and France to join them in sponsoring the founding conference of the United Nations. Agreement was reached on using the veto system of voting in the projected Security Council. The Soviet Union was promised an occupation zone in Korea in addition to recovery of all Soviet territories. The United States and Britain also agreed to recognize the independence of Mongolia, and to admit Ukraine and Belarus to the United Nations as full members. The new borders of Poland were recognized and it was agreed to recognize the new, Soviet installed, government of Poland.

As Soviet forces moved to encircle Berlin, American troops crossed the Rhine river. Adolf Hitler committed suicide and the commander of German forces in Berlin surrendered the city to the Soviets. An unconditional surrender to the Allies occurred and the war in Europe came to an end. It is said that the Allies won the war, but the burden of fighting had been borne, mainly, by the Soviet Union. It had also suffered the most. The war had actually been won, mainly, by the Soviet Union. *Hitler's racist dream, of the German nation becoming the "Lord of the Earth", had been completely shattered.* He had succeeded in committing mass murders of about twenty-five million Soviet citizens and millions of other Europeans. About a quarter million soldiers of the United States had died in fighting. For these "achievements", Hitler and his German nation had to pay with about fourteen million lives! Germany was divided into four zones of occupation soon after the surrender, as planned.

On April 5, 1945, the Soviet Union announced its decision to pull out of its neutrality pact with Japan. Meanwhile, the United States had been working on development of an Atomic bomb. The bomb was tested only once on July 16, 1945, in New Mexico. As the end of the Soviet-Japanese Neutrality Pact came closer, it was quite clear that the Soviet Union would try to push the Japanese out from all the Russian territories that they had occupied. Occupation, by the Soviet Union, of the Northern half of the Korean peninsula, above the 38th parallel, had already been agreed to. Despite heavy bombings, Japan showed no sign of a surrender, although most of its territorial gains had already been reversed. The fire-bombing of Tokyo in March, 1945 killed between 80,000 and 200,000 people and left more than a million homeless, but even this did not persuade Japan to surrender. The Truman administration in the United States felt that the Soviet Union would declare war on Japan, as it eventually did, and would demand control of a part of the territory of mainland Japan also, whenever a surrender did occur. Japan was an industrialized state and a huge amount of accumulated capital was at stake and the United States did not want to share this "prize". Further, the ruling class of the

US was already dreaming of Panamizing Japan and reaping huge amounts of capital in the future. It was, thus, decided to terrorize Japan into surrendering by dropping atomic bombs on its cities. The first bomb was dropped on Hiroshima city on August 6, 1945. The second bomb was dropped on Nagasaki city, on August 8, 1945. *More than one hundred thousand men, women and children died in these two cities, as soon as the bombs were dropped. More than two hundred thousand died eventually, as a result of exposure to nuclear radiation.* Terrorism had succeeded and Japan surrendered soon after the bombings. The surrender was to the United States only, denying the Soviet Union any part of the Japanese territory - as the US had desired. The United States has, since then, maintained the dubious distinction of being the first state in human history to engage in nuclear terrorism.

As agreed with its allies, the Soviet Union entered the Pacific Theater of the war within three months of the end of the war in Europe. The invasion of Manchuria began on August 9, 1945, exactly three months after the German surrender. It was launched simultaneously on three fronts as the Red Army advanced over the deserts and mountains from Mongolia, far from its resupply routes. The Japanese were caught by surprise in unfortified positions. At the same time, Soviet airborne units were used to seize airfields and city centers in advance of the land forces. In and around Korea, several Soviet amphibious landings were conducted, while the land advance also continued. In accordance with the agreement reached earlier with the US, Soviet forces stopped at the 38th parallel, leaving the Japanese still in control of the southern part of the peninsula. Later on, American forces landed at Inchon.

At the end of the war, there were about three million Japanese in China and Korea, most of whom were farmers in Manchukuo. They had been moved to Manchuria under a colonization plan. All the settlers and soldiers in Asia were repatriated back to Japan. Many political and military leaders were tried and convicted of war crimes, except the imperial family. The Japanese military had committed numerous atrocities against civilian and military personnel during the war, including large scale massacres of prisoners and civilians, large scale rapes of women, including rapes and murders of more than 20 thousand women in Nanjing alone. Thousands of women had been forced into sexual slavery, as "comfort women" for the Japanese military. Civilians had been burnt to death, or buried alive. Cruel biological experiments had been conducted on civilians and prisoners of war. However, those responsible for these abuses were not punished. Instead, the US offered immunity to those who had conducted biological experiments, in exchange for information about germ warfare.

The Second World War was a global military conflict lasting from 1939 to 1945 which involved most of the world's states, including all imperial powers, organized into two opposing military alliances: The Allies and the Axis. It was the most widespread war in history, involving more than 100 million military personnel and leading to mass murders of millions of prisoners and civilians, especially Jews, homosexuals and gypsies and the use of nuclear weapons for the first time in human history, it was the deadliest conflict in human history, with a loss of over fifty million human lives in Europe alone. In China 20-30 million people had died in the conflicts on its territory, during this period, and millions more before that.

Chapter - 7

Economics of Intensive Capitalism

In Europe, under intensive capitalism, wealth became more and more concentrated in the hands of tiny minorities, while the huge working classes were forced to live as wage-slaves. Population growth slowed down due to the intensive death rate of the working population by starvation wages and denial of medical care and the resulting decrease in birth rates. Wars between European imperial powers brought further death and destruction. Concurrently, intensive capitalism was transplanted into North America, Australia and Japan and, by the late nineteenth century, these regions were also caught up in the grip of full-scale intensive capitalism. This system has since moved from Europe and these regions, having evolved into a less inhumane system there, into Central Eurasia and South America and has become fully established in those territories. In this chapter we would analyze how socialized production is organized under this stage of socio-economic evolution of mankind – *to explain why extreme inequality, periodic economic crises, the drive towards creation of empires, an intensive death-rate and a slower birth-rate are its necessary and unavoidable consequences.*

The Production/Service Center Model

Most businesses start on a small scale, owned by one person or family, and engage in one kind of economic activity. With time, however, a business tends to grow in scale and gets involved in new related activities. Also, businesses may take over other businesses and thus get involved in a number of sectors. Thus, the range of economic activities that a corporation may be engaged in is unlimited. Activity in every sector of an economy involves setting up of activity centers, or production centers, which produce products and provide services of various kinds needed by a state's population, or abroad. We need to look at the organization of these production centers to understand what economic and social forces are involved in their establishment and operation and what consequences result from their operations.

For ease of understanding, we would first look at simple examples of production /service centers, without the complicated structures of today's factories, farms and businesses run by corporations owned by a large number of people. Let us assume that the production center is a factory and the owner of the factory does not contribute any effort, or labor, to its operation and only owns it, i.e., provides the capital for its set-up and operation, while a hired manager runs the factory, or the production center, with the help of some assistants who supervise the workers. If the owner of a factory, mine, or farm, also manages the production center, then, he may be considered a part-time worker, in addition to being the owner. If a business, or corporation, owns the production center, then there may be many owners. The only objective of the owner is to increase the capital that he has invested in the factory. This applies to multiple owners also, since their objective is to share in the capital accumulation that results from the operation of the production center.

As we would see, the economic fundamentals of processes involved in every production center are basically the same. Every production center, be it a factory, a mine, a business involved in distribution of goods, a bank, or another kind of financial institution, has four kinds of capital input as are indicated in Figure 7-01:

Fixed capital (C_FIXED) refers to land, buildings and machinery which constitute a production center. As we have discussed before, virgin land has no value. It, however, may have the value corresponding to the labor used in its improvement, or in constructing buildings on it for the factory. In general, land in industrial parks is sold for a price that is far above its real value. Land with its improvements, buildings and machinery are products previously produced and their value is the labor expended in creating them. These are forms of capital with "objectified labor", which gets incorporated into the products of the factory, in proportion to the number and kinds of products produced by the factory in a certain period of time, considering the life-span of each item of fixed capital.

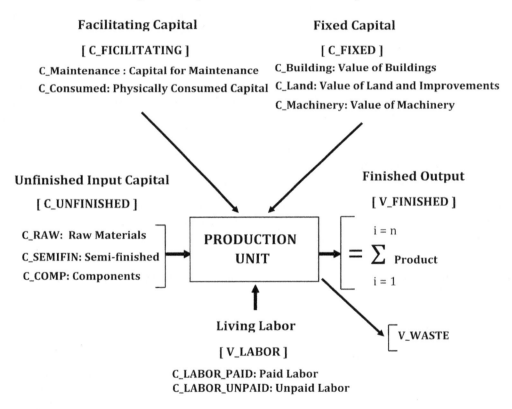

Facilitating Capital

[C_FICILITATING]

C_Maintenance : Capital for Maintenance
C_Consumed: Physically Consumed Capital

Fixed Capital

[C_FIXED]

C_Building: Value of Buildings
C_Land: Value of Land and Improvements
C_Machinery: Value of Machinery

Unfinished Input Capital

[C_UNFINISHED]

C_RAW: Raw Materials
C_SEMIFIN: Semi-finished
C_COMP: Components

PRODUCTION UNIT

Finished Output

[V_FINISHED]

$$= \sum_{i=1}^{i=n} \text{Product}$$

$$\begin{bmatrix} V\_WASTE \end{bmatrix}$$

Living Labor

[V_LABOR]

C_LABOR_PAID: Paid Labor
C_LABOR_UNPAID: Unpaid Labor

Figure 7-01: A Production/Service Center

Facilitating capital, (C_FACILITATIING) as the name implies, facilitates maintenance of the production process. It includes capital used in maintenance of plant and machinery and electrical energy, gas, water supply, etc., which are used to create the working environment of a factory, by heating, or air-conditioning buildings, or by providing drinking water for workers. Electrical energy, gas, lubricating oil and water may also be used in the production processes of the factory. Thus, this form of capital gets physically consumed during the production process and the labor, previously "objectified" in it, becomes part of the output of the production center.

During the production process, some of the raw materials and unfinished goods may be damaged and material used in facilitating the production process, like water, gas, electricity, oil, etc., may be wasted and not consumed as required by the production process. The cost of additional acquisition of these materials and services becomes part of the capital expended in the production process. Also, the disposal of waste, in most cases, becomes an additional cost for the production center. These costs may be considered part of the facilitating capital that is consumed, which also includes the cost of accounting and record keeping.

Unfinished capital (C_UNFINISHED) consists of raw materials and unfinished goods and components previously produced by other factories, farms, or mines. The items included in this form of capital are processed further to create the final products produced by the production center.

Living labor (V_LABOR) The machinery of the factory, or production center, is operated by workers who are hired by the owner, through the factory manager. Workers provide the *Labor* essential for the operation of the factory. This is living Labor which is used in creation of the final output of the factory and, thus, gets incorporated into its products. The workers of the factory operate the machinery of the factory and, using the raw materials and unfinished goods brought into the factory, create the finished products of the factory under the supervision of the factory manager and his management helpers.

In general, living labor is the main source of the increase in the capital invested in a production center. The income obtained by the sale of the total production of the production center is partly consumed by its workers (as wages) and, to some extent, by the owner. The net increase, after this consumption, is referred to as the *"Gross Profit"*.

All three forms of objectified capital, and the value added by living labor, are used for creation of the final products of the factory. Each item produced effectively includes the cost of a proportional part of Fixed, Facilitating and Unfinished capital. Each item, thus, includes the value of the total labor expended in its creation, including the non-living labor objectified in all forms of capital. *Thus, if all forms of capital were acquired at their exact value, the total capital expended, during the lifetime of the factory, would have almost the same total value as the total value of all the products produced, during the same time. The only difference would be due to accounting and wastage. If all the products of the factory are also sold at their value, then, the owner would be able to generate, virtually, no gain in his capital, i.e., there would be no profit.* This may not be clear at this point, but should become quite clear after we go through specific examples of various types of production centers and, then, revisit the production center in this general abstract form.

Under intensive capitalism, a factory is set up to generate not just profit, but *maximum profit*. Markets exist for all forms of input capital, labor and output. The *Law of the Jungle*, or the *Law of Supply and Demand*, applies to all inputs and outputs of the factory. The conflicts in

134

a factory are, firstly, between the owner and the suppliers of different forms of input capital and labor. The owner has to try to pay less than the value of input capital and labor. Further, he has to obtain the highest prices for products of his factory, compared to their value, so as to generate the *maximum profit that he desires*.

In the jungle called the market, those "animals" that are weak, are likely to lose in these conflicts. The question, then, becomes - which party is weak and which one has a stronger bargaining position in the set of relationships constituting a factory?

- The developers of land vs. the owner of the factory.

- The builders of the factory-building vs. the owner.

- Suppliers of machinery vs. the owner.

- Suppliers of water, gas, electricity, oil, etc., vs. the owner.

- Suppliers of raw materials vs. the owner.

- Suppliers of unfinished products and components vs. the owner.

- Workers providing their labor vs. the owner.

- Buyers of products of the factory vs. the owner.

Obviously, the developers of land and contractors constructing buildings are owners of their businesses and are in a similar strong position as the owner of this factory. They also own land and machinery used in land-development and building construction. Generally, they are not under financial pressure to come to agreements regarding price of land and cost of buildings. Similarly, suppliers of machinery, raw materials, unfinished items and components are in the same position. Utilities supplying water, gas, electricity, generally have fixed tariffs for supply of these services. They are not likely to make special concessions, under pressure, to those setting up new factories. Those supplying tools, fluids and chemicals for maintenance of machinery and other facilities at the factory are also in as strong a position as the owner of this factory. *What it means is that these forms of capital may be obtained by the owner of the factory at more or less prices than their values, but the differences between these prices and values are not likely to be big, under normal conditions of an economy.*

Now, that leaves the labor provided by the workers to be considered in the context of the social system in which the factory is set up. At the stage of intensive capitalism, when the factory is set up, society is generally going through a very painful transition from *advanced feudalism to intensive capitalism*. There are huge numbers of unemployed workers looking for jobs. Their needs and the needs of their families are urgent. Their immediate need for food and shelter forces them to work for any wages that may be offered by employers. The factory owner can offer much lower wages to these workers than the value of their labor, but the workers are too desperate to refuse to work on such wages. They are, thus, forced by market conditions to work for wages that are even below the requirements of subsistence. They end up becoming *wage-slaves*. Thus, the labor costs of the owner of the factory go down dramatically and he can make huge profits by, in effect, not paying for part of the labor that the workers provide. Thus, the *rate of intensive profit (RIP)* by the owner, is made possible by the glut in the labor market [1]

To clarify the processes involved in the generation of the high rates of intensive profit *(RIP)*, during intensive capitalism, we would analyze the workings of a few types of production centers, in the following sub-sections. Also, it should be noted that production in an industrialized state is of two kinds. Industries may produce products for consumption by the population or by other factories. These are known as consumer and producer goods.

In terms of consumer goods, the needs of the ruling class and the people are quite different. The people are desperate to survive and need food and medicine above everything else. They also need shelter, but are unable to afford to own their own houses. They may, instead, have to live in huts and make-shift housing, like card-board or mud houses, or rent living spaces. Renting is most common at this stage. The ruling class, in addition to food and clothing, needs luxury goods in general, but its total consumption is much less than the working class, because of its small size. In addition to consumer goods, the ruling class needs producer-goods for its factories, mines and other businesses, i.e., products like machine tools that are used in factories, mines and mechanized farms. The overall relationships of producer and consumer goods producing sectors of an economy would be examined after production centers of some common forms have been described.

A Factory Manufacturing Nails

Let us suppose that a factory has been set up to manufacture nails only. Not only that, but it manufactures nails of one size only, say only three inches long and a fixed thickness. Of course, such a factory, probably, cannot be found in real life! It would, however, make our example easy to understand. The inputs and output of such a factory are indicated in Figure 7-02. The components of Fixed Capital in this case are as indicated.

The factory would not be processing any unfinished items or components. It would use ingots of steel supplied by a steel mill, as raw material. The maintenance of the factory would include the maintenance of machinery and buildings. This would require tools and lubricating oil, etc. The factory's internal environment may be maintained by air conditioning and lighting, which would lead to consumption of electricity and gas. Machines may require water for cooling and workers may require it for cleaning-up and, also, for drinking. Suppose that electric machines are used, requiring electric power supply and one percent of the input steel is wasted in the production process. Suppose the factory employs one manager, one accountant, thirty shop-floor supervisors, twenty electricians, three hundred machine operators and two hundred unskilled workers, who are paid $2,000, $1,000, $500, $200, $100 and $50 per month, respectively. Also, suppose the buildings and machines of the factory have a useful life of 50 and 20 years, respectively. The buildings would lose 1/50 of their value in a year, due to depreciation. Similarly, the machines would lose 1/20 of their value in a year. Suppose, the land was purchased, with its improvements, for $200,000. For simplicity, let us assume it was obtained with a thirty-year loan from a bank, at six percent per year simple interest rate. Although the value of land is zero plus the value of improvements, in the real world of intensive capitalism no one can obtain land without paying a price. So, we have to assume it was obtained by using a loan from a bank. Thus, we can assume an annual cost of $12,000 per year as simple interest, or rent, of the land along with improvements like leveling and arrangements for electricity, water and gas supply. We are also supposing that the bank pays two percent simple interest to its depositors, i.e., $ 4,000 per year, which is the cost to the bank, based on the assumed value of the land and its improvements.

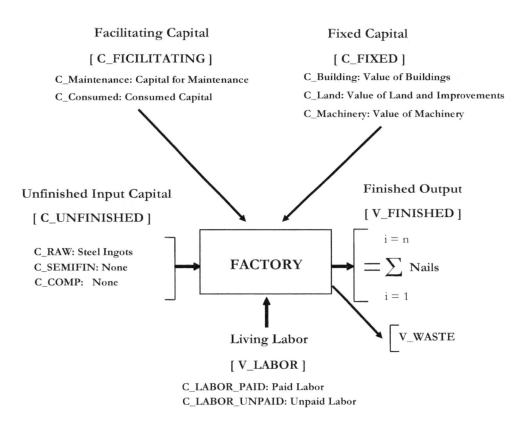

Figure 7-02: A Factory Manufacturing Nails

Suppose 300 ingots of steels are processed by the factory in a year, each yielding one million nails. Also, suppose the value of buildings, machines and each ingot of steel is $1,000,000, $800,000 and $1,000, respectively and the prices paid by the owner of the factory are $1,500,000, $1,200,000 and $1200 respectively.

If we assume that all components of facilitating capital are obtained at prices somewhat above their respective values, then, the accounting for *one year of operation of the factory,* may be done as follows, in terms of assumed prices and values. (*Values are enclosed in brackets*):

Depreciation/Replacement Cost of Fixed Capital:

Buildings: (50-year life) $30,000 ($20,000)   (X)

Machines: (20-Year life) $60,000 ($40,000)   (X)

Land: (Rent/Interest) $12,000 ($ 4,000)   (A)

Total: $102,000 ($64,000)

Unfinished Capital:

Cost of 300 ingots of steel: $360,000 ($300,000) (B)

Facilitating Capital:

Electricity: $26,000 ($21,000)   (C)

Gas: $12,000 ($11,000)   (D)

Water: $ 3,000 ($ 2,000)   (E)

Maintenance: $ 8,000 ($ 6,000)   (F)

Wastage and disposal: $ 3,000 ($ 2,000)   (G)

Total: $ 52,000 ($ 42,000)

Objectified Labor Cost:

Depreciation/Replacement of Fixed Capital: $102,000 ($ 64,000)

Unfinished Capital Cost: $360,000 ($300,000)

Facilitating Capital Cost: $ 52,000 ($ 42,000)

Total: $514,000 ($406,000)

Living Labor Cost

Manager (1): (2,000x1x12) $24,000 ($ 24,000)

Accountant (2): (1,000x2x12) $24,000 ($ 36,000)

Shop-floor Supervisors (30): (500x30x12) $180,000 ($ 360,000)

Electricians (20): (200x20x12) $ 48,000 ($ 96,000)

Machine operators (100): (100x100x12) $120,000 ($360,000)

Unskilled workers (200): (50x200x12) $120,000 ($480,000)

Total: $516,000

($1,356,000)

Total Cost for One Year

Objectified Labor Cost $514,000 ($ 406,000)

Living Labor Cost $516,000 ($1,356,000)

Total: = $1,030,000
($1,762,000)

Output for One Year

Nails produced 300,000,000

Cost/1000 Nails $1,030,000/300,000   = $3.43

Value/1000 Nails $1,762,000/300,000 = $5.87

Sales and Profit

Sale price/1000 Nails $6.30

Total Revenue $ 1,890,000

Profit/1000 Nails = $6.30 - $3.43 = $2.87

Total Intensive Profit in one year = $ 1,890,000-$ 1,030,000= $860,000

Rate of Intensive Profit (RIP, before tax) = $860,000/($1,030,000) *100 = 83.49%

Value of unpaid labor = $1,356,000 - $516,000 = $840,000

Rate of Exploitation of Workers (ROE) = 840,000/1,356,000 * 100   = 61.9%

Notice that the objectified labor costs include $90,000 as depreciation to replace the buildings and machines when they reach the end of their lives. Mr. X receives back all his recurring costs also. Thus, the initial capital invested by the owner of the company, i.e., Mr. X, remains unaltered at the end of the year. Waste and disposal costs are included in facilitating capital costs. We can see that the *rate of intensive profit* of the factory is about 83.49 percent. The rate of exploitation of labor is 61.9 percent, i.e., workers are not paid for 61.9 percent, or more than half, of their labor.

Several capitalists and different kinds of workers have actually participated in the creation of this factory and its operation. If we suppose Mr. X owned this imaginary factory and wanted to "make" maximum amount of money by using others, we would be neglecting several other beneficiaries who have also, indirectly, participated in this project. We have indicated the other beneficiaries as Mr. A, B, C, etc. Of course, the workers and their families have also benefited. Their wages, under intensive capitalism, are subsistence wages, but they have managed to stay alive! Mr. X, however has had a gross profit of $860,000 for the year, after paying off his suppliers, including Bank A and Mr. G, a total of $424,000. Also, he gets his payment of $90,000, as depreciation of buildings and machines of the factory. He retains this amount for the ultimate replacement of his investment in fixed capital. He retains the ownership of the factory and all its machinery and other facilities. For the operation of the factory, thus, there is no net cost to him.

Examining the costs and sale prices for the year, for the products and services supplied by others, we can see that Mr. "G" is not really a supplier. He is charging $3,000 for disposing the waste off. The bank, charging $ 12,000 per year for the loan for land acquisition, is indicated as "A". This requires a closer examination. The bank is giving this loan out of the money deposited by its customers, for which it pays an effective interest of four percent, amounting to $ 4,000 per year, which takes money creation by the bank into consideration. The bank, thus, has a yearly profit of $ 8,000 and Mr. X gets possession of the land, on which the factory has been established, from someone else, say Mr. Y. Now, Mr. Y must have obtained this land from another person by making a payment. Thus, the land would have passed through numerous hands before it is acquired by Mr. X. Since the original value of the land is zero, as it is a naturally occurring resource, then the question does arise as to why a price has to be paid for its acquisition. Of course, the improvements on this land do have some value. But the virgin land had no value originally. Someone in the distant past must have gotten possession of this land and maintained it by force. The land, then, must have passed through several hands, over numerous generations, till the practice of selling it came to be developed by some ancient people. Thus, the possession of the land is being sold, not the land itself. We are considering $ 12,000 payment as the yearly cost of possession of the land – we are calling it the "value" of the land, although it is only a pseudo-value. If we assume that the costs of all products and services provided by the suppliers to the factory are as shown in the table, then, we can summarize the gains from this factory, of all the capitalists besides Mr. X, as indicated:

The customers pay the full negotiated amount of $1,890,000 for the yearly output of the factory, based on the price of $ 6.30 per packet of one thousand nails. This amount is distributed three ways – An amount of $ 90,000 is retained by the owner as depreciation of fixed capital, so that the cost of operation of the "enterprise" for the owner remains zero. An amount of $424,000 goes to the suppliers "A-G" of facilitating and input unfinished capital, etc. This list also includes the bank "A" and waste disposal company "G". The suppliers end up with a total profit of $78,000. If their costs are higher than as indicated, then, their profits would

Capitalist	Costs($)	Sales($)	Profit/Loss($)
A	4,000	12,000	+ 8,000
B	300,000	360,000	+60,000
C	21,000	26,000	+ 5,000
D	11,000	12,000	+ 1,000
E	2,000	3,000	+1,000
F	6,000	8,000	+2,000
G	2,000	3,000	+1,000
TOTAL	346,000	424,000	+78,000

be proportionately lower. After paying wages of $ 516,000, the remaining amount of $860,000 is taken by the owner as his gross profit, with no further cost to him! The yearly operation of the factory and the resulting flow of capital is shown in Figure 7-03(a).

All the suppliers of the factory, including the bank "A" and Mr. "G" are, in effect, partners of the owner, Mr. "X", although they are bound by different agreements about what benefit they can derive from the operation of the factory. If we lump all the capital input together, except the value of labor, then a simplified diagram of the operation of the factory results. It is shown in Figure 7-03(b).

Figure 7-03 (a) View-1

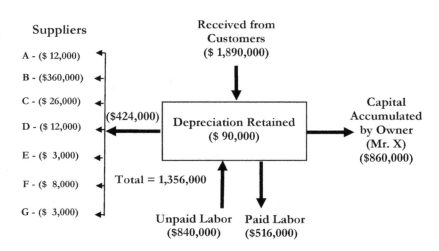

Figure 7-03 (b) View-2

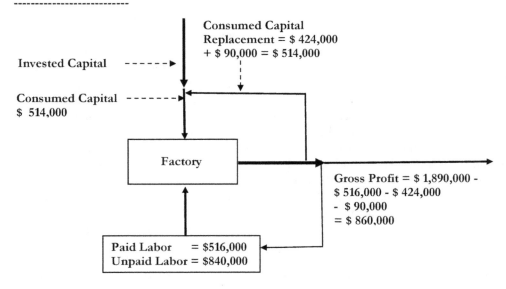

If we suppose that all the components of objectified capital are paid for at the asking prices of suppliers, as indicated, and all workers of the factory are paid the full value of their labor, then the results would be as follows:

Objectified labor cost = $ 514,000

Living Labor cost = $ 1,356,000

Total cost for the year = $ 514,000 + $1,356,000 = $ 1,870,000

Cost/1000 nails = $1,870,000/300,000 = $ 6.23

If, now, the nails are sold at the cost price, then, Mr. X would have no profit. He, however, would not lose any of his capital. His suppliers would, still, have a total profit of $78,000, based on the costs and prices of their products and supplies sold to this factory. These, of course, are not his objectives!

A Coal Mine

Let us look at the operation of a coal mining company that has acquired a mine from another company. The mine is established on land that has already been explored and a coal seam has been found in it. Tunnels have been dug to reach the "coal face" in the seam. The cost of land would include the cost of exploration by the other company, resulting in establishment of the mine. The cost of land, also, includes improvements, which, in a mine, consist of tunnels, basic rock supporting structures and temporary housing for workers.

The inputs and output of such a mine are indicated in Figure 7-04. The components of Fixed Capital in this case are the same as for the factory manufacturing nails. The factory would not be processing any unfinished items, or components. The coal would be mined from the coal seam, which is the only raw material for the mine. The coal is to be transported by mine-workers. The maintenance of the mine would include the maintenance of machinery and buildings. This would require tools and lubricating oil, etc. The mine's internal environment is maintained by fans circulating air and by lighting. This would lead to consumption of electricity. Workers would require drinking water. Mining machinery would require electric power supply. Suppose the mine employs one manager, one accountant, three mine supervisors, two electricians, thirty machine operators, and two hundred other skilled workers and one hundred unskilled workers, who are paid $2,000, $1,000, $500, $200, $100, $80 and $50 per month, respectively. Also, suppose the land with all its improvements and buildings has a useful life of 50 years, after which it would become totally useless, while the mining machinery has a useful life of 10 years. The coal would run out in 50 years. The improvements on land, including buildings, would lose 1/50 of their value in a year, due to depreciation. Similarly, the machines would lose 1/10 of their value in a year. Suppose, the land was purchased, with all its improvements, for $ 1,000,000, while it had cost the original company $800,000, including exploration costs. The value of the coal seam is, of course, zero. Suppose the mine produces 10,000 tons of coal in a year. Suppose the value of machines is $200,000 and their price paid by the owner of the mine was $210,000. Then, the accounting for one year of operation of the coal mine may be done as follows, in terms of prices, assuming prices for components of facilitating capital that are somewhat higher than their values, as in the previous example. (*Values are enclosed in brackets*):

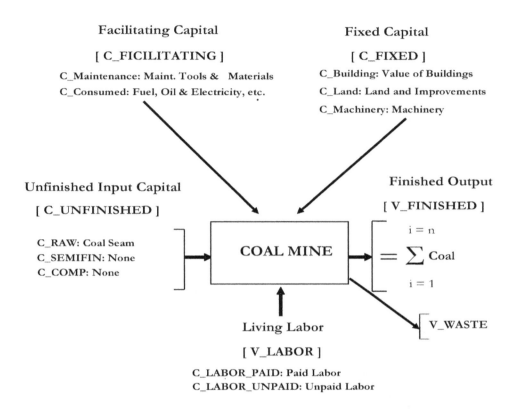

Facilitating Capital

[C_FICILITATING]

C_Maintenance: Maint. Tools & Materials
C_Consumed: Fuel, Oil & Electricity, etc.

Fixed Capital

[C_FIXED]

C_Building: Value of Buildings
C_Land: Land and Improvements
C_Machinery: Machinery

Unfinished Input Capital

[C_UNFINISHED]

C_RAW: Coal Seam
C_SEMIFIN: None
C_COMP: None

Finished Output

[V_FINISHED]

COAL MINE

$$= \sum_{i=1}^{i=n} \text{Coal}$$

V_WASTE

Living Labor

[V_LABOR]

C_LABOR_PAID: Paid Labor
C_LABOR_UNPAID: Unpaid Labor

Figure 7-04: A Coal Mine as a Production Center

Depreciation of Fixed Capital:

Machines:	$21,000 ($20,000)
Land and buildings:	$20,000 ($16,000)
Total:	$41,000 ($36,000)

Unfinished Capital: $ 0.00
Facilitating Capital:

Electricity:	$30,000 ($25,000)

Water:	$ 5,000 ($ 4,000)
Maintenance	$ 5,000 ($ 4,000)
Waste & Disposal	$ 5,000 ($ 1,000)
Total:	$45,000 ($34,000)

Total Objectified Labor Cost:

Depreciation of Fixed Capital:	$41,000 ($36,000)
Unfinished Capital:	$0 ($ 0)
Facilitating Capital	$45,000 ($34,000)
Total:	$ 86,000 ($70,000)

Living Labor Cost:

Manager (1): (2,000x1x12)	$24,000 ($24,000)
Accountant (1): (1,000x1x12)	$12,000 ($18,000)
Mining Supervisors (3): (500x3x12)	$18,000 ($27,000)
Electricians (2): (200x2x12)	$ 4,800 ($ 7,200)
Machine operators (30): (100x30x12)	$ 36,000 ($ 54,000)
Misc. skilled workers (200): (80*200*12)	$192,000 ($200,000)
Unskilled workers (100): (50*100*12)	$ 60,000 ($ 90,000)
Total:	$346,800 ($420,200)

Total Costs for One Year:

Objectified Labor Cost	$ 86,000 ($ 70,000)
Living Labor Cost	$346,800 ($420,200)
Total: =	$432,800 ($490,200)

Output for One Year:

Coal produced	10,000 Tons
Cost/Ton	$432,800/10,000 = $43.28
Value/Ton	$490,200/10,000 = $49.02

Sales, Labor Exploitation and Profit

Sale price/Ton $60.00

Total revenue $600,000

Total Intensive Profit = $600,000 - $432,800 = $167,200

Rate of Intensive Profit (RIP)=($60.00-$43.28)/43.28 *100 = 38.63%

Value of unpaid labor = $420,200 - $346,800 = $73,400

Rate of Exploitation of labor (ROE) = 73,400/420,200 * 100 = 17.46%

Notice that the objectified labor costs include $41,000 as depreciation to replace the machines and buildings when they reach the end of their lives. Thus, the capital invested by the owners of the company remains unaltered. From the above calculations, we can see that the rate of exploitation of labor *(ROE)* is 17.46 percent. It means that workers are not paid for about one-fifth of their labor – resulting in the *rate of intensive profit* of about 38.63 percent for the mining company

If we suppose that all the components of objectified capital are supplied at their value to the mining company, without any profit, then the rate of exploitation of the labor remains the same, i.e., 17.46%, but the rate of intensive profit increases to 43.95%, after removal of its loss of $16,000 to the suppliers of objectified capital.

If all the components of the objectified capital and living labor are obtained at value, i.e., $490,200, and coal is sold at its value of $49.02/ton, then the profit rate would be zero percent.

A Modern Agricultural Farm

Modern agricultural farms are like factories, using a high level of mechanization in their operations. These corporate farms operate on very large areas of land and may have industries based on their territory to process their agricultural products. In the United States and Brazil, these farms may have as much as 50,000 acres each, or more, of land under cultivation and use a large amount of machinery for cultivation, harvesting, storage and transportation of crops and, possibly, their industrial products.

These farms may use machinery like planters, tractors, earth-movers, thrashers, harvester combines, aircraft for spraying insecticides and trucks and other vehicles for transportation. These agrobusinesses hire a large number of workers to operate the agricultural machinery and also to man their offices, where sale and purchase of supplies for the farms and their production may be processed, along with record-keeping and accounting. The offices, also, use telecommunication equipment and computers.

The Fixed Capital of such an agrobusiness consists of land, buildings and machinery that have been mentioned already. The improvements on the land may include roads, tube-wells for drinking water or for irrigation, electric sub-stations for electric power supply from a local electric utility, gas connections for heating and cooking, animal sheds and feeding stations, crop storage silos for fodder storage for animals and living quarters for the owners and workers of the farm. The maintenance of the farm machinery requires maintenance staff and tools and supplies. In addition to electric energy and gas, consumable capital may consist of fertilizer, insecticides, etc. In addition to seeds, the input to a farm may consist of plant saplings and equipment for shipment of plants, like Christmas trees, and animals which the farm may be producing.

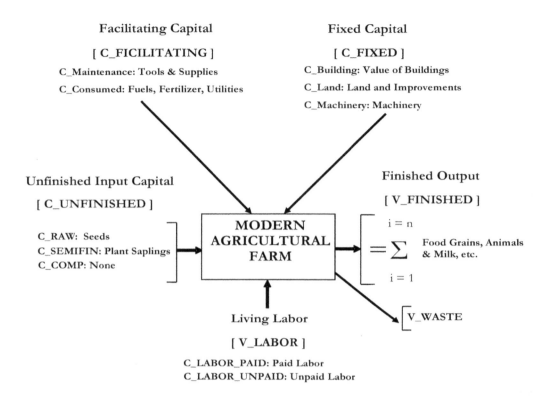

Figure 7-05: A Modern Agricultural Farm as a Production Center

We can see that all forms of capital are involved in these agrobusinesses. The model of a production center in the social system of intensive capitalism is exactly applicable to these farms, in all aspects, as indicated in Figure 7-05. Each form of capital has its market, including the output products of these farming corporation. Workers are the weakest participants in these markets. The labor market ensures that they receive much less than the value of their labor, thus ensuring a very high *RIP* for the owner/owners of the corporate agrobusinesses.

A Transportation Company

A transportation company may be involved in transport of people or minerals, fuel, chemicals and industrial products, etc. It provides a service for which it charges fees to its customers. A transport company may be involved in railroad transport, or may use aircraft transporting travelers, their luggage, or industrial products. If it is involved in road-transport, then it would use buses or trucks for this purpose. Although such a corporation would be involved in providing services, the components of its capital inputs are very similar to a factory involved in manufacturing.

In this case, the components of Fixed Capital and Facilitating Capital are as indicated in Figure 7-06. Workers of such an organization would include managers, accountants, ticket issuing clerks, cashiers, pilots, air-hosts and air-hostesses, drivers and unskilled workers like luggage handlers. The categories of workers would naturally depend on the kind of transportation company.

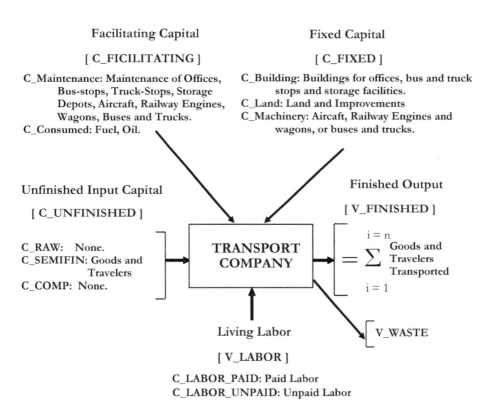

Figure 7-06: A Transportation Company as a Service Center

We have indicated goods and travelers as "Unfinished Capital" that enters the company. Obviously, virtually no capital is expended in acquiring these goods and travelers, unless we think about advertising costs. But the goods are received by the company and are transported to a different location where their value becomes higher because of the objectified and living labor used in transportation. The company charges the owners of goods for this cost. Transportation of goods may require short term, or long term, storage. This becomes an additional cost for the company. Similarly, travelers are transported to their desired locations and are required to pay for the value of objectified and living labor used in their transportation. Thus, travelers, as "Unfinished Capital" do not cost the company anything.

However, during transportation some seats for travelers may not be utilized, thus these become the source of extra costs to the company. Similarly, the un-utilized goods transportation capacity of its vehicles would create extra costs for the company. All costs of labor and other capital inputs are proportionally assigned to the goods and travelers transported. In this case labor and travelers are in a weak bargaining position vs. the management of the company and both end up losing, because of their unfavorable circumstances in the relevant markets. The travelers may be charged more than the value of their transportation and most categories of workers would be paid far less than the full value of their labor. Thus, the profits of the company, mainly, come out of the unpaid labor of its employees, though a part may be extracted from the travelers. The more the unpaid labor of its employees and extra transport charges, the greater the *rate of intensive profit (RIP)* and greater the *rate of exploitation (ROE)* of its workers.

A Trading Company

The operation of a company involved in trading is very similar to that of a transportation company. A trading company may consist of only one individual purchasing agricultural

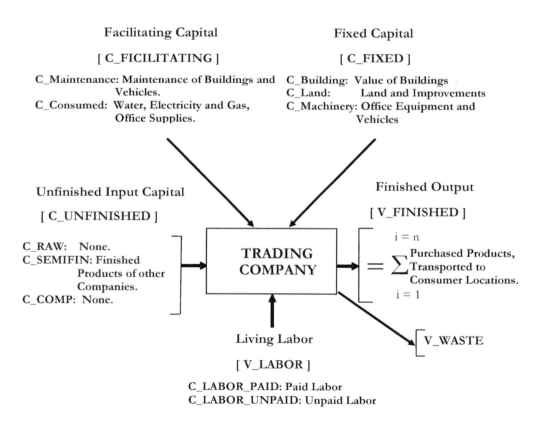

Figure 7-07: A Trading Company as a Service Center

products or industrial products of other companies, transporting them to his shop's location close to consumers, and selling those products at higher prices than the relevant costs. Large trading companies may have many stores distributed over a state, or over several states, all over the world. The company may bargain with producers and manufactures of various products, to obtain prices as low as possible. It may, then, transport those products to its store locations, to be offered for sale to customers.

Thus, the "unfinished" input capital of such a trading company consists of the products it purchases and its output consists of the same products. The operation of the company is obvious, as indicated in Figure 7-07. Even in this case, the weakest party in this organization are the workers providing the labor which constitutes the value added to the input products. The *RIP* of such a company is, thus, determined by how much of this labor is not paid for by the company. Further, it depends on how low are the prices that it is able to negotiate with its suppliers and how high the prices are that its customers have to pay.

A Bank

A bank does not produce any material product. It provides various kinds of services to its customers. It basically provides security to its customers who can deposit their money in their accounts at the bank, where it is safe if the deposits are insured against bank failure. The customers can also deposit their savings in savings-accounts, or accounts with withdrawal restrictions. Generally, the bank pays the depositors an interest on the amount deposited, the rate of interest depending on how restrictive the terms of withdrawal are. The more restrictive the terms of withdrawal are, the higher the rate of interest. Some banks also provide accounts for investment in stocks, bonds and mutual funds. A bank also provides loans for construction and renovation of houses or commercial buildings and for purchase of cars and boats. Some banks offer micro-finance loans to their customers for setting up small shops or small home-based businesses. A bank has qualified employees to handle the paper work and it also employs some unskilled workers. The source of profit of a bank is different from a factory in one respect. It charges its loan-customers a high rate of interest on the amounts loaned and pays its depositors a lower rate of interest. The difference between the two rates of interest, which the bankers refer to as the *spread*, is one major source of the profit of the bank.

The capital inputs and output of such a bank are indicated in Figure 7-08. The components of Fixed Capital and Facilitating Capital in this case are obvious. However, the Unfinished Input Capital and the Finished, or Output, Capital require explanation. Customers deposit their money in Checking and Savings accounts at the bank. The bank may, however, provide Investment accounts. The money in such accounts may be invested in the stock market. It can be invested in stocks, mutual funds, money-market funds, bond funds and exchange-traded funds (ETFs). These are capital markets. The bank, normally, does not share the risks involved in such investments, but does charge transaction fees for them. The mutual funds, etc., also charge fees, which may be reduced for the bank or its customers. The fees charged by the bank on investment accounts are another source of income for the bank. Normally, the deposits in the accounts at a bank far exceed the withdrawals. The bank invests these excess funds in the stock and bond markets or money markets of various kinds, for its own benefit. This is a big source of its income.

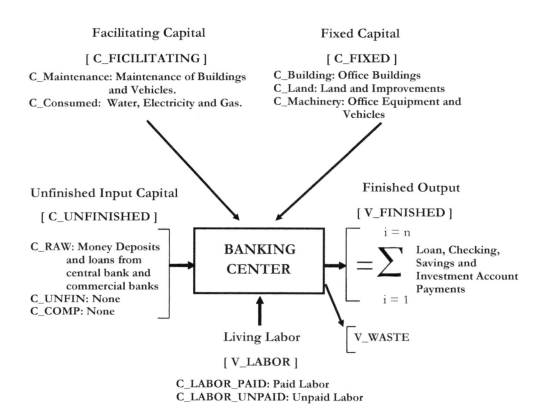

Figure 7-08: A Bank as a Service Center

As indicated, the bank issues loans of various kinds. A bank can, effectively, create money by loaning out funds that it does not have. Banks operating in the intensive capitalism of today have been forced to become much less abusive in this respect since the Great Depression in the 1930s. Now, all states have a central Reserve Bank, like the Federal Reserve Bank of the US, or the State Bank of Pakistan, that imposes a *Reserve Requirement* on banks, based on total funds deposited in their checking and some other accounts. This reserve requirement may be in the range of *2-15 percent*, i.e., a bank has to maintain this amount in the form of cash in its vaults or with the central bank. The banks can, thus, lend the remaining amount. However, as the loans are repaid, these payments, like all deposits in checking account, are reported to the central reserve bank and become part of the total amount to which the reserve requirement applies. Thus, banks can lend an amount several times the deposits in their accounts. Thus, money is created by these banks and the total money supply is increased/decreased by the central banks by changing the reserve requirements and adjusting interest rates. The central banks also have a *"discount rate"* for lending to major banks

150

and financial institutions. The *"federal funds rate"* is used by banks in the US for overnight inter-bank lending. In the US, all loan interest rates are generally pegged to the *prime rate*, which banks use for loans to their major customers, or institutions. All these interest rates are related to the interest rate set by the central bank of a state. Thus, money loaned by the commercial banks may be made "easy money" by lowering the central bank rate. The opposite happens if a central bank raises the interest rate, making it more expensive and difficult for companies and individuals to borrow money from commercial banks. Loans issued by commercial banks are increased further by the "easy money" policies of central banks. Thus, their income from loans is not just based on the "spread", but is much more than that.

If the social system is at the *intensive capitalism* stage, then, most of the deposits at a bank are in accounts maintained by professionals, business owners and owners of various kinds of capital. The workers, who only have subsistence living and have no surplus cash, cannot afford to open such accounts. The deposits in a bank are the main source of income for the bank. The loans are mainly made to those who own cars, houses and businesses, i.e., the members of the ruling class. These individuals have the capital to offer as collateral. We can list the costs and sources of income of a bank, during a specific period, as follows:

Costs:

- Depreciation of buildings and office equipment. The price of land may also change. Land, generally, appreciates with time.

- Interest paid to customers, on accounts maintained at the bank.

- Facilitating capital costs, including the cost of maintenance of buildings, vehicles, etc., and the cost of electricity, gas, water, office supplies, etc., consumed during the period and wastage of these resources.

- Payments to employees for various kinds of labor.

Sources of Income:

- Fees charged to customers for opening and maintenance of checking, savings and investment accounts.

- Interest from loans issued. This is generally the main source.

- Income from money deposits invested in stock, money and bond markets.

- Unpaid labor of, mainly, unskilled and low-skilled workers.

An Insurance Company

The risk of injury and death and loss of belongings is a part of life. People can die young due to accidents, leaving their spouses and children without any source of support. Belongings of a family can be lost due to floods, fire, theft or robbery. A manufacturing company may lose part of its products, due to accidents in transportation. Even capital invested in machinery and equipment may be lost by a company due to accidents like the air-crash of an aircraft owned by a transport company, or loss of office equipment due to fire in a bank. An insurance company promises to protect its customers against such risks, provided they pay a premium for this service.

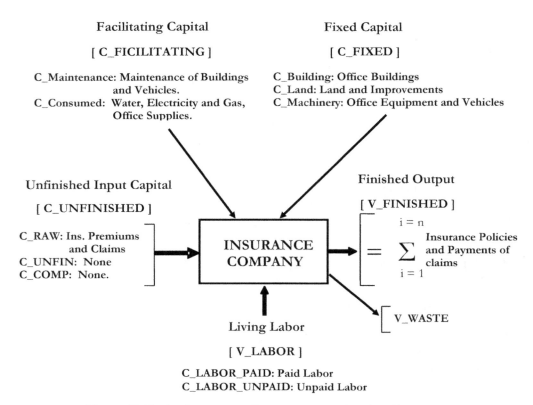

Facilitating Capital

[C_FICILITATING]

C_Maintenance: Maintenance of Buildings
and Vehicles.
C_Consumed: Water, Electricity and Gas,
Office Supplies.

Fixed Capital

[C_FIXED]

C_Building: Office Buildings
C_Land: Land and Improvements
C_Machinery: Office Equipment and Vehicles

Unfinished Input Capital

[C_UNFINISHED]

C_RAW: Ins. Premiums
and Claims
C_UNFIN: None
C_COMP: None.

**INSURANCE
COMPANY**

Finished Output

[V_FINISHED]

$$= \sum_{i=1}^{i=n}$$ Insurance Policies and Payments of claims

V_WASTE

Living Labor

[V_LABOR]

C_LABOR_PAID: Paid Labor
C_LABOR_UNPAID: Unpaid Labor

Figure 7-09: An Insurance Company as a Service Center

The inputs and output of such an insurance company are indicated in Figure 7-09. The components of Fixed and Facilitating capital are self-explanatory. There is no unfinished input capital. The company may offer several different types of policies for life, disability, business and home insurance, etc. It may also insure products under shipment, against losses due to breakage, etc. The customers are required to make premium payments to keep these policies in force. The premiums are based on the probabilities of losses of various kinds. For example, life insurance is based on life expectancy for healthy individuals at various stages of their lives, when they decide to obtain such insurance policies. The statistics of life expectancy are tabulated in the form of what are called *"actuarial tables"*. The premiums are calculated on the basis of such tables, after adding a margin of expected profit. The policies are marketed by insurance agents. The required premium rates of these policies are set to include these "commissions" paid to the insurance agents as costs.

If the social system is, purely, at the *intensive capitalism* stage, then all the insurance policies are obtained by professionals and business owners to insure their lives, property and businesses. The workers, who only have subsistence living and have no surplus cash, cannot afford them. Thus, the costs and sources of income of an insurance company may be listed as follows:

Costs:

- Depreciation of buildings and office equipment and the depreciation, or appreciation, of land.

- Payment of claims.

- Facilitating capital costs, including the cost of maintenance of buildings, vehicles, etc., and the cost of electricity, gas, water, office supplies, etc., consumed during the period and wastage of these resources.

- Payments to insurance agents and employees for various kinds of labor.

Sources of Income:

- Unpaid labor of office workers.

- Unpaid labor of insurance policy sale agents.

- Premiums of policies issued to customers.

A Primitive Agricultural Farm

In a feudal society, agriculture farms are not mechanized and do not have access to modern technologies of production. Still, the capital inputs to such a farm are basically the same as described in previous sections. Figure 7-10 indicates the inputs and outputs of such a farm. If society is in transition to *intensive capitalism*, then, as has been described in the previous chapter, there is an excess of workers on land. They could have been working as share-croppers before, but their share may shrink further as the power of the landlord keeps growing. If they were working as serfs, the small plots of land allotted to them may be forcibly taken over by their landlords or by others with larger land holdings and, therefore, possessing greater power.

In a feudal society at the stage of *advanced feudalism*, the peasants, serfs, or share-croppers, are totally dependent on the farms and their landlords. Their families live on the farms in make-shift housing, like mud-houses, without any education or medical facilities for their children or themselves. Their food is grown by them on the farms and they, generally, raise chickens, cows, sheep and goats, which also provide part of the food for them and their families. Since they have no other skills except the ability to sow and produce crops on land, their dependence on the farm and the landlord is extreme.

Initially, the machinery of the farm may consist of only ploughs and simple implements, requiring very little maintenance. Maintenance is provided by skilled workers in the relevant village of the landlord. Water supply is generally arranged by the landlord. The unfinished input capital consists of seeds only, unless the landlord decides to set up a fruit orchard on the land, or starts an animal farm, in which case plant saplings may be supplied by a plant nursery and some animals may be purchased by the owner. In more advanced mechanized farms, fertilizers and insecticides may also be supplied by the landlord. Food grains and, possibly, fruits are the outputs of the farm. If animals are also being raised, they may also form the output of the farm, and may be disposed-off by the owner. Such farms are the standard production centers during the stage of *feudalism* and they remain the same, or get worse in terms of abuse of workers, during the initial stages of *intensive capitalism*.

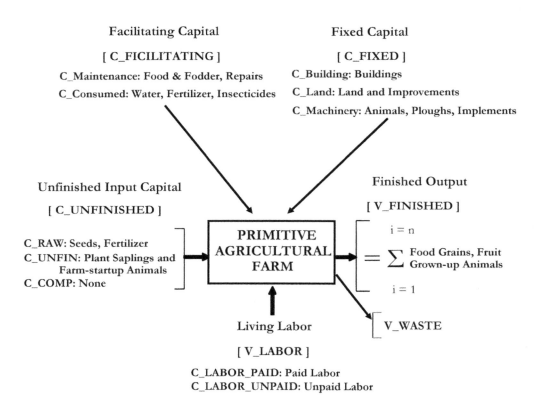

Facilitating Capital

[C_FICILITATING]

C_Maintenance: Food & Fodder, Repairs

C_Consumed: Water, Fertilizer, Insecticides

Fixed Capital

[C_FIXED]

C_Building: Buildings

C_Land: Land and Improvements

C_Machinery: Animals, Ploughs, Implements

Unfinished Input Capital

[C_UNFINISHED]

C_RAW: Seeds, Fertilizer

C_UNFIN: Plant Saplings and Farm-startup Animals

C_COMP: None

PRIMITIVE AGRICULTURAL FARM

Finished Output

[V_FINISHED]

$= \sum_{i=1}^{i=n}$ Food Grains, Fruit Grown-up Animals

V_WASTE

Living Labor

[V_LABOR]

C_LABOR_PAID: Paid Labor

C_LABOR_UNPAID: Unpaid Labor

Figure 7-10: A Primitive Agricultural Farm

With further development of technology, mechanization of agriculture has become faster and has devastating effects on agricultural workers and their families. As agriculture is mechanized, peasants working on the land become surplus much faster and are evicted by their landlords. As machines, like tractors and thrashers are brought in, their fear of eviction keeps growing till eviction becomes a reality. Thus, the peasants, as a class, lose any bargaining power they may have had previously.

It is not difficult to imagine why the farm-workers, or share-croppers, would receive much less than what they deserve on the basis of the value of their labor. They are the only ones who work on the land. Only the landlord owns the land, the ownership having been obtained by force, or payment of a price for someone else's initial forcible occupation, as discussed before. Yet, the landlord may take half, or one-third, of the produce of the land as *"his share"*. Landlord's share of a crop is basically the rent of the land in kind. It may include the cost of seeds, fertilizers and insecticides being returned to him with a profit, according to a contract. It is well known, that under these conditions of servitude, the landlords also sexually abuse the families of the workers on the land and the workers are forced, by their circumstances, to bear these indignities in addition to economic exploitation.

A University

Students enter a university as its unfinished input capital and leave the university as qualified engineers, doctors and scientist, etc. Thus, a university changes untrained, low-skilled, workers into high-skilled workers, i.e., it creates higher-level human capital as its output.

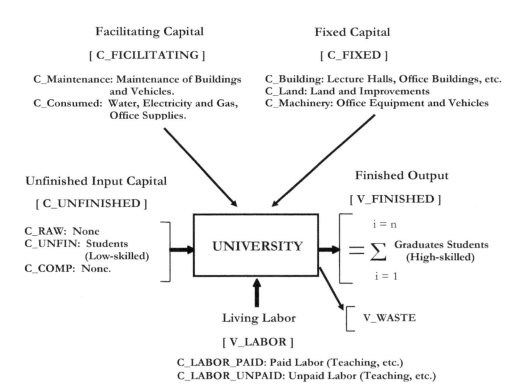

Facilitating Capital

[C_FICILITATING]

C_Maintenance: Maintenance of Buildings and Vehicles.
C_Consumed: Water, Electricity and Gas, Office Supplies.

Fixed Capital

[C_FIXED]

C_Building: Lecture Halls, Office Buildings, etc.
C_Land: Land and Improvements
C_Machinery: Office Equipment and Vehicles

Unfinished Input Capital

[C_UNFINISHED]

C_RAW: None
C_UNFIN: Students (Low-skilled)
C_COMP: None.

Finished Output

[V_FINISHED]

UNIVERSITY

$$= \sum_{i=1}^{i=n}$$ Graduates Students (High-skilled)

V_WASTE

Living Labor

[V_LABOR]

C_LABOR_PAID: Paid Labor (Teaching, etc.)
C_LABOR_UNPAID: Unpaid Labor (Teaching, etc.)

Figure 7-11: A University as a Service Center

The living labor in a university is provided, basically, by its teachers and professors who impart knowledge to its students. Additional labor is, of course, needed for maintenance of the facilities of the university. Under intensive capitalism, most of the students entering universities belong to the ruling class. At a later stage of the system, some others of relatively lower income levels may also be able to get in.

A university is a service center, not a production center. The function of all educational institutions is the same. Thus, what applies to a university, also applies to a school, a college, or a vocational training institution. Figure 7-11 shows the capital inputs and output of a university, as a service center.

A Hospital

Sick individuals enter a hospital as its "input human capital" and leave it as healthy individuals if they survive after treatment and hospitalization. Thus, a hospital protects human capital and tends to produce "higher valued" human capital. Doctors and nurses provide the labor to treat patients.

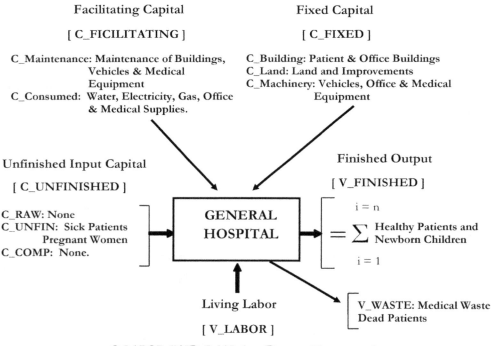

Figure 7-12: A General Hospital as a Service Center

Thus, a hospital may be viewed as a service center for its patients. However, at the stage of intensive capitalism, hospitals, mostly, serve the ruling class and not the people of a state. This situation changes as society develops. At a certain stage, individuals with lower incomes may be able to afford medical treatment in a hospital, but at the peak of intensive capitalism, this is generally not possible. Figure 7-12 shows the capital inputs and output of a hospital.

Summary and Conclusions

Production in a social system at the stage of *intensive capitalism* consists of production of goods and services at many types of production/service centers. Some such production/service centers have been described in the previous sections of this chapter. The list of types of production centers can be very long. It depends on numerous production processes of specific products and services. Some of these production processes may require less skilled labor. Others require highly skilled labor, e.g., software developers, scientists and researchers. At the initial stage of intensive capitalism, however, the labor force is mainly unskilled. The operation of these centers is identical in terms of basic input and output of capital, as indicated in Figure 7-01.

Input to any production center consists of two basic forms – objectified capital/labor and living labor. Objectified capital consists of three basic components – fixed, facilitating and unfinished input capital. Combination of objectified capital and the value of living labor results in finished output of the company, factory, mine, agricultural farm, or a bank, etc. In Figure 7-01, the flow of real capital is shown, i.e., capital not in money form. The output is sold in the relevant economic markets and its proceeds, in money form, are used for payment of agreed wages to labor, after retention of part of the capital as depreciation and to replace the objectified capital consumed by the company during a certain period. Another component is retained to replace the unfinished input capital. Part of the remaining proceeds are consumed by the owner/owners. The net capital gain after these deductions is used in calculation of the gross profit of the company, before taxes.

Components of objectified capital may be obtained by a center at prices that are lower or higher than their values. However, because of the glut of workers in the labor market under intensive capitalism, the workers do not have the bargaining power to demand wages for the value of their labor (V_LABOR), and do not get paid for all the labor they provide. Effectively, they get paid for part of their labor and have to provide additional unpaid labor to the owners of the production center. Their unpaid labor is the main source of the huge rate of intensive profit (RIP) for the owners of the production/service center, as a result. Under intensive capitalism, wages of the workers may not be sufficient even for their subsistence and subsistence of their families.

All the components of capital are used in production of the finished capital (V_*FINISHED*) consisting of, generally, a large number of products. Assuming that only one type of product, or service, is produced by a production center, the value incorporated in each produced item may be denoted as V_*Product*. Components of fixed, facilitating and unfinished input capital are indicated in Figure 7-01. Thus:

$$C\_FIXED = C\_LAND + C\_BUILDING + C\_MACHINERY$$

The capital required for operation of the center may be referred to as running capital (C_RUN). It consists of three components - facilitating capital, unfinished capital and labor.

$$C\_RUN = C\_FACILITATING + C\_UNFINISHED + V\_LABOR$$

If a production center produces "n" products in a certain period of time, say a year, month or a day, then, depending on the expected life of each kind of objectified capital, the proportional part of that form of fixed capital is used up in that period. Thus, each product is created by consumption of nth part of each form of capital used during the period, as input to the production process:

Fixed capital used during the period: C_FIXED_{Period}

Running capital used during the period: C_RUN_{Period}

Thus,

$$C\_FINISHED_{Period} = C\_FIXED_{Period} + C\_RUN_{Period}$$

$$V\_PRODUCT = (V\_FINISHED_{Period})/n$$

We can categorize the capital input to the production center in another way and simplify the notation for a period as follows.

Objectified Labor for Period (O) = $C\_FIXED_{Period} + C\_FACILITATING_{Period} + C\_UNFINISHED_{Period}$

Paid Labor for Period = P_{Period}

Unpaid Labor for Period = U_{Period}

Then, for a period, in symbolic form:

Rate of Intensive Profit, (RIP) = $U/(O+P) * 100\,\%$

Rate of Exploitation of Workers, (ROE) = $U/(U+P) * 100\,\%$

The output of any production center, during a certain period, contains objectified labor, paid labor and unpaid labor for the period. *The unpaid labor is the only source of the profit obtained by the owners of the production center.* However, part of the profit may be transferred to the suppliers of objectified capital, or may be received from them, depending on the terms of the transactions between the production center and its suppliers. *The more the unpaid part of labor, the greater is the rate of intensive profit (RIP). However, the RIP may be affected by fire and other accidents during production and distribution.*

If all the input is paid for at its value, then "U" would be zero for the period, yielding zero percent profit for the members of the ruling class involved, i.e., the owner/owners of the production/service center (or "company") and other production or service centers, acting as suppliers of fixed, unfinished and facilitating capital, including any banks "financing" such capital. The rate of exploitation of workers would also be zero.

Two Basic Sectors of an Economy

An economy normally has numerous sectors of production, like agriculture, machine-tool production and aircraft, military equipment and textile production, etc. Each sector of the economy may be viewed as one production center, or one company with many production

158

centers, so that the model of the sector as a whole, becomes the same as the model for one production center. All the capital inputs and outputs of such a sector model would be the same as for an individual factory. The same applies to the model of the whole economy of a state. However, one detail needs to be looked at. The industries in a state economy can be divided into two basic kinds - those engaged in manufacture of *consumer goods and services* and those engaged in production of *producer goods and services*.

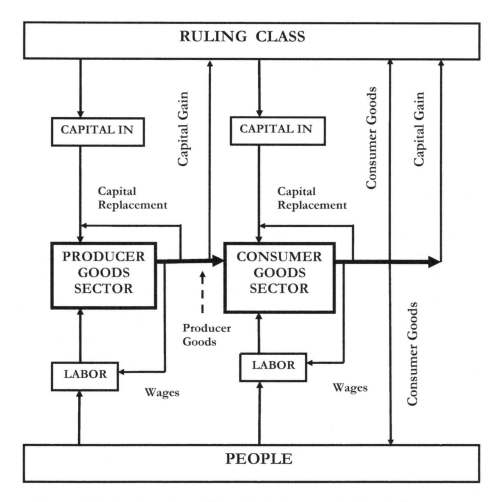

Figure 7-13: An Overview of Two Basic Sectors of a State's Economy

Producer goods are machines and other equipment used in manufacture of other producer or consumer goods. They are not consumed, as such, by the general population. They are used by other producer or consumer goods manufacturing factories and companies. However, every company, or factory, no matter of which kind, pays wages to its workers.

Similarly, companies engaged in providing services can be divided into two categories – those providing services to other companies and those providing services to consumers among the people. Companies may require services like transport, banking, insurance, design and consultancy and other consumers may need transport, education and health care, etc. The insurance policies, investment accounts, stock-market, bond-market and money-market operations of financial institutions basically affect the ruling class only. Also, lending mainly affects the ruling class, since members of this class own all buildings, factories, machines and houses, etc., in an extreme case of intensive capitalism. They and their companies obtain these loans for new acquisitions or improvements to existing businesses, buildings and houses. Some members of the ruling class gain by such transactions through these financial institutions, others lose. But the gains of some are basically balanced by the losses of others within the ruling class. Loans, if any, obtained by workers are sought to protect themselves against starvation and disease. Mostly, such loans are denied because of the absence of *collateral*. The working people, however, do need to use the services of transport companies. Education and health-care facilities, to the extent they exist and are affordable for working families, may also be used by them. Generally, such facilities do not exist, or are too costly, for workers to take advantage of, under the earlier stages of intensive capitalism.

The two sectors of the economy of a state are shown in Figure 7-13. One is a producer goods production and services sector and the other is a consumer goods production and services sector. All the capital inputs to each company, except the capital used for payment of wages to labor, can be combined and indicated as one input. Thus, all the capital consumed by a sector over a period, say, of one year, except capital used for payment of wages, is directly, provided by members of the ruling class. Provisions are, however, made for payment of wages from the sale of the output of the sector. Wages, thus, correspond to the paid labor provided by the workers. Similarly, replacement of depreciation of fixed capital and replacement of all the other forms of objectified capital consumed during a period is retained from the sale of the output. In Figure 7-13, wages, depreciation of fixed capital and replacement of other objectified labor capital are *not indicated as money flows*, but as *flows of equivalent* capital or goods. To understand the processes indicated in the diagram, these may, however, be imagined in terms of money interchangeably.

After accounting for replacement of consumed objectified capital and wages out of the output of each sector, the remaining output corresponds to the unpaid labor provided by the working class. From sale of all the output of each sector, all the capital corresponding to unpaid living labor is retained by the ruling class and is its gross "Capital Gain", as indicated, for each sector of the economy. It is partly used by the ruling class to meet its consumption needs, including the maintenance of its government over the people. The remaining unpaid living labor forms the net capital gain over one year.

In the diagram, "CAPITAL IN" refers to the total capital consumed by a sector, out of the capital invested by the ruling class. This includes all the Unfinished Capital, Facilitating Capital and Fixed Capital consumed by the sector in a year. "Capital Replacement" refers to replacement of all this consumed capital, including depreciation of fixed capital. In a system of *intensive capitalism*, wages of the working class are completely used up in obtaining the basic necessities of life. The gain, or capital accumulation in both sectors of an economy, occurs due to the unjust and abusive operation of the labor market. This capital accumulation is not clearly indicated in this diagram. It would be dealt with in detail, later on. However, we can see that virtually all the labor is provided by the people or the working class. As corporations develop in the later stages of intensive capitalism, the managers and owners of such corporations may also contribute to the production process and may be considered as part-time workers. For the economy as a whole, their labor contributions to the production processes, however, remain relatively insignificant and effectively negligible.

The output of companies, producing consumer goods, is sold to both the members of the ruling class and the people. All the luxury goods are, naturally, consumed by the ruling class. The people, with starvation wages, are not capable of purchasing any luxury goods. In fact, all their wages are used up in purchase of essential items necessary for survival and subsistence - food, medicine, clothing and shelter. Generally, their wages are not sufficient to meet even these needs. Further, *since their demand for these essentials is naturally urgent, it cannot be deferred or postponed, i.e., it is not "elastic". Thus, workers generally have to pay much more than the value of these essentials.*

A State as a Production Center

By combining the producer goods and services sector of an economy with the consumer goods and services sector, we can arrive at the diagram of a state's economy as shown in Figure 7-14. In this diagram, "Objectified Capital Consumed" is part of that invested by individuals, companies and the government of the state, in a certain period of time, say, a year. Since the government is established and operated by the ruling class, just like a non-profit corporation, all the capital invested, under the specific conditions of intensive capitalism, belongs to the ruling class. We are not considering the capital owned and invested during the stage of advanced feudalism, some of which would generally survive till the later stages of this new developing socio-economic system, within the old conditions of the society of a state. The paid and unpaid components of living labor and the replacement of consumed objectified capital should be quite obvious by now, since it is based on the same concepts as a production center or a sector of state-economy.

The total output of the state for a period of one year, including both goods and services, is referred to as the *Gross Domestic Product (GDP)*. However, classical economists calculate it in terms of prices. Thus, if for a period of one year:

O = *Total objectified-labor capital consumed*

P = *Value of paid-labor input*

U = *Value of unpaid-labor input*

Then, we have,

Total capital input during the year = $O+P$

Total value of output during the year (GDP) = $O+P+U$

Rate of Gross Intensive Profit (RIP) % = $U/(O+P) * 100$

Rate of Exploitation of Workers (ROE) % = $U/(U+P) * 100$

Some of the production in a state is consumed by the population of the state. The working class consumes the production of basic necessities based on its wages. If consumption by the ruling class, including the expense on the maintenance of its government, is denoted by "c" Then, the increase in the capital, hoarded by the ruling class or its capital gain in a year, would be:

Capital gain = U - c

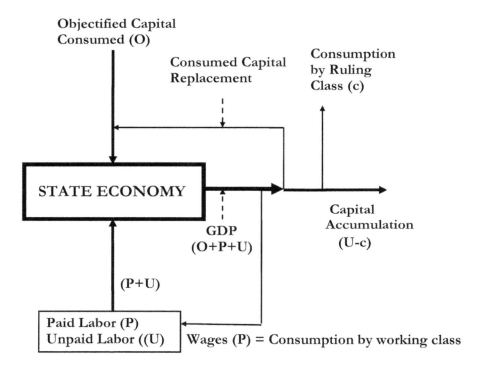

Figure 7-14: The State as a Production Center

Further, the following details need to be clarified here:

- There are markets for fixed, facilitating and unfinished-input capital goods. Some production centers may be able to obtain these forms of capital at prices below their value; others would obtain them at prices that are above their value. But since all production centers are owned by the ruling class, the gain of one individual or company is the loss of another, within that class. Thus, "O", the total effective objectified hoarded capital, utilized in a state's economic activity during a year, is re-extracted by the ruling class from the production for that year. It does not have any

162

net effect on the total capital accumulation in a state over a year, although economic activity cannot take place without it. Only, the unpaid labor is accumulated as capital during the year, after meeting the consumption needs of the ruling classes.

- The proportion of the value of labor, that would be paid to workers, would depend on labor market conditions. If there is a glut of workers, they would end up being paid much less than the value of their labor. If there is a shortage of labor, wages would tend to rise and workers would get paid for a higher proportion of their labor. In general, as described previously, workers get paid much less than the value of their labor in a system of intensive capitalism. Even when machines are introduced into factories and mines, etc., the higher skilled workers are not paid much higher wages because of the high unemployment in the working class. Thus, their rate of exploitation (ROE) would be very high. As we shall see, as the social system evolves into its next stage, the rate of exploitation does tend to go down.

- The system functions on exploitation of the working class. If workers are paid wages corresponding to the value of their labor, there would be no unpaid labor and, thus, no exploitation of labor. We refer to the rate of intensive profit (RIP), since some of the profits obtained by the owners of production centers are "socially necessary" to meet the costs of the social system's operation in an ideal state. These costs are social costs incurred by an *ideal state* in providing physical and social security to the population and ensuring its well-being. *We refer to any capital gain above the level that meets these requirements, as "loot" based on exploitation.* What level of profit by production centers, or a state's economy, is necessary to meet the *"socially necessary"* functions of the ideal state? This depends on a detailed understanding of how an ideal state would function. We would describe the requirements of an ideal state later, in Chapter-20. The profit is termed "loot" if it is not used by a state, or its institutions, for the functions that are required of an ideal state and its society – i.e. what is required for the security and well-being of its population, but becomes part of the hoarded capital of the ruling class, instead.

Capital Accumulation under Intensive Capitalism

A state that has reached the stage of *intensive capitalism* in its evolution may have part of its economy still operating under the conditions of the previous stage of *feudalism,* since it is in transition towards the peak of *intensive capitalism.* Thus, institutions of the two stages of evolution may co-exist. As we have seen, the operation of primitive farms and other production centers in the two social systems are very similar. The small "factories" making shoes, earthen utensils, etc., or the workings of a black-smith's workshop, manned by the black-smith and his "helpers", or a cobbler's shop, etc., are very similar to large factories and production centers during intensive capitalism. The scale of these institutions is, of course, very different. However, the purpose is the same. During the transition, the old primitive farms would become larger as the landlords take over the lands of the small land-holders. The farms also become mechanized and are likely to become much larger by further acquisitions of adjoining farms, so that machinery can be economically used for their operation. The peasants working on these farms would be evicted and would have to move to the cities in search of jobs, to survive. At the beginning of this transition, the landlords would be owners of capital and they would focus on mechanization of their farms and then would set up factories in the cities. The jobs created in the cities would not be sufficient to absorb the surplus labor pushed into the cities and a glut of desperate workers would result there. The purpose of all corporations, and smaller businesses engaged in economic activities in a state, at the stage of intensive capitalism, is the maximization of the rate of capital accumulation, and the rate *of intensive profit (RIP)*. Thus, the whole purpose of the economy of the state is maximum capital accumulation by the ruling class, under the existing market conditions. The model of the domestic economy of the state may, thus, become as shown in Figure 7-15.

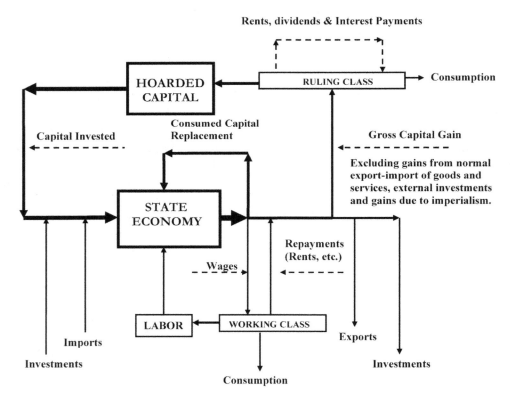

Figure 7-15: Capital Accumulation under Intensive Capitalism

The ruling class invests some of its hoarded capital on mechanization of its farms and some is invested in setting up new industries. The farms and factories use this capital for hiring workers also. The workers provide the labor to increase the value of products and thus increase the capital invested. The ruling class markets the products of its production centers

164

and pays wages to the workers. These wages of the workers are completely used up, as repayments to the ruling class, in acquiring means of subsistence - food, clothing and shelter. The ruling class replaces the objectified capital consumed during a period of, say, a year, from the output that has been produced by the economy of a state. Part of this is the "depreciation allowance" to replace its fixed capital consumed during the production process. The rest of the objectified capital replaced is the facilitating and the unfinished capital invested in the economy that has been consumed in a year. Some of the labor value/capital provided by the working class is returned to it as wages. Out of the total capital consumed during the production processes, the unpaid labor provided by the working class remains as the gross capital gain to the ruling class.

It must be pointed out that the markets of all components of fixed, facilitating and unfinished input capital do not affect the ruling class as a whole. In each market, a factory or company, may purchase goods at prices above or below their value, but this would not affect the total capital of the ruling class. In each case, the gain of one owner of capital is the loss of another and the loss of one owner of capital is the gain of another, within the same ruling class. The net gain or loss to the ruling class, due to these transfers or transactions, is zero.

Similarly, individuals, or companies, may obtain loans to finance the construction of a factory, or factories, and to set up new businesses, or to purchase the consumable capital goods or the input unfinished goods, or even to pay wages to employees. Also, rents may be incurred by individuals of the ruling class and companies. The terms of the loans, normally, guarantee a rate of interest to the lending bank or company. Thus, the lender, effectively, becomes a partner of the owner/owners of the factory or company. The lender does not share all the risks of such a business, however. *As the principle of the loans is returned to the lenders, there is no effect on the hoarded capital of the ruling class. The interest paid to the lenders, does not increase the capital gain of the ruling class, since it comes out of the debtor's total capital gain during a period. Both the lender and the loan recipient are members of the ruling class. Also, members of the ruling class may gift property to individuals of their class or offer it to companies engaged in specific sectors, to encourage their work. All such transactions are within the ruling class and do not affect its total hoarded capital.*

Out of the gross capital gain, some of the production of the state is directly consumed by the ruling class itself and some is consumed by its government. The rest of the generated *capital gain* is added to its hoarded capital. *This circulation of capital continues, while the conditions of the working class do not change and its members keep on dying because of lack of food, shelter and medical care.* At the same time, the capital hoarded by the ruling class keeps on growing at an increasing pace. There are arrangements for loans and investments services, offered by the financial institutions to the ruling class, which reduce it to a set of individuals who remain busy in counting "their" financial assets "earned" by "their hard work"!

It needs to be pointed out that the government of the state is controlled and, effectively, owned by the ruling class. Government may levy taxes and give benefits to the corporations or individuals of the ruling class. These transactions are within the ruling class. Thus, the government may be considered as belonging to the ruling class and does not need to be shown as separate from it. The government may also impose taxes on the people, or may give benefits to the people, though rarely. These taxes, if any, are considered part of repayments by the people out of their wages. Similarly, benefits given to the people are considered part of wages, as shown in the diagram.

A state and its corporations have trade relations with other states. Export and import of products results. Also, investments may be made by state-owned institutions, the members of the ruling class, or its corporations, into such institutions of another state and vice versa. These trade and investment activities result in transfer of capital from one state to another and, thus, affect the capital accumulation in both those states. We would analyze these trade and investment relations in Chapter-12.

Characteristics of Intensive Capitalism

Growth of Corporations

During the nineteenth and twentieth centuries, corporations grew in size and economic influence. Some of these corporations became trans-state corporations, with their operations spread over many states. General Electric was one such corporation. This is an inherent characteristic of intensive capitalism.

Introduction of Machinery and Automation

Under intensive capitalism, in each sector of the economy, new machinery and more and more automation is introduced to increase the productivity of labor, i.e., *the output per worker employed*. As automation and mechanization grow in one sector, the RIP of that sector tends to increase more and more slowly. This, results in transfer of new investment to new sectors of the economy, especially the newly established sectors, which have come into being due to development of new technologies, and where higher RIP can be expected. Marx has referred to this phenomenon as the *"law of the falling tendency of the rate of profit"*. To understand, more clearly, why this happens, let us examine the rate of profit of a production center, or an economy of a state, as a whole, as we have derived in previous sections of this chapter. The formula, we thus derived, also applies to any sector of an economy.

Rate of Intensive Profit (RIP), $\% = U/(O+P)*100$

Thus, the rate of profit depends on unpaid living labor. As technology develops and more and more machinery and automation are introduced into a company, or factory, the proportion of objectified labor (O) increases in the total labor used in it, while the proportion of living labor (U+P) decreases. The introduction of additional machinery, and the resulting increase in automation, is dictated by the desire for higher profits by increasing the total throughput of the factory or company. It is also dictated by the competition between different factories, or companies, producing the same or similar products, since each enterprise owner desires to increase his market share and total profits. The owner, also, wants not to offer any wage increases to workers he retains for operation of machinery. Under labor market conditions of intensive capitalism, i.e., large numbers of unemployed workers, he is likely to succeed in maintaining low labor costs. The productivity of the company increases, however, and the company is able to sell more *and increase its total profits, while its profit rate for each product has gone down, in general.* As more and more companies enter the same sector of production, the total amount they produce increases tremendously, till the demand for those products begins to be saturated. It happens rather quickly under conditions of intensive capitalism, since the workers laid off by companies make the unemployment and misery of the working class even worse – thus decreasing demand for many types of products. This causes the prices of products in the relevant sector of an economy to go down further. ***Thus, in general, the introduction of more machinery, more-sophisticated machinery and automation of production processes, tends to reduce profit rates of companies involved, although the total profits may increase dramatically due to acquisition of bigger market share during the earlier stage of introduction of new technology in the relevant sector.***

Imperialism

As the economy of a state, as a whole, develops into a highly mechanized and automated economy, with its agriculture, manufacturing and services sectors and its rate of intensive profit (RIP) begins to decrease in most sectors, the insatiable greed of the ruling class forces it to seek other states to increase its hoarded capital by using lower wage workers. It, then, pushes its government to seek colonies and set up empires to exploit and abuse other peoples of those states, whose standard of living is lower, who are willing to work for lower wages and who are vulnerable to conquest, and whom its state then seeks to subjugate. *Thus, colonization and imperialism are the direct and natural tendencies of a society, and state, which has reached the level of intensive capitalism.* This tendency exists even in states at the stage of primitive, or advanced, feudalism, but its intense growth under intensive capitalism has been dramatic. This is what happened in European states and the US and these states became imperial states.

At that time, the whole world was divided into empires of those imperial powers. Thus, inter-state trade at that stage was really inter-empire trade. The trade within each empire was tightly controlled by the relevant imperial power. Trade was just another means of abusing the subjugated peoples by means of controlled market subjugation, including the use of currency of the imperial power as the reserve currency with arbitrarily set exchange rates. This stage of *intensive capitalism* lasted in US, Western Europe, Australia, and Japan from the start of the French Revolution in 1789, till the end of the Second World War in 1945. Parts of East Asia, Central Eurasia and South America are going through that stage *of intensive capitalism* at this time.

Inter-State Trade

Inter-state trade is one way that the ruling class finds to satisfy its ever-increasing craving for a higher rate of capital accumulation. Since the advent of intensive capitalism after the French Revolution, inter-state trade has, thus, gone through an explosive growth. This growth in trade has been accompanied by terribly destructive wars between imperial powers for control of subjugated peoples and the "right" to trade with them under extremely favorable terms for themselves, and extremely unfavorable and abusive terms for the subjugated peoples, nations and states. Since at the time of intensive capitalism in Europe, US and Japan, the other states were not even nominally independent, the trade between the imperial powers and their possessions was, obviously, highly abusive. We would analyze inter-state trade in Chapter-12, which corresponds to the period of globalization after decolonization and emergence of somewhat independent states in the non-imperial world. In that chapter, the abusive nature of inter-state trade, under "normal conditions" of intensive capitalism, benevolent capitalism and globalization, would also become clear.

Social Inequality

As more and more capital is accumulated in the hands of the ruling class of owners of business and industry, while the working class is driven more and more towards subsistence living, *the gap between the standards of living of the two classes continues to grow*. The exploitation of the working people leads to massive unemployment, social tensions, strikes and demonstrations against the excesses of the rulers. It leads to social explosions, big and small, like those which occurred in Europe in the nineteenth century. Such social explosions can lead to violence, blood-shed and, even, a revolution, i.e., replacement of the ruling class by another.

Drop in Population Growth

The massive migration of agricultural workers, from countryside to cities, and the subsistence wages paid to workers, lead to an intensively high death rate and falling birth rates, due to massive starvation and lack of medical care for the working class. This is the phenomenon referred to as "social murder" by Friedrich Engels. We refer to it as *"Intensive Death Rate"* to indicate its correspondence to intensive capitalism.

The "Boom" & "Bust" Phenomenon

Periods of "boom" and "bust", or "business cycles", which the modern economic system goes through, had been known to occur for a long time [2]. Depressions, or recessions, occurred in the US and Europe in 1873-1896 and 1920-21. The "Great Depression" was an especially severe occurrence of this phenomenon.

During the early stages of development of industry, there was some confusion as to why recessions occurred. But, as intensive capitalism, developed, the processes it goes through became quite clear. Now, it is quite well-understood why and how recessions occur leading to a contraction of the economy, how recovery comes about, or can be brought about, so that the system begins another period of growth - till it, inevitably, collapses again. We would describe the processes underway in an economy characterized by intensive capitalism. The processes remain, basically, the same even in the next stage of the evolution of capitalism [3]. In fact, the system was going through the beginnings of these evolutionary changes in the US economy during the Roosevelt administrations, when the Great Depression occurred.

When the economy of a state is in recession, under the system of intensive capitalism, and has reached the bottom, i.e., it is no longer contracting substantially, there is a high rate of unemployment and wages may have fallen much below subsistence level. Large numbers of people may be starving under such conditions, because of their low incomes, or no income at all. Thus, the buying power of the people would have fallen to a very low level. The ruling class would have money to spend at this time, out of its hoarded capital, but the total buying power of the class would not be high as regards consumer goods, because the number of individuals in this class is very small compared to the working class. Thus, the demand for goods and services would be very low at this stage. Also, capitalists would not be investing in new or existing factories since they would not expect to sell their products and hope for a high Rate of Intensive Profit (RIP).

INCREASING RIP

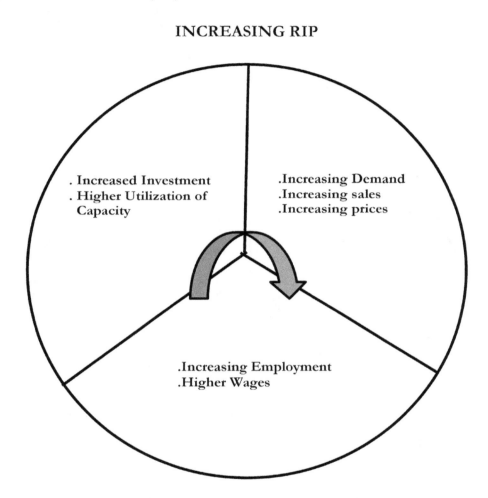

Figure 7-16: The Wheel of Blind Greed

Only producers in the basic consumer industries would still be able to sell their products, since essential consumer items like food, clothing, medicines, etc., would still be needed. However, the demand for these goods, and services like transport, would also have decreased –because the people would be focused on survival only. Thus, at this stage, demand would have reached a minimum and unemployment would be at a maximum and investment would have come to a virtual halt.

In the era of intensive capitalism, before the 1930s, governments used to make "easy money" available by lowering interest rates and increasing money supply, when the economy was in a deep recession or depression. The easy money policies make it possible to get new loans, or refinance old loans at lower interest rates and have been continued till now, supplemented by direct government intervention. The increased money supply helps corporations and businesses in their operations. Banks, for example, can issue more loans to businesses and owners of factories. The owners may hire more workers in their businesses, as a result.

DECREASING RIP

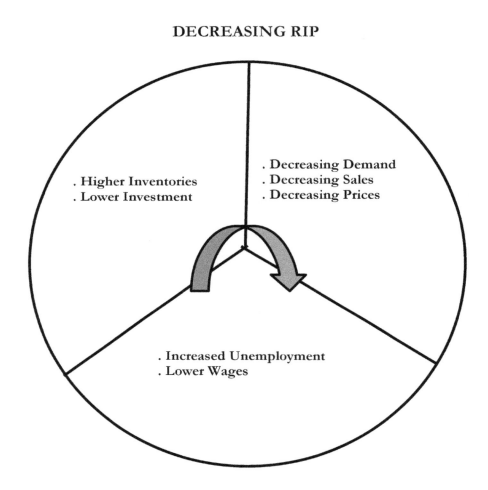

Figure 7-17: The Wheel of Blind Terror

The increase in employment causes more money to be spent by workers. This increases sales and prices begin to increase. The increase in prices and sales, encourage businesses to hire more workers to increase production. Thus, inventories begin to decrease and the investment increase may create demand for producer goods. Thus, unemployment begins to drop and workers who have jobs, have money to spend. In time, the wage levels begin to rise. The increased demand for producer goods and higher income of the working class increases the demand for products in general. Thus, sales begin to grow and, in time, prices of products also begin to increase. Thus, by making easy money available, governments put the *"Wheel of Blind Greed"*, as indicated in Figure 7-16, in motion, although, in the beginning, it turns very slowly. Easy money alone cannot quickly solve the problem, especially under intensive capitalism when workers have no savings or capital. As employment increases, total wages also increase. Thus, the working class has more money to spend and demand for basic consumer goods starts to grow. The resulting sales cause the ruling class to invest more. The excitement of higher and higher rate of intensive profit (RIP) causes the investment rate to keep on rising. Thus, the rate of growth of the economy keeps on increasing.

A stage ultimately comes when, the unemployment rate has fallen to a low level, the capacity utilization of factories has risen to a high level and the economy reaches a state of high production. This is the point, which we refer to as the *"Vista Point"*, when the economic conditions are ideal if they could be stabilized and the rate of growth of the economy could be maintained. But, **since blind greed is in control of the economy and all economic activity, this cannot be done. At the "Vista Point", the owners of factories and mines, etc., are very excited at the prospect of still greater profits in the future. Thus, the economy always overshoots the Vista Point.** Investment keeps on increasing, causing overproduction of goods and services. Also due to greatly increased demand for more and more workers, wages of workers employed by industry begin to rise. To deal with this "problem", businesses raise the prices of their products and try to resist the wage demands of their workers. Ultimately, this leads to a slowdown of demand growth. Because of over-production and the slowing rise in incomes of workers, leading to slowing demand growth, the higher prices cannot be maintained. When some businesses notice that their products cannot be sold, or cannot be sold as fast as before, or their prices are no longer acceptable to consumers and their inventories have begun to grow, they reduce or stop further investment and start laying workers off. Thus, the *Wheel of Blind Greed* suddenly becomes the *Wheel of Blind Terror* shown in Figure 7-17, as businesses, and their owners, are terrorized by the prospect of losing the capital they have accumulated by "their" hard work!

As more and more workers are laid off, the unemployment rate starts increasing. Wages fall and total income of the working class starts to decrease. The lack of investment and the fall in wages cause the total demand for goods and services to decrease, causing sales and prices to decrease. This causes the investment to decrease or stop and more and more workers get laid off. As the wheel turns, the economy sinks deeper and deeper into recession, even if financial institutions do not fail. Soon the bottom is reached again, completing a full cycle of growth and recession.

As can be seen from the behavior of an economy under the conditions of intensive capital accumulation, the economy goes through the business cycle because the *unlimited and blind greed* of the ruling class is in control of the economy. The economy starts growing when money and demand is artificially created by the government and greed motivates the ruling class to start employing workers and start investing again. Even if the government does not create money at a high rate to "stimulate" the greed in the economy, the economy does start to grow after it has been depressed for a long time. This is what used to happen before the Great Depression, when generally no action was taken, except increase in "easy money" supply, as an economy reached its bottom. However, in such cases, the people had to suffer a lot more, because of very high unemployment rates over a long time.

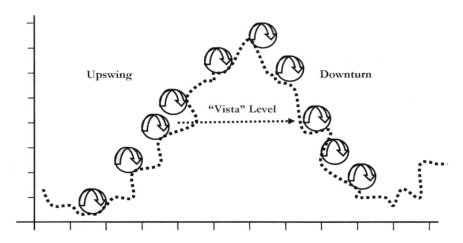

Figure 7-18: The Business Cycle

During the growth phase, when the economy reaches the *"Vista Point"*, it cannot be stabilized at a moderate but constant rate of growth because of the desire for higher rates of capital accumulation. Only, lack of demand for goods and services stops the economy from growing. The crisis of overproduction does not cause the economy to slow down, but instead, due to the fear of the loss of the accumulated capital by the ruling class, it causes it to *collapse quickly*. Thus, the business cycle has remained an essential part of the system of *intensive capitalism*. Marx had forecast that this phenomenon of "Economic Crises" would get worse with time, and it did.

What we have described is the core phenomenon that triggers a recession as part of a business cycle. A recession, however, can also be triggered by bank failures and a resulting financial crisis, which brings a sudden stop to all investment resulting in massive layoffs – thus kick-starting the core processes of a recession.

The Keynesian Solution

A British economist observed the ravages of the Great Depression and suggested a scheme of management of the business cycle. He noticed that there is a *"multiplier effect"* at work in increases in investment and demand. Continuation of both investment and demand at some moderate pace is critical to continuation of economic activity, without violent fluctuations.

"The law of capitalist production which really lies at the basis of the supposed 'natural law of population' can be reduced simply to this: the relation between capital accumulation and the rate of wages is nothing other than the relation between the unpaid labor which has been transformed into capital and the additional paid labor necessary to set in motion this additional capital. It is therefore in no way a relation between two magnitudes which are mutually independent, i.e. between the magnitude of the capital and the numbers of the working population; it is rather, at bottom, only the relation between the unpaid and the paid labor of the same working population. If the quantity of unpaid labor supplied by the working class and accumulated by the capitalist class increases so rapidly that its transformation into capital requires an extraordinary addition of paid labor, then wages rise and, all other circumstances remaining equal, the unpaid labor diminishes in proportion. But as soon as this diminution touches the point at which the [unpaid] labor that nourishes capital is no longer supplied in normal quantity, a reaction sets in: a smaller part of revenue is capitalized, accumulation slows down, and the rising movement of wages comes up against an obstacle. The rise of wages is therefore confined within limits that not only leave intact the foundations of the capitalist system, but also secure its reproduction on an increasing scale. The law of capitalist accumulation, mystified by the economists into a supposed law of nature, in fact expresses the situation that the very nature of accumulation excludes every diminution in the degree of exploitation of labor, and every rise in the price of labor, which could seriously imperil the continual reproduction, on an ever larger scale, of the capital relation."

Karl, Marx (1867), <http://www.marxists.org/archive/marx/works/1867/*Capital, volume-I, chapter 25, section-I* /index.htm>, accessed October 16, 2019.

When investment is increased in some business by hiring new employees or acquiring new plant and equipment, this has two effects. It creates demand in the business, for producer goods of other businesses, which may increase their investment in turn and cause increases in demand for producer goods of other industries, and so on. Also, the investment creates jobs and leads to hiring of workers, or an increase in the wages of existing workers, who may have to work over-time. The workers, thus, have an increase in their income, which they quickly spend to meet their urgent needs. The spending by the workers, then, creates demand for consumer goods of other industries, which in turn hire more workers, or make existing workers work over-time. This, then, increases the income of the working class further, leading to a further increase in demand. *Thus, the initial investment gets multiplied manifold. Similarly, the increase in demand also gets multiplied manifold.* If money is placed in the hands of workers directly, by government giving grants or reduction in taxes, this has the same *multiplier effect on demand.* The decreases in investment and demand have the opposite multiplier effect. Keynes, thus, proposed changes in both monetary and fiscal policy, to manage a recession He observed that an economy cannot come out of a recession or depression unless the *"aggregate demand"* for goods and services is increased. By aggregate demand he meant the total demand for goods and services by the people, the ruling class, its corporations and the government. This could be done by lowering interest rates and increasing money supply, but

173

this process could be strengthened by government systematically increasing demand by its budgetary, or fiscal, policies. The government could, for example, lower taxes on the people, leave more money in their pockets, give them small sums of money as *"stimulus payments"*, or increase the un-employment and social security payments, if any. These measures would tend to drive the government budget in the direction of a deficit and put money in the hands of the people who would readily spend it on readily consumable items. This would generate demand for the suppliers of those products, reducing their inventories. The demand would grow manifold due to the multiplier effect, causing the employers to start or increase hiring of workers, who would now have wage-income to spend, thus generating more demand. This would, ultimately, lift the whole economy out of recession or depression.

… By a continuing process of inflation, governments can confiscate, secretly and unobserved, an important part of the wealth of their citizens. By this method they not only confiscate, but they confiscate arbitrarily; and, while the process impoverishes many, it actually enriches some.[4]

… But this long run is a misleading guide to current affairs. In the long run we are all dead.[5]

… The decadent international but individualistic capitalism in the hands of which we found ourselves after the war is not a success. It is not intelligent. It is not beautiful. It is not just. It is not virtuous. And it doesn't deliver the goods. In short we dislike it, and we are beginning to despise it. But when we wonder what to put in its place, we are extremely perplexed.[6]

… I do not know which makes a man more conservative — to know nothing but the present, or nothing but the past.[7]

… Capitalism is the astounding belief that the most wickedest of men will do the most wickedest of things for the greatest good of everyone.[8]

John Maynard Keynes: Quotes retrieved on January 14, 2020, from en.wikiquote.org, https://creativecommons.org/licenses/by-sa/3.0/legalcode

---

Keynes also proposed the opposite policies in the upswing phase of the business cycle, or the growth period of an economy. Thus, if the economy has had a reasonable rate of growth and the unemployment is low, the money supply can be reduced and the interest rates on loans can be increased. These measures would discourage investments by the ruling class and would increase mortgage payments for new home buyers. Similarly, the government's budget can be *moved towards a surplus*, even if a surplus is not achieved. This can be done by increasing taxes on the people and businesses and reducing benefits for social security recipients, if any. Increases in taxes on businesses would tend to slow down investment. Increases in taxes on the people, would reduce their disposable income, thus reducing money in their pockets, leading to reduction in demand and spending. The reduction of any social security payments, like unemployment benefits, would also lead to reduction in spending and demand.

The measures outlined above, have to be taken in measured and graduated steps so that extremes of unemployment and depression, and overproduction due to the "boom" in production growth, are avoided. The efforts to slow down the booming economy are meant to keep it growing as near the "Vista Point" as possible. But, since the economy is bound to overshoot the "Vista Point", these measures help in achieving a *"soft landing"* when the inevitable "bust" or "crash" arrives, i.e., these measure ensure that the coming recession is not severe and the resulting unemployment, repossessions of houses due to payment defaults, defaults on personal loans and decreases in demand in general would not lead to large-scale individual and business bankruptcies and failures of industrial corporations and banks, etc.[9]

In short, Keynes had suggested that only governments could manage the business cycle by taking action both in their *fiscal policy*, i.e., their budgetary measures, and *monetary policy* - whether in the form of taxation, industrial regulation, public works, social insurance, social-welfare services, deficit spending, or changes in money supply and interest rates. When asked to specify what kind of spending by government would take an economy out of recession, his reply was - *"build pyramids"*!

In the beginning, there was strong opposition to the proposals that Keynes put forward for management of a business cycle. It was not surprising. What Keynes was suggesting can be summarized as follows:

- When the growth, or 'boom" is underway, the government should monitor the investment, employment and inflation. It should raise taxes and/or reduce expenses as rate of expansion of the economy increases - to reduce aggregate demand. It may also raise interest rates and reduce money supply with the same objective, if inflation develops.

- When the recession, or the "bust" period, arrives, the government should increase expenditures, including making "stimulus payments to citizens, and should reduce taxes to make more money available in the hands of the population - to increase aggregate demand.

Objections to the Keynesian scheme continued for quite some time. When this scheme of management was actually tried by governments and it yielded the desired results, the criticism decreased.

Chapter - 8

Benevolent Capitalism

Even before the Second World War began, the United States had developed the biggest economy in the world. Mainland US had been untouched by the First World War and, as the Second World War started, it saw an opportunity to achieve global economic, military and political dominance. The goals of the war were stated clearly in 1941, before the United States entered the war. It planned to gain unhindered access to the markets and raw materials of all states of the world and, for this, intended to have freedom of navigation in all oceans. It intended to create a stable global market dominated and controlled by it through global financial institutions, to be created to ensure unrestricted trade relations. To create these conditions, it intended to create the United Nations - a global organization to ensure prevention and control of conflicts between states and their "collective security" - *under its hegemony*. To co-opt the Soviet Union and its demands for freedom for all nations, de-colonization of all imperial possessions was implied, though not clearly stated. However, the hidden agenda of Panamization of imperial powers and of captive market subjugation of the whole world was not spelled out.

During 1943, discussions had started about postwar relations with the Soviet Union, as defeat of Hitler's Germany became imminent. Three post-war objectives seemed reasonable from the point of view of the US and its allies. The first objective was to get the United Nations and its associated agencies into operation. The second was to work out methods of collaboration with the Soviet Union; this was seen as a prerequisite for the smooth and successful operation of the United Nations. It was also clear that only the US was in a position to deal with the political problems arising out of the chaos of a world destroyed by the two world wars. Discussions were already underway for creation of the United Nations and global financial institutions.

When the leaders of the three allied powers met at Potsdam, near Berlin, in the middle of 1945, it became clear that the world was splitting up into two spheres of influence - those of the United States and the Soviet Union. The sphere of influence of the Soviet Union was also likely to expand due to the increasingly likely takeover of China by its Communist Party. Under Stalin's paranoid dictatorship Soviet forces had established, or were in the process of establishing, new governments and administrations of communist parties in all states under soviet control. Stalin was in no mood to allow any freedom of self-government to those administrations either. Stalin's style of military enforcement of his rule was quite obvious even at that time. It was then realized that Roosevelt's "four policemen" could not jointly enforce world peace - and the concept was, quietly, buried in this meeting! Instead, a policy of "containment" of the Soviet Union was formulated after the Potsdam meeting, as proposed by George Kennan, the US ambassador in Moscow.

As was discussed in Chapter-7, below-subsistence living of the working class and the accompanying *Intensive Death Rate* are the main characteristics of a social system at the stage of *intensive capitalism*. In Europe, intensive capitalism came to an end after the Second World War. This system had taken about three hundred years to become dominant in Europe, since its beginnings in the early eighteenth century. France was the first state, where this social system became dominant after the French Revolution. After the Second World War, the relationship between the ruling classes and the people of Europe began to change. A shortage of labor was developing and also the power and influence of the Soviet Union was increasing and the ruling classes of West European states increasingly felt threatened as state after state in Eastern Europe fell under Soviet control. Under these conditions, the people were able to demand better working conditions and wages and the ruling class felt compelled to make concessions to the people. The same happened in North America, especially in the United States. These changes were implemented in the US, Western Europe, Japan and Australia during 1945-1955. Thus, the economic activity in these regions became less abusive, although its organization remained, basically, the same. We refer to the new stage of political and social development of Europe and North America, as *Benevolent Capitalism*.

A revolution occurred in China soon after the war ended. When the Soviet Union also became a nuclear power in 1949 and then the Korean peninsula exploded into war, the United States radically increased its military budget and expanded its military reach. The policy of "containment" of the Soviet Union and its allies, including China, was also radically upgraded. In these circumstances, the Panamization of Western Europe, Japan, South Korea and Taiwan was implemented. We would look at the new global institutions of the United Nations (UNO), International Monetary Fund (IMF) and the World Bank in the following sections. The Chinese Revolution, the Korean War, the policy of "containment" and the process of "Panamization" in Europe and Asia, would also be discussed.

Chinese Revolution

During the fighting against the Japanese, as described in Chapter-6, the alliance of CPC and KMT was an alliance forced by the Soviet Union. The two parties remained hostile towards each other, while fighting against the Japanese also. The PLA developed the concept of guerrilla war and engaged in this new kind of warfare against the Japanese. In 1941, clashes between the Communist and KMT forces intensified and Chiang Kai-shek again demanded that the PLA evacuate certain territories. The PLA complied but was ambushed by KMT forces during its evacuation, causing hundreds of thousands of deaths. This ended the Second United Front between the two parties. However, the guerilla war had won strong popular support for the CPC within the Japanese-occupied areas

By the end of the War with Japan, the balance of power in China's civil war had shifted in favor of the CPC and their main force, the PLA, grew into a huge organization. CPC had created many "Liberated Areas" containing one-fourth of the country's territory and one-third of its population. Many important towns and cities were included in these areas. Moreover, the Soviet Union turned over all of its captured Japanese weapons and a substantial amount of its own supplies to the CPC, who received control of Northeastern China, also, from it. The CPC's ultimate source of support, among the people, was its land reform policy. The

CPC continued to make the irresistible promise in the countryside to the massive number of landless and starving Chinese peasants that by fighting for the CPC they would be able to take farmland from their landlords. This strategy enabled the CPC to obtain a huge supply of manpower for combat and logistic support, despite suffering heavy casualties throughout the civil war.

In the last months of the Second World War in East Asia, Soviet forces launched a huge offensive against the Japanese in Manchuria, along the Chinese-Mongolian border. The Soviet Union succeeded in occupying Manchuria completely by the end of the war. Under the terms of the Japanese unconditional surrender dictated by the United States, Japanese troops were ordered to surrender to KMT troops and not to the CPC. KMT troops were then airlifted by the United States to occupy key cities in North China, while the countryside was already dominated by the CPC. President Truman was worried about the CPC military taking over even the major cities in the North and decided to delay the requirement of the immediate surrender of Japanese forces, so that the KMT would achieve full control over those cities and the US forces could do the same about control of Chinese seaports.[1]

The United States strongly supported the KMT forces. Strategic sites were guarded by the US marines. The US had equipped and trained hundreds of thousands of Chiang's troops. At this point, those troops were transported by US forces, to occupy areas in Manchuria. Chiang's men also seized most of the banks, factories and commercial properties vacated by the Japanese Army. Thus, the KMT had come completely under US control by this time. Peace negotiations between the two sides did not yield any agreement and the truce fell apart in the middle of 1946. Then the full-fledged *"War of Liberation"* began between the two Chinese armies, starting with Chiang Kai-shek's large-scale attacks on CPC-controlled territories.

The PLA, then, adopted a gorilla war strategy. They would suddenly attack KMT forces from all sides, relying on their intelligence sources among the people. If and when KMT reinforcements would begin to arrive, they would disperse in all directions without offering any conventional resistance. Thus, they avoided open confrontations with the forces of the KMT, and were prepared to abandon territory while causing maximum damage to Chiang's forces. In most cases, the surrounding countryside and small towns had come under CPC influence. These tactics were very successful and, within a year, the balance of power became favorable to the CPC. The PLA crossed the Yangtze River and captured Chiang's capital, Nanjing. Ultimately, the KMT retreated to Taiwan under American protection. By late 1949, the People's Liberation Army was pursuing the remaining KMT forces in southern China, some of whom retreated to Burma under guidance of the CIA. They continued to offer some resistance for some time, by carrying out military operations on Chinese territory including Tibet, till the PLA became fully in control of the mainland.

On October 1, 1949, Mao Zedong proclaimed the *People's Republic of China* with its capital at Beijing. At the time of Chiang Kai-shek's defeat and withdrawal, Taiwan was, nominally, a Japanese controlled territory. In December 1949, Chiang proclaimed Taipei, Taiwan, the temporary capital of the *Republic of China* and continued to claim his regime as the sole legitimate government of China. Things changed further with the onset of the Korean War in

June 1950. At this point, China had freed itself from the domination of both the US and Japan. Only, Taiwan remained in the hands of the US-backed Chiang Kai-Shek regime. United States upgraded its policy of "containment" by tripling its military budget and increasing its economic support for all states bordering the Soviet Union and China. The US Seventh Fleet was ordered into the Taiwan Straits to prevent the PLA from attacking the Chiang regime. Also, the United States increased supply of all kinds of military equipment to the ROC for its land and air forces. Economic and financial relations between the US and the Taiwanese ruling classes were upgraded. The Panamization of Taiwan was, thus, taken up as another extension of the desired global empire dominated by the US ruling class.

The Korean War

The Soviet Union also became a nuclear power in 1949, having exploded an atomic bomb. It was, however, much weaker than the United States in nuclear capability, i.e., it did not have atomic bombs comparable to the United States, in numbers and power, or yield. It, thus, decided not to immobilize its active military forces till it felt secure from the nuclear threat posed by the United States. Soviet forces in Eastern Europe were also not withdrawn to Soviet territory. The war had come to an end, but the military tensions were very high at that time, as the two super powers, formerly nominal allies, had assumed threatening postures towards each other.

The Korean Peninsula had been part of the Japanese Empire since 1910. A guerilla war had been waged by the Communist Party of Korea. The leadership of this war was in the hands of a Korean named Kim IL Sung. The guerillas were initially based in Manchuria, but were driven out to Siberia by the Japanese. Some of them had joined the Soviet Red Army and had fought along with it. Because of this fighting against the Japanese, the Communist Party of Korea, and its leader, were very popular in Korea. At the Yalta and Potsdam Conferences, it had been decided that Korea would be divided into two occupation zones, North and South of the 38th parallel and the Japanese would have to surrender to the two allied powers in the respective zones. The Soviet forces entering Korea were able to establish control in the peninsula's northern area. In accordance with the agreement with the United States, Soviet forces stopped at the 38th parallel, leaving the Japanese still in control of the southern part of the peninsula. Sometime later, American forces, also, landed at Inchon. Following the Japanese surrender in September 1945, Korea was divided into North and South Korea along the 38th parallel, as planned. The population of North Korea was only nine million out of the total population of about thirty million, but most mines and industries were in the North, while the South had mainly agricultural territory. A "socialist" regime, on the Soviet model, was quickly set up in the North. The Soviet Union trained the regime's armed forces, supplied them with military equipment and, then, withdrew from Korea in 1948, leaving Kim IL Sung and his regime in control. Kim's government carried out land reforms, taking over large estates and giving ownership of pieces of agricultural land to peasants. Industries and mines were, similarly, taken over by the state. These measures increased Kim's popularity throughout North and South Korea. The North Korean state was renamed the Democratic People's Republic of Korea (DPRK).

A provisional government had been established in the South by the Communist Party of Korea. Peoples Committees had been set up on the lines of the Soviets in the Soviet Union. Thus, the US forces faced popular resistance right from the time of their arrival. The US command established control by restoring to power the key Japanese colonial administrators and their Korean collaborators; and refused to recognize the provisional government of the short-lived People's Republic of Korea (PRK) backed by the Communist Party of Korea. These policies provoked civil uprisings. A commission was set up by the Soviet and US Commands to create a unified state of Korea, but it could not function because of disagreements between the two powers. When the commission decided that the country would become independent after a five-year trusteeship of the two occupying powers the Korean population revolted in the South. Some people protested and some rose up in arms. To suppress this resistance, the US command banned strikes and outlawed the PRK Revolutionary Government and the PRK People's Committees. This led to more protests and violence. Civil disorder spread throughout the country. Strikes by railroad and other workers started and the students attacked the police and the police attacked them. Many students, policemen, government officials and landlords were killed and the US command declared martial law. This increased the unpopularity of the US occupation authorities. Even the ruling class of South Korea opposed the trusteeship plan. The United States, then, decided to hold elections in its occupation zone. A demonstration commemorating Korean resistance to Japanese rule ended up becoming an uprising resulting in deaths of tens of thousands of citizens at the hands of the South Korean Army. The elections were also marred by terrorism and sabotage resulting in hundreds of deaths. US-educated Syngman Rhee was made the President of South Korea after he "won" US-controlled "elections", while the US continued to suppress all opposition to "its" South Korea. This state was re-named the Republic of Korea (ROK). Rhee's régime continued the expulsion of communists and leftists from politics. In frustration they withdrew to hilly areas and began to prepare for a guerrilla war against the ROK. In late 1948, South Korean leftist soldiers rebelled against the ROK government and their rebellion was violently suppressed. In 1949, most of the US troops were withdrawn, mostly to Japan, from where they could be quickly redeployed to Korea if necessary. This superficial withdrawal was aimed at cooling the nationalistic resentments of the Koreans. Killings of South Koreans continued at the hands of troops of the ROK. However, communist "marauding bands" were blamed for this violence.

Thus, one nation was divided into two states and this unnatural division still persists. During this period, the United States was preoccupied with the Panamization of Europe and Japan. Korea and Taiwan had been given lower priority, due to its military budget constraints. With the Communist victory in China, Taiwan was also given more attention. In South Korea, the United States trained the local Army and Police. The Army of the ROK was about the same size as that of the DPRK. However, in accordance with a long-term plan of Panamization, it was not given heavy military equipment. It was also, naturally, led by a US general of its "Advisory Group" in South Korea. As guerilla fighting erupted in the ROK, its Army got busy dealing with it. Protests by communists, their allies - students, labor organizations, etc., continued, and were suppressed by force. Tensions continued to grow throughout the Korean peninsula, despite the "withdrawal" of US forces.

After the victory of the Communist Party of China, the Soviet Union wanted the People's Republic of China (PRC) to take over the seat reserved for China in the Security Council of the UN. United States opposed it and wanted its puppet Chiang Kai-Shek regime to retain the seat. In 1949, the Soviet Union conducted its first atomic bomb test and became a nuclear power, like the United States. In the middle of 1950, the Soviet Union boycotted the Security Council in protest against the US insistence of denying China's rights in the UN. While the Soviet Union was boycotting the Security Council, DPRK decided to liberate all the Korean territory and launched an attack across the 38th parallel.

Thus, a civil war started in Korea, pitting the two regimes against each other. The Soviet Union had approved Kim's plan of "unification of Korea within three days"! However, it did not commit to provide combat troops. In negotiation between China and the DPRK, China committed to provide large numbers of Chinese troop, if it became necessary. DPRK and its allies did not expect the US to intervene, since most of the US troops had been previously withdrawn to Japan. As about 90,000 troops with about 150 tank moved South at high speed, the forces of the ROK disintegrated and fled. Both the ROK and the US were taken by surprise. The United States took advantage of the Soviet boycott of the Security Council, to get a resolution passed, recommending that "the members of the United Nations furnish such assistance to the Republic of Korea as may be necessary to repel the armed attack and to restore international peace and security to the area."

Soon, the United States responded with its air and naval forces and, later, with its land forces, under the UN flag. Other countries were asked to contribute military forces, at least of a token size, for this "military action by the United Nations". Several states did send their forces. However, the United States provided about 90% of the non-Korean troops. The initial assault by the forces of the DPRK had a strong momentum and the US and ROK forces were quickly driven to the very South of the Korean peninsula. In about one month, only the port of Pusan and a perimeter around it remained in control of the United States. There was a very bloody struggle between the two sides as the North Koreans threw everything they had at the US forces in an effort to gain complete control over Korean territory, but failed to do so. More and more US forces entered the Pusan perimeter and, after consolidating their hold, the US Commander, General MacArthur, launched an amphibious assault on the port of Inchon near Seoul, completely bypassing the forces of the Korean People's Army (KPA) in the South and cutting off their supply lines. US forces in the Pusan perimeter, then, broke out of this bridgehead and pushed the KPA forces towards the North, ultimately recapturing all the territory lost to the KPA.

It was, then, quite clear that the US goal, now, was to conquer all of Korea as the US forces launched an attack on the DPRK and these forces captured most of its territory, including its capital, Pyongyang. US forces reached almost the end of the DPRK territory in the North, reaching, in some places, the Yalu River that forms the border between China and Korea. When the US forces, reached close to the Yalu River, the Chinese responded with a huge counter-attack, initially involving about 300,000 troops. The Chinese forces were referred to, by the Chinese government, as the People's Volunteer Army (PVA). China had made this decision after consultation with the Soviet Union.

Tactics of the Chinese forces were quite unexpected, since the United States had not faced guerilla fighting before. The PVA troops moved mainly during the night to avoid attacks by the US Air Force. They also avoided regular roads and mountain paths and attacked the US forces from two or three directions at the same time. US aerial reconnaissance had difficulty sighting PVA units in daytime, because their march minimized aerial detection. The PVA marched during the night and prepared for the next night's advance at the end of their march, deploying elaborate camouflage techniques. While marching, soldiers remained motionless if an aircraft appeared, until it flew away. Thus, China had made a very difficult decision to defend the DPRK, although its troops were not equipped to fight, virtually without air cover, against the well-equipped land and air forces of the United States. This decision changed the attitude of the Soviet Union also. Within a few weeks, the Soviet Union started providing air cover to the Chinese troops and also increased its aid to China.

The US forces were pushed back down the peninsula. Seoul changed hands once again, but the US forces, after their retreat, were able to hold the Chinese back about 70 miles South of Seoul. Then, General Ridgway led the US forces in a slow advance northward, inflicting heavy casualties on the PVA and the KPA and the US forces re-recaptured Seoul. This was the fourth and final time it changed hands! Patrols of the US and allied forces, then, started crossing the 38th parallel. Both sides were exhausted and neither side was able or willing to launch a big attack with the forces available to it. Negotiations for a cease-fire were started but they dragged on for a long time.

General MacArthur had been constantly pushing Washington to remove the restrictions on his forces. His public hawkish statements contradicted the instructions of his superiors including the President of the United State. MacArthur was indicating that he was thinking of launching nuclear attacks on cities in North Korea, China and the Soviet Union, but President Truman did not approve of such attacks because of their possible consequences. He fired MacArthur for insubordination, because the general had been publicly challenging him for months. In his biography, he says:

"I fired him because he wouldn't respect the authority of the president. That's the answer to that. I didn't fire him because he was a dumb son of a bitch, although he was, but that's not against the law for generals. If it was, half to three-quarters of them would be in jail." [2]

Merle Miller, Quoting President Truman, *in Plain Speaking: An Oral Biography of Harry S Truman*. Reprinted by permission of the copyright holder, Ms. Hanley.

In early 1953, President Eisenhower took over from President Truman. Negotiations for a cease-fire were continued, but these were, now, accompanied by threats of nuclear attacks on Chinese cities. It was decided to get ready to launch a nuclear attack on China, if the Chinese started to fight again. The US secured a specific promise from the British government, under Winston Churchill, that it would support the US in extending military operations against China if necessary. At that time, the US considered Chiang's puppet regime as the sole legitimate government of China. *Thus, in the middle of the year, the CIA informed Nehru, the Indian Prime Minister, through a senior official, that the United States was getting ready to launch a nuclear attack on China and intended to do so if the leadership of the Peoples' Republic did not agree to a truce.* This message was planted deliberately in India so that it would get to the government in Beijing. That is exactly what happened.

The government of the People's Republic did not want a wider war, nor could it face a nuclear attack. It made the only reasonable decision it could make and accepted a truce in Korea. Much later, President Eisenhower confirmed the events that finally led to the truce in Korea. He thought that the threat had brought the Chinese under control, but it had, also, made the Chinese leadership deeply aware of the nuclear danger in the future. The leadership decided to speed up its own nuclear weapons program and prepared to face such an attack by building huge underground shelters for its civilian population and by reorganizing its military to face such an attack with guerilla warfare.[3]

Nuclear terrorism succeeded again and, finally, the fighting ended in mid-July 1953, when the armistice agreement was signed. More than thirty thousand US troops had died in this war and about a million each of Chinese and Koreans. The agreement restored the de-facto border between the Koreas near the 38th Parallel, by creation of a buffer Demilitarized Zone (DMZ).

The New World Order

The new world order, that the ruling class of the United States desired to create, required that all European imperial powers and Japan hand over control of their conquered territories to the local ruling classes, thus providing easy access for the US, to their resources. This was referred to as de-colonization. Violence and conflicts were to be expected in this process. Creation of a global organization for control of military conflicts and dispute resolution was, thus, part of the US plan. The United Nations was set up for this purpose. Also, a system of controlled-market subjugation had to be set up, so that the United States could maximize the exploitation of the rest of the world, especially the newly "de-colonized" territories. The process of de-colonization was started through the Trusteeship Council of the United Nations. Post-war reconstruction of Europe, and other areas, later on, was handled by the Marshal Plan and the World Bank. International Monetary Fund (more accurately referred to as the "Imperial Monetary Fund") was set up to increase and stabilize global trade on terms extremely favorable to the United States. The European imperial powers were Panamized through the Marshal Plan and NATO. Japan, Korea and Taiwan were also Panamized, by financial and other arrangements between the ruling class of the US and the local ruling classes of those states and military arrangements between governments of those classes. Thus, in about twenty years, the United States successfully imposed a new imperial order on most of the world, except those states in which communist parties had taken control. The Soviet Union, China and other states ruled by their communist parties, were subjected to the policy of *"containment"* We would look at elements of this new world order in the following sections.

The Bretton Woods System

In the 19th and early 20th centuries gold played a key role in international monetary transactions. At that time, the British pound was, effectively, the reserve currency, because of the size of the British economy and the size of the British Empire. The international value of currencies was determined by the price of gold in those currencies and gold was used to settle international accounts. The gold standard maintained fixed exchange rates that were seen as desirable because they reduced the risks in inter-state trade. Imbalances in international trade were theoretically rectified automatically by the gold standard. A country with a trade deficit would have depleted gold reserves and would thus have to reduce its money supply, by buying gold with its local currency. The resulting fall in demand for foreign products would reduce imports and the lowering of prices would boost exports; thus, the deficit would be rectified. Any country experiencing inflation would also lose gold to its trading partners and, therefore, would have a decrease in the amount of money available for the local population to spend. This decrease in the amount of money would act to reduce the inflationary pressure.

The allies of the United States were economically exhausted by the war and needed US assistance to rebuild their economies and to finance their inter-state trade. Before the war, the French and the British had realized that they could no longer compete with US industries in an open marketplace. During the 1930s, the British had created their own economic bloc to shut out US goods. Britain did not want to surrender its system of *captive market subjugation* to the United States so it resisted the "free access" clause in the Atlantic Charter, but the US was determined to obtain its access to the British Empire. The combined value of British and US trade was about half of the entire world's trade, at that time. For the US to open global markets, it first had to acquire full access to the resources of the British Empire. While Britain had economically dominated the 19th century, the US intended to take over this role.

The US was clearly the most powerful state at that time. So, ultimately, it was able to impose its will on the other imperial powers, including Britain. A devastated Britain had little choice. Two world wars had destroyed the country's principal industries that paid for the imports of half of its food and nearly all its raw materials except coal. The British had no choice but to ask for aid. Not until the United States signed an agreement to grant Britain aid of billions of dollars, did the British parliament ratify the Bretton Woods Agreements. Similarly, de Gaulle, at that point the leader of the French Empire, was forced by economic condition of France after the war, to ask the US for large US dollar loans. Most of the requests were granted; in return France promised to curtail government subsidies and currency devaluation that had given its exporters advantages in the world market. After the war, many US planners believed that the fundamental causes of the two world wars lay in economic discrimination and trade warfare. Specifically, they had in mind the trade and exchange controls of Nazi Germany and the imperial preference system practiced by Britain, by which members of the British Empire were given special trading status. France and the US, also, had such policies. It was believed that unrestricted free trade led to peace between imperial powers. On the other hand, high tariffs, trade barriers, and "unfair" economic competition led to war.

The experience of the Great Depression was on the minds of the planners at Bretton Woods. They wanted to avoid a repeat of the disaster of the 1930s, when the insistence by creditor states on the repayment of war debts by their allies and reparations by their adversaries, led to a breakdown of the international financial system and a worldwide economic depression. The so-called "beggar thy neighbor" policies that emerged as the crisis continued saw some trading states using concurrent competitive currency devaluations in an attempt to increase their exports and lower imports. Also, bilateral trading blocks and other haphazard barriers and restrictions on international trade and investment had caused severe damage. They not only slowed the recovery of global inter-state and inter-empire trade, but also hindered the cross-border flow of capital and investment. Thus, at Bretton Woods, preventing a repetition of this process of concurrent competitive devaluations was the most important objective.

The new global economic system required a globally acceptable currency for investment, trade, and payments. However, there was no global institution in existence at that time to issue currency and manage its use. Meanwhile the Soviet Union was fast increasing its gold production, giving it an advantage in the continuation of the gold standard. The US wanted to eliminate or reduce this advantage. Although the gold standard had met the requirement of a global currency in the past, the US had other ambitious plans. To meet US ambitions, a system of fixed exchange rates was set up, to be managed by the newly created global institutions using the US dollar as a reserve currency. Thus, the chief features of the Bretton Woods system were an obligation for each country to adopt a monetary policy that maintained the exchange rate by tying its currency to the US dollar and the ability of the IMF to bridge temporary imbalances of payments. The political basis for the Bretton Woods system was in the combination of two key conditions - the shared experiences of the Great Depression, and the concentration of economic power in a small number of states (United States, Britain, France, Italy, Canada, Australia) which was further enhanced by the exclusion

State	Currency	1946 Rate	1948 Rate	1949 Rate	Current Currency	Current Rate
Belgium	Franc	43.830		50.000	Euro	0.910
Canada	Dollar	1.000		1.100	Dollar	1.330
France	Franc	119.100		350.000	Euro	0.910
Germany	Mark		3.330	4.200	Euro	0.910
Italy	Lira			625.000	Euro	0.910
Japan	Yen			360.000	Yen	107.300
Britain	Pound	0.248		0.357	Pound	0.810

Figure 8-01: Recent Exchange Rates and those Fixed in 1946, 1948 & 1949.

of a number of important states (China, Germany, Japan and the Soviet Union) due to the war. A high level of agreement among the powerful on the goals and means of international economic management facilitated the decisions reached by the Bretton Woods Conference. Its foundation was based on a shared belief in intensive capitalism. However, the governments of the imperial powers differed on the role of the state in their economies.

What ultimately emerged was the system of *"pegged rates"*. Members were required to peg their currencies in terms of the reserve currency and to maintain exchange rates within a band of plus or minus 1% of parity, by intervening in their foreign exchange markets. The initial pegged rates along with some changes in 1948 and 1949, are shown in Figure 8-01. The recent approximate exchange rates are also shown in it. The US dollar took over the role that gold had played under the gold standard in the international financial system. The United States, also, agreed to link the dollar to gold at the rate of $35 per ounce of gold. At this rate, foreign governments and central banks were allowed to exchange dollars for gold. Thus a "gold exchange standard" was adopted. All European states that had been involved in World War II were highly in debt and had transferred large amounts of gold to the United States, a fact that contributed to the appreciation of the US dollar. This resulted in the financial supremacy of the United States, as the pegged rates were negotiated on the basis of the exchange rates prevalent at the time - *putting other imperial powers and, especially, former imperial possessions at an immense disadvantage*. Further, the US remained free to create as much money as it wished at any time, through its federal reserve bank. This artificially created money could be easily exchanged for valuable products of the rest of the world - thus the US achieved an extreme imperial financial control over all of mankind, except the states ruled by their communist parties! However, as Figure 8-01 shows, the currencies of most large European economies were devalued between 1946 and 1949, thus strengthening the US dollar and giving it a further advantage over other imperial powers and the Soviet Union. The $35 per ounce price was maintained to keep the price of gold low enough so that the Soviet advantage would remain "under control".

The imperial powers also agreed that the global economic system required government intervention. In the aftermath of the Great Depression, public management of the economy had emerged as a primary activity of governments in the imperial states. Employment, stability, and growth were now important subjects of public policy. In turn, the role of government in the state economy had become associated with the assumption by the state of the responsibility for ensuring a degree of economic wellbeing for its citizens, since the Great Depression had given rise to a popular demand for government intervention in the economy. Also, the system of management of the business cycle proposed by Keynes had a great effect.

Bretton Woods, then, created a system of triangular trade: The United States would use the dollars, created by its central bank, the "Federal Reserve", *at virtually no cost*, to trade with a *target set of states* now subjected to *collective captive market subjugation* by all the imperial powers together. The target set of states were the former possessions of one or another imperial power, to which the relevant imperial power had given right of access to the United States. For example, South Asia was such a possession of Britain and Syria that of France. Thus, the United States was, now, able to trade with these target states, at a tremendous "profit", expanding its industry and acquiring raw materials from those developing states. It would use the surplus so "earned" to invest in economies of European imperial powers and Japan and grant loans to them to rebuild their economies and make the United States the market for their products. This would also allow those powers to purchase products from their former imperial possessions at almost the same exchange rates they had imposed on them before. This reinforced the role of the United States as the *global imperial super power* dominating the west European states and Japan and indirectly dominating their "separate" imperial possessions.

Mainland United States had not been touched by the war, while Europe and the Soviet Union were devastated by it. Thus, after the war, the US became the world's leading industrial, monetary, and military power. Being at the center of the global market gave the United States unprecedented freedom of action in pursuing its imperial goals. A trade surplus made it easier to keep armies abroad and to invest outside, especially in Panamized Europe, Japan and Korea. It also had the power to decide when and how to intervene in global crises. As a result of the new global monetary system, exporting to the US became the primary economic goal of the ruling classes of developing and developed economies, putting the United States on the way to creation of a global empire firmly controlled by its ruling class.

International Monetary Fund (IMF)

IMF started its financial operations in 1947. It was expected to advise countries on policies affecting their monetary systems. For this purpose, two rival plans were proposed by the United States and Britain. British proposals would have established a world reserve currency administered by a central bank vested with the power of creating money and with the authority to take actions on a large scale. In case of balance of payments imbalances, Keynes recommended that both debtors and creditors should change their policies. As outlined by Keynes, countries with payment surpluses should increase their imports from the deficit countries and thereby create a foreign trade equilibrium. Thus, Keynes was sensitive to the problem that placing too much of the burden on a deficit country would cause difficulties for it. But the United States, as a likely creditor, did not accept this plan. It saw an imbalance as a problem only of the deficit country. Because of the overwhelming economic and military power of the United States the participants at Bretton Woods had to accept the US plan.

A system of subscriptions and quotas was set up for the IMF. This created a pool of state currencies and gold subscribed by each state, the gold being contributed in dollars. IMF was not to act as a world central bank capable of creating money. It was required to manage the trade deficits of states in such a way that they would not produce concurrent currency devaluations which may trigger a big decline in their imports. When joining the IMF, members were assigned "quotas" reflecting their relative economic power, and, as a sort of credit deposit, were obliged to pay a "subscription" of an amount in proportion to the quota. 25% of the subscription was to be paid in gold or currency convertible into gold (i.e., the US dollar) and 75% in the member's own currency. The IMF was to use this money to grant loans to member countries with financial difficulties. Each member was entitled to withdraw 25% of its quota immediately in case of payment problems. If this sum was insufficient, a state was also able to request foreign currency loans. The maximum amount of such a loan was determined by the size of its quota. Members were required to pay back debts within a period of less than five years. The IMF set up rules and procedures to keep a country from going too deeply into debt year after year. The IMF was to monitor the economies of its members. It continues to do so even now.

The IMF, also, originally provided for adjustment of the exchange rate by a state by up to 10%, by an international agreement. Such a devaluation was expected to be rare and would have tended to restore equilibrium in its trade by expanding its exports and reducing imports. This, however, was allowed only if there was a "fundamental disequilibrium", as determined by its members. Members were allocated voting rights in proportion to their quotas. Since the United States was contributing the most, US dominance was ensured. It held one-third of all IMF quotas in the beginning. This was enough for it to veto all changes to the IMF charter. Further, the IMF was based in Washington, D.C., and staffed mainly by US economists. It regularly exchanged personnel with the US Treasury, as if it was a department of the government of the United States. All this ensured US domination of this permanent institution.

International Bank for Reconstruction and Development (IBRD)

The International Bank for Reconstruction and Development (IBRD) was also created by the Bretton Woods planners. Originally, the IBRD was expected to make use of its own funds to issue loans or to underwrite private loans to ensure reconstruction of infrastructure and a speedy postwar recovery of Europe. When the Marshal Plan was launched, the IBRD shifted its focus to other parts of the world. Until the late 1960s, its loans were earmarked for the construction of seaports, highway systems, and power plants. Later on, it also began to finance projects aimed at poverty alleviation, achievement of universal primary education and control of major contagious diseases, etc., thus enabling the US in negotiating political concessions.

Collapse of the Pegged Rate System

At the end of the Second World War, the US had more than half of global gold reserves. After the war, inter-state trade increased rapidly, but the mining and production of gold could not keep pace with it. Also, the total amount of gold available for maintenance of the reserves of various states could not keep growing with the growth of inter-state trade because the open market for gold diverted a large quantity of gold to other uses by individuals and manufacturing companies. The same happened to the gold in possession of the US Federal Reserve Bank, because the balance of trade of the US began to shift and became negative in the early 1950s. Further, the Soviet Union emerged as the biggest gold producer. This gave the Soviet Union an advantage as compared to other major economic powers. This was something that the US and other imperial powers had not expected and were worried about. $35/ounce was the price meant to be used for transfers between central banks. As the open market price of gold began to differ from the $35/ounce price (It is currently about $ 1300/ounce!) fixed under the Bretton Woods scheme, efforts were made to maintain the free market price for gold near the $35 per ounce official price. This was necessary because as the difference of the two prices increased, it tempted states to buy gold at the official price and sell it on the open market, while trying to solve their internal economic problems.

In 1967, the IMF introduced "special drawing rights" (SDRs), as a kind of new currency. The SDRs were set as equal to one US dollar, but were not usable for transactions other than between central banks and the IMF. The purpose of the introduction of the SDR system was to prevent states from buying pegged gold and selling it at the higher free market price. Central banks of member states were required to accept holding more SDRs than their previous required contributions. However, interest was to be charged, or credited, to each state based on its SDR holding. The use of SDRs as paper gold seemed to offer a way to effectively change the IMF into a central bank, instead of the US.

As the balance of payments of the US developed deficits, it began to erode confidence in the dollar as the reserve currency. As European states and Japan recovered from the effects of the war, their economies began to grow fast. Per capita income of their populations also came closer to that of the US and their total reserves ultimately exceeded the gold stock of the US. This was the source of the loss of confidence. As the Vietnam War heated up, the US tried not to pay for it and its social support programs through its tax revenue. This caused an increase in the flow of gold to foreign countries to pay for the military expenditures involved and led to further deterioration of the US balance of trade position. By 1970 the US held less than one-fifth of international reserves, but could not maintain even those reserves because of the desire to maintain the fixed exchange rates and the obligation to convert dollars into gold on demand. Also the convertibility of the Western European currencies and the Japanese yen had facilitated a huge expansion of international financial transactions, but the high levels of money creation by the Federal Reserve of the US, had not only created inflation in its own economy, but because of the fixed exchange rates, this inflation had continued to be exported to the rest of the world. One possible solution to the twin problems of the falling gold stocks in the US and the exported inflation was to raise the official price of gold in terms of the US dollar and all other currencies, but such an increase would have benefited the Soviet Union and other gold producing countries and the US government did not want that to happen. Ultimately, in mid-1971, Nixon "closed the gold window", making the dollar unconvertible to gold directly. This brought the Bretton Woods system to a virtual end. This action, referred to as the *Nixon shock*, created the situation in which the U. S. dollar remained a reserve currency for some states, while many currencies, like the British Pound, became free floating. By the early 1980s, all major economies, outside the Soviet Block, were using floating currencies.

When the delinking of the US dollar and gold happened, the exchange rates of various currencies had become acceptable to the ruling classes and their governments in virtually all states not ruled by their communist parties. As fluctuations in exchange rates began to occur, the US continued to create money and borrow more from the US corporations, other global corporations, other governments and oligarchs of the ruling classes of the world of capitalism and feudalism, by issuing "treasury notes" through its Federal Reserve Bank – especially, the ruling oligarchs of Bantustans, like Saudi Arabia, and Panamized states became the creditors. Since these ruling classes, or the global "elite", had financial links with the US economy, they themselves, and the governments dominated by them and their corporations, continued to lend money to the US. This has continued till now. Even China has invested in the US treasury notes to manage the exchange rate of is currency, the Yuan, and to forestall aggressive US political moves. After 1974, following the dramatic increases in oil prices, the ruling tribe of Saudi Arabia agreed to sell oil in US dollars only. Under Saudi and US pressure, this requirement was agreed to by OPEC also. This further strengthened the position of the US dollar as the semi-global reserve currency even after the financial power losses due to the Nixon shock. The total US debt has, now, reached about twenty-three trillion dollars. The value of the US dollar has come down, but the US continues to borrow more and more money from abroad and continues with its imperial adventures to grab more and more resources of other states. These adventures are, effectively, financed by the "elite" owners of capital in the US and the rest of the world, as mentioned before, without much cost to the US government!

The United Nations

The United Nations was established immediately after the World War, replacing the League of Nations, as part of the US design for the new world order. Originally, it had 51 members. The number has grown to about 200 now. The UNO charter states that the purposes of the UN are - the maintenance of international peace and security; the development of friendly relations among states; and the achievement of cooperation in solving international economic, social, cultural, and humanitarian problems. It expresses a strong hope for the equality of all people and the expansion of basic freedoms. The principal organs of the UN were the General Assembly, the Security Council, the Economic and Social Council, the International Court of Justice, and the Secretariat. The International Monetary Fund (IMF) and the World Bank function as Organizations associated with the UNO, although they are not formally part of it. All UN administrative functions are handled by the Secretariat, with the secretary-general at its head.

The General Assembly

The only UN body provided by the charter in which all member states are represented is the General Assembly. The General Assembly was designed to be a deliberative body dealing chiefly with general questions of a political, social, or economic character. It meets in a regular annual session and special sessions are sometimes held. It has several committees set up to deal with specific matters categorized as political, security, economic, humanitarian, etc.

The Security Council

The Security Council was set up as an institution with primary responsibility for preserving peace. Unlike the General Assembly, it was given power to enforce measures and was organized as a small and effective executive institution. Also, unlike the General Assembly, the Security Council in theory functions continuously at the seat of the UN. The council has 15 members. Five permanent members are - China, France, Britain, the United States, and Russia. The 10 (originally six) nonpermanent members are elected for two-year terms by the General Assembly. A proposal was made to add five new permanent members without veto powers, but no action has been taken on it. There are two systems of voting in the Security Council. On procedural matters the affirmative vote of any nine members is necessary, but on substantive matters the majority of the nine members must include all the five permanent members.

Under the charter, the Security Council may take measures on any danger to world peace. It may act upon complaint of a member or of a non-member, on notification by the secretary-general or by the General Assembly, or on its own. In general, the council considers matters of two kinds. The first is "disputes" that might endanger peace. Here the council is limited to making recommendations to the parties after it has exhausted other methods of reaching a solution. In the case of more serious matters, such as "threats to peace," "breaches of peace," and "acts of aggression," the council may take enforcement actions. These may range from full or partial break of economic or diplomatic relations to military operations of any level considered necessary. By the terms of the charter, the UN was forbidden to intervene in matters "which are essentially domestic," but this limitation was not intended to hinder Security Council measures intended to prevent threats to peace. *The charter was intentionally kept ambiguous regarding domestic issues which could also be considered as threats to peace. This kept an opening for a deliberate intervention in domestic issues of states, if the US and its allies considered them dangerous and threatening for, "international peace and security".*

Performance

In practice the UN has not evolved as was expected by the United States. Originally it was composed largely of its allies - the "peace-loving" states, who were coming together under American "leadership", to prevent future aggression and to promote "humanitarian purposes". Close cooperation among members was expected, especially in the Security Council. The conflict between the United States and the Soviet Union did not allow this to happen. The charter had envisaged a regular military force available to the Security Council and directed the creation of the Military Staff Committee to make appropriate plans. The committee - consisting of the chiefs of staff of the Big Five - was unable to reach agreement. The United States did not recognize the People's Republic of China for a long time. It insisted that only the regime in Taiwan represented all of China and, thus, only Taiwan had to be given the privileges meant to be given to China. Thus, no regular forces were established. The charter anticipated that regional security agreements would supplement the overall UN system. But the United States created such comprehensive alliances as the North Atlantic Treaty Organization (NATO), the Organization of American States, the Central Treaty Organization (CENTO) and the Southeast Asia Treaty Organization (SEATO), as part of its policy of "Containment of Communism". The Soviet Union, in response, created the Warsaw Treaty Organization. The proposed military role of the United Nations was, thus, taken over by these Organizations controlled by the two mutually hostile super powers. The United Nations, however, continued to play a limited role, within constraints imposed by the "cold war" between the United States and the Soviet Union, in the emerging conflicts in Asia, Africa and Latin America. Because of continuing opposition from the Soviet Union and frequent use of its veto in the Security Council, the United Nations had largely become a forum for public debate only. With disintegration of the Soviet Union, the US has achieved somewhat dominant role in the UN. Now, it can obtain the consent of the other four permanent members of the Security Council relatively easily. If it cannot obtain the consent, it acts unilaterally and the Security Council cannot act against its wishes, because of its veto power. This is how the US had acted even before. The UN can take action only for matters on which the permanent members of the Security Council are in full agreement.

Panamization of Europe
Marshal Plan 1948-52

After the Second World War, the Bretton Woods arrangements were largely adhered to and ratified by the participating governments. It was expected that monetary reserves of states, supplemented with necessary IMF credits, would finance any temporary balance of payments deficits. But this did not prove sufficient to get Europe out of its state of devastation. The United States was running huge balance of trade surpluses, and the US reserves were immense and growing, while Europe badly needed help to finance its reconstruction. Without help from the United States, Europe faced serious economic, social and political deterioration. Infant mortality rates had increased and millions of orphans wandered in the streets of burnt-out buildings of cities of Europe. Millions of people had become homeless. In addition, Communist parties were making dramatic progress among the people of Europe. However, the IMF could make loans only for current account deficits and not for capital investment or reconstruction purposes.

The reduction of gold and dollar reserves in European states, made it difficult to import essential items for use by existing industrial facilities. Also, food shortages had demoralized the working people. These shortages of basic industrial raw materials and food were hampering production. Agricultural and industrial production had gone down by about one quarter, while exports went down by half. This economic crisis was making already serious political problems much worse. Britain, because of the drain on its resources, decided to withdraw British forces from Greece. Greece was going through a civil war that had worsened the deterioration of its economy. When the British government informed President Truman that Britain could no longer sustain the burden of supporting Greece in its struggle against its internal guerrilla insurgency, President Truman promptly extended economic and military help to it. Conditions in Germany were the worst in Western Europe, because of the terrible destruction it had been subjected to during the war. Intelligence agencies were warning that widespread poverty was promoting popular discontent, because of which support for communist parties was growing fast. The US felt that Soviet Union stood to gain politically from these conditions. The fear was that the deep economic crisis would lead to victory of the Communist parties in several European countries.

Because of the deteriorating conditions in Europe, the US set up a program of economic support for all European states on a regional basis. This program was named "The European Recovery Program" and was also referred to as the "Marshall Plan". It was to provide large-scale financial and economic help for rebuilding Europe largely through grants rather than loans. Financial aid was provided to Greek, so that it could suppress its internal insurgency. Large scale grants were extended to other states. Economic regulations and restrictions on Germany and Italy were scrapped. This financial support was designed to increase European exports and it was expected that the resulting European recovery would ultimately benefit the United States by widening markets for its exports. The main objective of the plan was the regional integration of US-controlled Europe, including Germany. The total cost to the US was equal to about two percent of its GDP. The program included public and private collaboration and had the support of ruling classes of Europe and the US. Thus, business organizations, labor unions, farm associations and other private groups were involved in addition to government organizations. A network of cooperation was, thus, created across Western Europe

A high degree of economic progress and stability returned to Western Europe, due to the Marshal Plan. Inflation came under control and inter-state trade recovered to a great extent by 1952. During this Plan period, Western Europe's aggregate GDP increased by about a third. Germany recovered dramatically, because a large part of its highly skilled human capital had survived the devastation of war. The resulting powerful European economy was, thus, expected to confront and contain the economic and political power of the Soviet Union. The plan, ultimately, led to the European Union and the Euro Zone of the present. The ruling classes of individual European states were effectively amalgamated into one, which fused into the ruling class of the United States. The economic interests of these ruling classes, thus, became highly intertwined. Thus, not only the economic crises in Europe had been dealt with, but also, the fundamental objective of unification of the interests of the European and North American ruling classes had been achieved. This was the foundation on which the Panamization of Europe could proceed further without hindrance.

North Atlantic Treaty Organization (NATO)

While the work on the Marshal Plan was in progress, the North Atlantic Alliance (NATO) was created in 1949 as part of a Panamization plan for Europe. This military alliance had three intertwined purposes - suppression of the revival of military rivalries in Europe, achievement of US military dominance on the continent, and encouragement of European economic and political integration. All this, combined with the fusion of the ruling classes of Europe and the US, added up to Panamization of Europe, which not only served to expand US imperial control onto Europe, but was also part of the plan of "containment" of the Soviet Union. This Treaty laid down that an armed attack against one or more of its members, would be considered an attack against all of them and that following such an attack, each member would take "such action as it deems necessary, including the use of armed force" in response. It also laid the foundation for cooperation in military preparedness among its members, and provided for non-military cooperation among them. NATO, however, did not have a command structure for coordination of military operations. When the Soviet Union also began to develop nuclear weapons and the Korean War broke out, such a consolidated command structure was created and Headquarters of the Alliance were set up near Paris. US General Dwight D. Eisenhower was appointed its first Supreme Commander. Lord Ismay of the United Kingdom became its first Secretary General.

As a result of the Marshal Plan and establishment of NATO, political stability gradually returned to Western Europe. New Allies joined the Alliance. When the three Western-controlled occupation zones of Germany were combined into one state and West Germany became a member of NATO, the Soviet Union and its Eastern European allies formed the Warsaw Pact in 1955. A conflict between the two alliances was narrowly avoided over Cuba - during the 1962 missile crises. US involvement in Vietnam escalated during the 1960s. In 1966, France withdrew from NATO's integrated military command structure. As a result, the NATO's Headquarters were moved to Brussels.

The original US and NATO nuclear strategy was "Massive Retaliation", which meant that in response to a Soviet attack, the Western allies would respond with full scale nuclear attack. This had forced the Soviet Union to station large numbers of conventional forces in Eastern Europe. In the wake of the Cuban Missile crises, it was realized that every political or military move by the Soviet Union did not call for and could not be met with massive retaliation onto the Soviet Union. Thus, US President John F. Kennedy formulated a strategy of "Flexible Response". This strategy enhanced NATO's conventional defense posture by offering military responses short of a full nuclear exchange in the event of a conflict.

By the 1960s, the complete Panamization of Europe had been achieved. A unified economy had been created in most of Europe, the Common Market forming its core. The process of further integration of European and North American economies was also being systematically continued. This had created a loosely unified ruling class in Europe, dominated by the ruling class of the United States. In addition, the armed forces of European powers had been organized into one unified command structure dominated by the United States, through its conventional and nuclear forces based in Europe, led by a General of the US army. However, the Panamization of Europe was different from that of Porto Rico, Korea or Taiwan. In Europe, several Imperial Powers had been Panamized. Their imperial territories were not transferred to the United States as direct imperial possessions. As these imperial powers were forced to transfer their controlled territories to the local ruling classes, they were allowed to retain the links they had established with the local ruling classes and their newly-created armed forces. Thus, the European imperial powers, basically, became sub-imperial powers, dominated by the United States.

The Conference on Security and Co-operation in Europe was convened in 1973. Two years later, the Conference led to the negotiation of the "Helsinki Final Act", as a political agreement. The Act bound its signatories – including the Soviet Union and members of the Warsaw Pact – to respect the fundamental rights of their citizens, including the freedom of thought, conscience, or belief. A period of relaxation of tensions followed - referred to as "détente". The 1979 introduction of Soviet troops into Afghanistan and the Soviet deployment of SS-20 ballistic missiles in Europe led to the suspension of détente.

Concessions to the Working Classes in Europe

While the economic reconstruction of Europe and its economic integration were in progress and the United States was establishing its hegemony over Europe, the European states were also pushed to make concessions to their people whose standards of living had dramatically fallen due to the two devastating world wars. All European states gradually allowed formation and promotion of labor unions and laws were promulgated to manage collective bargaining by workers. Unemployment, healthcare and disability insurance were introduced. Wages were allowed to rise, the education systems were expanded to cover the whole population of each state and provisions were made for support of families who could not afford the cost of education of their children, on their own. Also, a social safety net gradually arose, between 1945 and 1970, covering support for all handicapped individuals and their families. All this was done at considerable cost to the ruling classes, but the fear of communism and the Soviet Union forced the ruling classes to make these concessions to their people, since the further advance of the Soviet System would have meant the complete elimination of all capital accumulation for the ruling classes. Thus, the Soviet Union, by its very existence brought about a dramatic change in the social system of Western and Central Europe.

Panamization of Japan

During the Second World War, Japan basically lost all the territory it acquired after 1894. In addition, the island of Okinawa, along with some smaller islands close to it, were taken over by the United States. The US set up large military bases in Okinawa. A territorial dispute started with the Soviet Union regarding the possession of three Islands, in the Kurile chain. This dispute has not been resolved till now. The remains of Japan's war machine were destroyed, and war crime trials were held. Several hundred military officers committed suicide right after Japan surrendered, and many hundreds more were executed for committing war crimes, including the Japanese Prime Minister during the war. The Emperor, however, was not declared a war criminal in accordance with the US plan of Panamization.

After the war, the occupation administration of Japan was headed by General MacArthur. Initially, about four hundred thousand troops were stationed across Japan, mostly around the capital. The number decreased with time. MacArthur kept much of the local Japanese government intact and did not attempt to micromanage it, preferring to rule in the way the British had run India for decades before the war. MacArthur kept the bureaucrats and technocrats who had always run Japan and retained the country's most other institutions. Rigid censorship was imposed on all media. A purge of alleged militarists and ultranationalists was conducted. The Japanese Prime Minister during the war and some top generals were executed after trial for war crimes. The war crimes against the people of Japan, especially nuclear terrorism, were of course never mentioned.

MacArthur was given extensive responsibilities beyond the Japanese home islands, the most significant of which was the repatriation of hundreds of thousands of Japanese troops from areas they held at the end of the war. He ordered the recall of all Japanese diplomatic personnel abroad and broke all diplomatic ties between Japan and other states. These relations were initially managed by the US occupation authorities. Policies were established that continued, rather than dismantled, the business conglomerates that had long dominated the Japanese economy. Instead these big conglomerates were made to go public and to engage in cross-investment with US Corporations. Capital flows between the ruling classes of the two states were, thus, pushed and cross-ownership of the Corporations of the two states was achieved progressively. Some progressive measures were also adopted within Japan. These included the decentralization of the education system and the police forces and down-sizing of large land holdings.

A four-power Allied agreement (between the United States, the United Kingdom, the Soviet Union and China) had called for a commission to formulate a new Japanese constitution. To avoid Soviet participation, a "Constituent Assembly", consisting of US generals and their assistants, secretly formulated a new constitution for Japan. It allowed the Emperor to remain as a symbolic head of state, with no executive powers, but forced Japan to completely surrender its rights to defend itself or to maintain armed forces for this purpose. It also included some progressive provisions guaranteeing fundament human rights, including giving women the right to vote. The new constitution was imposed on Japan with effect from May 3, 1947, although it was approved earlier.

A peace treaty between the rulers of the United States and Japan was negotiated and signed by the two governments in 1951. Officially, it was claimed that the US occupation had ended, although the US forces continued to occupy their bases in Japan, while Japan was not allowed to maintain any military organization. The Korean War helped in the recovery of the Japanese economy. The living standards of the population began to rise. Also, through capital flows in both directions, the interests of the two ruling classes had begun to converge and become intertwined. The position of the ruling Liberal Democratic Party (LDP) improved and it achieved dominance against the leftist opposition. When the Korean War was coming to an end, it was felt that the occupation of Japan and Korea was becoming too costly. The US, then, pressured Japan into establishing a "self-defense force" despite the provisions of the constitution it had itself imposed on Japan. The Self Defense Force was established in 1954, accompanied by large public demonstrations and protests. At the same time, the US made available to Japan, the area covered by its former *"Greater East Asia Co-Prosperity Sphere"*, minus mainland China, for exploitation under the hegemony of the US. Thus, Japan, like the European "allies" of the United States, became a sub-imperial power and the planned Japanese co-prosperity sphere actually became an *"American, Greater East Asia Co-Prosperity Sphere"*!

Bilateral talks on revising the 1951 security pact began in 1959, and the new Treaty of Mutual Cooperation and Security was signed in Washington in 1960. When the pact was submitted to the Japanese parliament, the Diet, for ratification, it became the subject of heated and bitter debate over the relationship between the United States and Japan. There was considerable violence as the leftist opposition made efforts to prevent its passage. It was finally approved by the House of Representatives. Japan's Socialist Party deputies boycotted the lower house session and tried to prevent the LDP deputies from entering the chamber. The police removed them by force. Massive demonstrations and rioting by students and trade unions also occurred at this time. This prevented a scheduled visit to Japan by President Dwight D. Eisenhower and, also, precipitated the resignation of its Prime Minister. The treaty was, however, passed by the upper house of the Japanese parliament. The approval of the treaty occurred by default, because of the boycott by the opposition.

Article-1 of the treaty established that each country would seek to resolve any international disputes peacefully. The treaty also gave prominence to the United Nations in dealing with aggression. Article 5 dealt with an armed attack by a third party. It required that the United Nations Security Council be involved. Under the treaty, both parties assumed an obligation to maintain and develop their capacities to resist armed attack in common and to act together in case of armed attack on territories under Japanese administration.

Article 6 of the treaty provides for stationing of US forces in Japan. It specified the provision of facilities and areas for their use. This treaty initially involved a military aid program that provided for Japan's acquisition of funds, weapon systems, and services for the state's essential defense. Although Japan no longer received any aid from the United States by the 1960s, the agreement continued to serve as the basis for purchase and licensing agreements ensuring compatibility and interoperability of the weapon-systems of the two states. It also served as the basis for the release of classified technical data to Japan, to facilitate the use of the fast-developing Japanese industry for its participation in manufacture of weapons for the United States.

While the economic integration of the Japanese economy with the states of the "Greater East Asia Co-Prosperity Sphere" was in progress, under the hegemony of the United States, Japan and its other partners, e.g., South Korea and Taiwan, were also pushed to make concessions to their working classes. These concessions were similar to those which had to be made by the European states under the US hegemony - collective bargaining, healthcare, insurance against unemployment and disability were allowed. Public education was expanded to cover all children and support for poor families was provided as part of a wide social network of protection for the people. The purpose of these concessions by the ruling classes of these states, under pressure of the US, was to reduce the attraction and influence of communism among the working classes.

The economies of the two states are, now, highly integrated via trade in goods and services. They are large markets for each other's exports and important sources of imports. More importantly, Japan and the United States are closely connected via capital flows. Japan is a major foreign source of financing of the US "national debt", i.e., state debt, and will likely remain so for the foreseeable future. Japan is also a significant source of foreign private portfolio and direct investment in the United States and the United States is similarly the source of much of the foreign investment in Japan. However, with the reintroduction of the market economy in China, Japan has gone through a major restructuring of its economy. Its GDP growth has slowed down and many Japanese companies have moved their manufacturing to China. Okinawa has been returned to Japanese control. But the US bases remain and the dominance of the US military over Japanese military forces persists, as is required for maintenance of the Panamized status of Japan.

In 2015, under the US policy of "Pivot to Asia", the Japanese constitution was revised to enable Japan's military to engage in collective defense with its allies, making Japan's military into a more "normal" military like those of other sovereign states. Thus, hegemony of the United States over Japan is thorough and complete. In short, Japan has been fully Panamized. The same applies to South Korea, where the same structure of imperial control exists at this time. However, a slow de-Panamization of Taiwan is in progress at this time, because of the steady increase in the economic, military and political power of the People's Republic of China.

The Global Class War
Containment

In 1947, the US separated its Air Force from its Army, and made it independent. It also created the Central Intelligence Agency. In 1948, the Marshall Plan came into being and the negotiations leading to the North Atlantic Treaty Organization (NATO) were initiated. At the same time, restoration of the economies of both Germany and Japan was taken in hand and these states were made part of the community of states Panamized by the United States. At this time the United States was the only state possessing atomic weapons. Under this nuclear threat, the Soviet Union did not fully demobilize its forces after the end of the war and most of these forces remained stationed in Eastern Europe, rendering a US nuclear attack on the Soviet Union unfeasible.

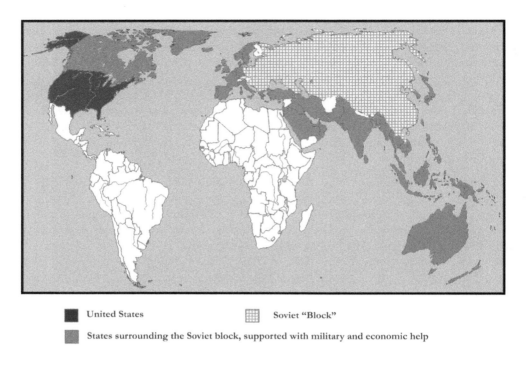

■ United States   ▦ Soviet "Block"

■ States surrounding the Soviet block, supported with military and economic help

Figure 8-02: Containment

After the war, the governments established in Easter European states, were established on Stalin's orders. Most of those governments still had popular support, because of the image of the Soviet Union as the liberator of the states and the policies initiated by the new governments. For example, Gomulka's government was quite popular in Poland. However, because of the rigid control exercised by paranoid Stalin on the policies of those governments and the continued heavy deployment of Soviet forces, the support for them disappeared ultimately. Even those governments resented the Soviet hegemony and, specially, the assassinations of their members, who were not following orders, or "scheming" against Stalin only in his own paranoid imagination.

The detonation of a Soviet nuclear device in late August 1949 and the establishment of the People's Republic of China, as a result of the Chinese Revolution, shifted the balance of power towards the Soviet bloc. As a result, the United States decided to develop the next-generation weapons based on nuclear fusion. A thorough secret study (referred to as NSC-68) of the global political and military situation was conducted and its recommendations were eventually approved by the US President in September 1950, after the North Korean forces entered South Korea. Thus, "containment" of the Soviet Union became the main objective of US foreign policy.

The US could no longer continue to rely on the threat of nuclear retaliation alone against the Soviet conventional forces, since the Soviet Union could also respond with nuclear weapons and inflict unacceptable damage to the United States. The US had to increase its conventional forces also. The study recommended a huge several-fold increase in the military budget. It was felt that, by the year 1954, the threat of first use of nuclear weapons against the Soviet Union would no longer be a credible threat. It not only recommended a conventional and nuclear military build-up, but also proposed a massive increase in military and economic assistance to all states bordering the "socialist block", i.e., all states from Turkey to Vietnam on the periphery of the Soviet Union and China - and other states behind this *"first line of defense"* which was also referred to as a *"bulwark against Communism"*. Finally, assistance to European and East Asian allies of the US was also recommended - to build up their military forces also, to reduce the cost to the United States.

As a result of this huge build-up of US military forces, bases were established in several states bordering the Soviet Union, Eastern European states and China. These included, in addition to NATO bases in Europe and Canada, bases in Turkey, Iran, Pakistan, Thailand, Taiwan and the Philippines. The US navy was expanded and introduced into all oceans. The US air force covered the whole world except the "Soviet Block". Economic assistance was extended to Turkey, Iran, Pakistan, Afghanistan, India, Thailand, Philippines, Taiwan, South Korea, in addition to Japan and European states allied with the US. In addition to NATO and the Security Treaty with Japan, the United States signed treaties to set up the *Central Treaty Organization (CENTO)* and the *South East Asia Treaty Organization (SEATO)* and started providing their member states with military assistance. Thus, the Soviet Union, China and their allies were encircled on all sides by US-allied states and their military forces, backed by US military forces based in those states.

The economic assistance to the encircling states stressed population control, setting up of hydroelectric dams and power stations and expansion of irrigation systems to settle the increasing rural populations of these states, on the assumption that the increasing urban populations of these impoverished states could be won over by the communist movements in these states if the people's standards of living could not be improved and large scale unemployment developed. Mikhail Gorbachev referred to "containment" as "cordon sanitaire" and described the intervention of Western imperial powers into the Soviet civil war and other actions by the US, including the threat of a nuclear attack after the Second World War, as follows:

"Ever since the October Revolution, we have been under permanent threat of potential aggression. Try getting in our shoes and see for yourself. A civil war with foreign forces involved, intervention by fourteen states, an economic blockade and cordon sanitaire, no diplomatic recognition (by the US up to 1933), armed provocations in the East and, finally, a devastating and bloody war against

fascism which came from the West. Nor can we forget the plans for an atomic attack on the Soviet Union by the American military and the National Security Council. We also ask why the West was the first to set up a military alliance, NATO, and is always the first to develop new weapon systems." [4]

As part of the policy of containment, the US also took some defensive measures. These *"defensive measures"* were meant to defend the US ruling class against the people of the US, by making concessions to them. Roosevelt had already set up the Federal Deposit Insurance Corporation (FDIC) to protect bank deposits of the people. The *Social Security* system was set up and a health insurance scheme was introduced for retirees, known as Medicare. Many other measures were undertaken to provide a social safety net for the working people. Taken together these measures were meant to co-opt the Communist Party of the United States, which was demanding these protections for the working class. Simultaneously, a witch-hunt was conducted in the media and government offices for suspected members of the Communist Party and their sympathizers. *Persecution of those people followed, but virtually the whole program of the Communist Party of the United States was implemented by the Democratic Party in the Roosevelt, Truman and following administrations.* Similarly, the governments of European allies of the US were encouraged to implement similar measures to reduce communist influence among their voters. These concessions dramatically altered the conditions under which the workers of Europe had lived till then. Labor unions were encouraged and laws regarding collective bargaining were modified in favor of the workers. The social security systems that developed in Europe, Canada, or Australia, ultimately, were much more liberal than the system in the US. These measures had, by the 1970s, totally altered the system of *Intensive Capitalism.* We refer to the new system of capital accumulation as *Benevolent Capitalism.*

De-Colonization and Wars of National Liberation

As the policy of containment was being implemented, holes in the encirclement of the Soviet Union and the People's Republic of China (PRC) began to develop. The Communist Party of Vietnam had started its struggle against the French colonizers in 1944. This changed into an armed struggle and guerilla fighting on the lines of the communist party of the PRC. In 1954, after negotiations, the French withdrew from Indo-china and "French Indo-China" was divided into three national states. Vietnam was further divided into North and South Vietnam, with a DMZ dividing the two Vietnamese states, on the lines of the Korean division. The United States replaced France as the imperial power in Indochina.

The United States had decided that the classical policies of the imperial and colonial powers were not sustainable in the face of Soviet determination to help the conquered peoples in freeing themselves of the imperial yoke. New, more deceptive, policies were required to meet this "threat". It was proposed to hand over each territory to its *local ruling class* and to withdraw all direct political and military control. This was in line with the US position in negotiations with Britain and other imperial powers, much before the US entered the Second World War. The process of de-colonization was to be handled by the Trusteeship Council of the UN, but was not handed over to this Organization in most cases, in an effort to keep the Soviet Union out. Britain did agree to give "independence" to its imperial possessions and started this process with British India, which was split up into seven states - three Buddhist majority states, Burma (now Myanmar), Sri Lanka and Bhutan: two Hindu majority states, India and Nepal; and two Muslim-majority states, Pakistan and the Republic of Maldives. The British objective was to avoid creating states with large minorities, since they were afraid that the minorities would be taken over by the respective communist parties. China was being taken over by its communist party, at that time. Britain also granted independence to its other imperial possessions in Asia, Africa and the Americas. British colonies of Canada, Australia, and New Zealand were given more independent status, although the British Monarch is retained by these colonies as their symbolic head of state. South Africa was handed over to its white minority in 1961. The white minority had already set up a racist regime, based on policies of "Apartheid", after 1948. Similarly, Rhodesia was split up into two states - Northern Rhodesia was renamed Zambia after independence and Southern Rhodesia was handed over to its tiny white minority. In both South Africa and Southern Rhodesia, liberation movements developed and were helped by the Soviet Union and China. The natives prevailed in both colonies and the Union of South Africa and Zimbabwe became independent states with white European minorities. Both the white racist regimes had been dismantled. Arab states of Egypt, Sudan, South Yemen, Oman, Iraq, Libya, became independent. Britain had, however, previously separated a part of Iraq and created a Bantustan controlled by the Sabah tribe and large deposits of oil had been discovered there. It was named Kuwait. Seven Bantustans in the Persian Gulf were united into the United Arab Emirates ("United Arab Bantustans"). Similarly, the Bantustans of Saudi Arabia and Oman were declared "independent". British East Africa was split up into Kenya, Uganda and Tanganyika (later renamed Tanzania when the off-coast island of Zanzibar united with it.) Ghana and Nigeria were also given independence. Nigeria has survived the, mainly French, effort to separate its Southern part into "Biafra", because of its oil discoveries. The Soviet Union helped Nigeria in preserving its unity. The colony of Hong Kong was not returned to China till 1997. Colonies of Malaya, Singapore, Sabah and Sarawak were united into "Malaysia" because Indonesia, which had already won its independence from Holland, was claiming the Northern part of the Borneo Island. Malaysia exists even now, although the island of Singapore has seceded from it. As the British dismantled their Empire, they also staged a systematic withdrawal of their military forces - East of Singapore, East of Eden and, ultimately, East of Suez. Britain retains its naval and air bases in Cyprus and the Spanish island of Gibraltar. It has yet to stage an East of Gibraltar withdrawal, which it is unlikely to do till all the Bantustans it has created on the Arabian Peninsula are dismantled and the Panamization of the Middle East comes to an end. This is not likely to happen soon.

In 1952, the Egyptian King, Farooq, had been overthrown by his military chief. After consolidating his power, Colonel Nassir, wanted to build a dam on the river Nile to generate electric power and increase the area under cultivation in Egypt. The imperial powers refused to help Egypt. In 1956, Nassir decided to nationalize the Suez Canal, which had been built by Britain, to pay for the construction of the dam. This led to an invasion of Egypt by Britain, France and Israel. The Sinai Peninsula was captured by Israel within a few days and bombings of Cairo were carried out by Britain in an attempt to assassinate Nassir. The Soviet Union, at this point, issued threats to the imperial powers that it would take unspecified military action if the Egyptian territory was not returned. Israel and the other imperial powers complied. Then, the Soviet Union agreed to finance the construction of the Aswan dam. It had, thus, managed to penetrate the walls of the system of "containment". This led to creation of good-will for the Soviet Union in the Arab mind. This good-will continued to grow as the Soviet Union got more and more involved with the liberation of mankind from imperial domination in Asia, Africa and Latin America.

France, also, reluctantly dismantled its Empire. It had withdrawn from Indo-China and had "granted" independence to its colonies in West and North Africa and Djibouti, except Algeria which was declared "French". People of Algeria organized a Liberation Front and, mainly, with the help of the Soviet Union, forced France to withdraw from its newly discovered "France" in North Africa. France separated the Christian-majority area of Syria and gave it independence as "Lebanon". Rest of Syria was also granted independence. Spain had already "lost" most of its colonies in South America. Morocco also became independent, but Spain retained control of "Spanish" Sahara, an area South of Morocco, because of its potash deposits, which it continued to exploit. The people of the area organized the Polisario Liberation Front, with Algerian help, to resist the Spanish control. Morocco claims this territory. A plebiscite was promised to determine the status of this territory, but could not happen because of disagreements about whether it would be about autonomy or independence. The Belgians had violently abused the people of the Congo, cutting their hands and feet as punishments for "disobedience" and committing other unspeakable atrocities. Around 1960, as the demand for freedom grew among the people of their colony, they suddenly decided to leave and create a separate "state" of Katanga in Southern part of Copper-rich Congo. A civil war ultimately developed and led to a high level of violence. The US, ultimately, succeeded in getting the popular leader of the Congolese, Patrice Lumumba, assassinated and had a military dictatorship established, headed ultimately by Mobuto, the commander-in chief of the army. This dictatorship lasted for about forty years, when Lumumba's followers regained power and renamed the state, as the Democratic Republic of the Congo. This state continues to face foreign interventions, because of its mineral resources, and the consequent violence continues. In Ethiopia, its Emperor, Haile Selassie, "the Lion of Juda", was overthrown by a Liberation Front and the "Lion" was, appropriately, imprisoned in a zoo! He was later placed under house arrest. Then civil war started between different groups. It ended when, Eretria was allowed to secede. Portugal was the only imperial power that refused to grant independence to its colonies in Africa. Its main colony, Brazil, had already become independent. The Liberation movements in their African colonies ultimately drove the Portuguese out. Angola and Mozambique faced a higher level of violence because South Africa tried to replace the role of the Portuguese in an effort to suppress the activities of the African National Congress (ANC) of South Africa and the Liberation movements of Southern Rhodesia, Angola and Mozambique, which had joined hands with the ANC. Finally, people of all the four "front line" colonies succeeded in freeing themselves, with the help of the Soviet Union and China.

Thus, the process of de-colonization was the result of a global conflict. The Soviet Union helped the subjugated peoples in taking up arms and freeing themselves of the imperial yoke. The ruling classes of the imperial powers had to retreat. Most of them, however, decided to retreat in an orderly fashion, handing over power to the local ruling classes in the hope that these local members of the ruling classes would be able to keep their nations under economic subjugation and provide the imperial powers with the capital that they desired, though the rate of such capital extraction could only be sustained at a reduced level. The conflict was truly global. The Soviet Union and the United States were not fighting directly, but their war continued in the subjugated regions of Asia, Africa and Latin America. Some of these sub-conflicts are described in the following sub-sections.

Vietnam

Guerilla fighting had erupted in South Vietnam in the late fifties, when the French left and the US took over the "responsibility" for Vietnam. During early 1960s, during the presidency of John Kennedy, American forces were introduced into South Vietnam, first as "advisors" to the puppet regime of South Vietnam, and, later, as regular units engaged in fighting. The level of violence continued to grow as the guerilla resistance increased and the US forces were built up. The US increased economic help as ports, air ports and roads were constructed and new systems of communication, education and health were set up. It also created a large South Vietnamese Army, in accordance with its usual plan of Panamization. It was, of course, never mentioned that the impoverished Vietnamese had done no harm to the United States or its people.

After the Korean War, some US soldiers had decided to live on in Korea and China, rather than return to the US. They also issued statements sympathetic to the Koreans and the Chinese. The American politicians could not understand why American soldiers would have this dramatic change of mind. The politicians and the ruling class of the US could not see that they had conducted mass murders of the people of the two nations, leading to millions of deaths. The people of the two nations had not done anything against the US or its people. The soldiers could see that, but the ruling class of the US could not. Thus, it was "discovered" that China had perfected a new technique of *"brain washing"* - they could wash the brains of US soldiers and make them hostile to the US! The ruling class of the US, thus, began to believe its own lies! Something similar was discovered in Vietnam. When the resistance to the US occupation grew, the US was not considered at fault - the North Vietnamese had *"infiltrated"* into their own nation in the South! It was forgotten that all the Vietnamese, in North and South Vietnam, are one nation! First, they were attacked and occupied by the French and then by the US. The US forces were destroying their homes, spraying poisonous chemicals on their families and villages and on agricultural land and forests around them – at times, even, committing atrocities against men, women and children. When President Johnson wanted to bomb North Vietnam, it was suddenly "discovered" that the North Vietnamese had attacked American naval vessels, thus, providing a justification for more tonnage of bombs to be dropped on the country than was dropped in the whole Second World War! Thus, anti-communism had risen to a level of psychosis. The whole ruling class of the US had become *psychotic*. It had begun to "see" things that did not really exist!

McNamara started giving the same arguments for continuation of the war, as had been given for the Korean War. "Communism" had to be resisted there and then; otherwise more states were likely to fall like dominos. The people everywhere were deliberately misinformed. They were not told that communism, basically, meant living like a community where nobody would abuse another - no matter how bad the implementation of this principle might be. They were not told that the real threat was to the ruling classes of imperial powers and their imperial possessions, since communism meant that members of those classes of capitalists would be dispossessed of their hoarded capital. If the ruling classes of all the imperial possessions, like Vietnam, were eliminated by state take-over of "their" factories, farms, etc., then the loot and plunder of those areas would come to an end – destroying part of the system of imperialism and capital accumulation, on which the very existence of the US ruling class depended. Thus, the war continued and violence continued to grow. The terrain of Vietnam was extremely suitable for guerilla fighting. McNamara continued to derive great pleasure from the "favorable" body counts of people being killed on both sides. A cease fire was negotiated during the presidency of Richard Nixon, but broke down soon after. The South Vietnamese Army collapsed and US troops were withdrawn from Vietnam in 1975. The secret bombings and land attacks on Laos and Cambodia also came to an end. The US had killed more than one million Vietnamese in this conflict. A larger number had been rendered disabled. Abnormal babies continue to be born in Vietnam, even now, due to the poison sprayed by the US forces. The US also lost lives of more than 40,000 soldiers in Vietnam.[5]

Vietnam and Laos were taken over by their communist parties, but an organization named the Khmer Rouge seized power in Cambodia. The Khmer Rouge were extremely violent, primitive, men who had been fed an insane variety of half-baked "socialism". They went on a rampage emptying cities of their residents and killing everyone who could be suspected of having some level of education, including everyone wearing glasses! They also began to attack Vietnam and Laos, in the name of the Khmer Empire of the distant past. The Vietnamese invaded Cambodia and set up a regime friendly towards them, consisting of elements of the Cambodian army. The US immediately started providing military and financial support to the Khmer mass murderers in the belief that "the enemy of your enemy is your friend"! When another organization had been set up under the control of Cambodian Prince Sihanouk, the US support was gradually shifted to that organization. About two million Cambodians had, meanwhile, died in this "new" war, partly due to US "help".

The situation in Indo-China has, since, stabilized. On insistence of the US, China launched an attack on the Northern parts of Vietnam when it refused to dismantle the regime it had set up in Cambodia. Vietnam, then, complied and Cambodia came under the control of a local regime which did not include the Khmer forces. The Khmer rouge launched a guerilla war against the new government but failed to dislodge it. Ultimately, when the guerilla fighting had come to an end, the leader of the Khmer rouge retreated into Thailand, where he died and his body was unceremoniously thrown on a pile of wood and old furniture and burnt to ashes - a befitting end to that delusional mass murderer.

The US had used cluster bombs during its "secret" bombing of Laos. The cluster bombs consisted of large numbers of "bomblets" which were meant to kill people. A large number of such bomblets remain scattered in the countryside and continue to cause deaths of civilians including children. President Obama visited the country in 2016 and promised to provide help for detection and removal of these vicious remnants of an undeclared war against the people of Laos.

Cuba

In 1953, Nikita Khrushchev took over the leadership of the Communist Party of the Soviet Union, after Stalin's death. Among the first steps he took, he unveiled the excesses of the Stalin regime and its mass murders of Soviet and East European citizens, especially the members of the Soviet Communist Party, and other communist parties within the Soviet area of influence. Stalin's grave was dug up and his remains were transferred to his native Georgia. Thus, a debate started within the Communist Party of the Soviet Union. Stalin's collaborators were arrested and the head of the secret police, Beira, was executed, since he had carried out the witch hunts, organized the farcical trials and actual murders during the *Great Terror*. Because of these debates, especially when many in the Communist Party were still under the spell of Stalinism and the people of the Soviet Union were mostly unaware of the excesses of their "great leader", the position of Nikita Khrushchev was not very strong. He had to show that he too was a great leader in order to consolidate his position in the regime. He made several trips abroad, especially to Egypt and Indonesia, to impress his critics at home. He ordered the construction of the Berlin wall to prevent citizens of East Germany from leaving for West Germany. When China showed signs of independence, Khrushchev moved Soviet forces to the Soviet-China border and entered into exchanges of diatribes with the Chinese leadership. Mao stated several times that the United States was a *"paper tiger"* and the Soviet Union was afraid of this paper tiger. Khrushchev had replied that the Chinese should know that the "paper tiger" had *nuclear teeth!* In this effort to dominate the communist party, he made a blunder in Cuba and was removed from his position of leadership by a troika of leaders - Kosygin, Brezhnev and Podgorny. This blunder is known as the *Cuban missile crises*.

Cuba had been grabbed by the United States from Spain, during the American-Spanish war of 1898. However, unlike Porto Rico, it had been converted into a Bantustan. The Batista family had been put into power and an agreement had been imposed on Cuba to allow introduction of US forces if and when the US considered it necessary. Also, territory on the Guantanamo bay had been taken over by the US, for construction of a naval base. With time the US became very unpopular in Cuba, as the people of Cuba continued to live in poverty while the Batista family lived in luxury ostentatiously and Cuba became a playground for criminals from mainland US.

In 1958, the Batista regime was overthrown by a small group of revolutionaries led by Fidel Castro. The US did not respond militarily at first, since the revolutionaries did not seem to have any connection with the Soviet Union or any communist party. The Soviet Union did provide economic, and later, military support to the new regime. Cuba, under Castro, started building up its military strength and organized the people in its support. Systematically, businesses and farms were taken over by the state. When it became clear that the Castro regime was actually the regime of a communist party, the United States organized a group of exiles and US citizens into an armed forced and the CIA tried to overthrow the regime. As the CIA-controlled forces landed on what was known as the Bay of Pigs, they were met by armed Cubans, who had been trained to defend Cuba in case of a surprise attack. The invaders were either killed or had to surrender and this invasion came to be known as the Bay of Pigs Fiasco.[6]

The Soviet Union, in time, became more and more involved in Cuba, and, at one point, began to move missiles and other weaponry into Cuba. It had already installed some short-range nuclear missiles in Cuba and a brigade of Soviet troops, who had been stationed in Cuba as "trainers", had taken control of those missiles. The Soviet Union was in the process of installing medium-range missiles which would have had New York and San Francisco within their range, when these activities were discovered by the CIA. The bases for the missiles were already under construction. In response, the United States put a "quarantine" around Cuba, and ordered the Soviet Union to stop shipping missiles to Cuba, and although it did not say so, it threatened to invade Cuba otherwise. Squadrons of bomber and fighter aircraft had been moved into Florida. Nikita Khrushchev of the Soviet Union reacted by warning the United States that if Cuba was invaded, the Soviet Union could also launch a nuclear attack. There was a real danger of nuclear war, since the US did not know about the short-range nuclear missiles in possession of the Soviet troops on the island. These missiles were operational and could have been used in a conflict. Khrushchev had to give in ultimately. In negotiations with the United States, his regime demanded the dismantlement of US missiles in Turkey and an undertaking by the US not to invade Cuba. The US agreed and the Cuban missile crises came to an end. It also brought the career of Khrushchev in the Communist Party of the Soviet Union to an end. His critics had rightly accused him of gambling with the security of the Soviet Union.

Israel and Colonization of Palestine

While de-colonizing its Empire, Britain also helped in the creation of a new Jewish colony in Palestine, which caused immense suffering to the people of Palestine and led to violence during and following its creation; and continues to do so, till now.

Judaism is the original belief system that emerged in the Egyptian civilization about four thousand years ago. This was the time when society had evolved only into very primitive feudalism, with rulers who were considered to be Kings and Gods at the same time. The primitive Egyptian civilization had a religion which considered the rulers, known as the pharaohs, as semi-gods. They were supposed to have human qualities as well as God-like qualities. They lived like human beings and ultimately died like human beings, but were believed to live on after death in their graves - the pyramids. This was the first civilization which developed the concept of life after death. Because of this belief, the pharaohs, once in power, would spend much effort in constructing the pyramids, so that they would have a safe and comfortable place to live on, after their death. In this slave-owning, primitive, feudal society, thousands of slaves would die to create these structures for the pharaoh's "life" after death!

Apparently, Judaism emerged among the tribes enslaved by the pharaohs, as a new religion based on the concept of an invisible God. This new belief system challenged the native religion of the Nile civilization and, after a long conflict with Pharaoh's armies, according to the Jewish mythology, the Jews ended up migrating to Palestine. With the passage of time, about two thousand years ago, a new version of this belief system emerged, as a result of a violent political conflict within the Jewish community and Imperial Rome. That is how Christianity was born. A yet new version of the belief system emerged in Arabia about one and a half thousand years ago. It is referred to as Islam. Thus, the original belief system, which emerged in Palestine and its surrounding territories, has evolved into three religions - Judaism, Christianity and Islam. Further, over time, these three religions have developed several sects each.

Many Jews left Palestine, as a result of violent conflicts there. According to Jewish mythology, the ancestors of all the present-day Jews were driven out of Palestine by the Roman rulers. Their numbers, however, did grow as members of other tribes, mostly East European, intermarried with them and adopted their belief system. However, in terms of number of believers, Judaism could not grow much, because of the long-standing practice of discouraging such intermarriages with "gentiles", who were considered non-believers and individuals belonging to hostile tribes. Also, not being able to engage in agriculture, the Jews developed financial and business skills, because of which they were able to financially exploit local "host" populations wherever they went, without blending in them. They, thus, always earned the hostility of the local people everywhere, because of the exploitative existence of a large number of individuals of this community. Further, the more religious zealots among them continued to desire their "return" to Palestine in accordance with their mythical beliefs, despite the obvious fact that most of them had acquired a European genetic heritage.

In 1896, a Jewish leader, named Theodore Herzl, proposed a plan to create a separate Jewish state. The idea was to colonize either Argentina or Palestine with the help and cooperation of one of the existing imperial powers and set up a state for all Jews of the world. Negotiations were conducted with several imperial powers, to obtain territory for the state. The Ottoman Emperor, the Russian Tsar and the British were approached. Britain was willing to participate in the plan, so it was proposed to set up a Jewish Company, in London, to manage the project, under the supervision of a Jewish "Society". Britain, offered the territory of Uganda to the followers of Theodore Herzl, or "Zionists", for this purpose, but the Zionist movement ended up favoring the colonization of Palestine, to create a "homeland" for the Jews.

> .... **"Hence it is that even Jews faithfully repeat the cry of the Anti-Semites: "We depend for sustenance on the nations who are our hosts, and if we had no hosts to support us we should die of starvation." .....Even if we were a nation of entrepreneurs--such as absurdly exaggerated accounts make us out to be-- we should not require another nation to live on. We do not depend on the circulation of old commodities, because we produce new ones."**

> .... **"Let the sovereignty be granted us over a portion of the globe large enough to satisfy the rightful requirements of a nation; the rest we shall manage for ourselves. Should the Powers declare themselves willing to admit our sovereignty over a neutral piece of land, then the (Jewish) Society will enter into negotiations for the possession of this land. Here two territories come under consideration, Palestine and Argentine. In both countries important experiments in colonization have been made, though on the mistaken principle of a gradual**

infiltration of Jews. An infiltration is bound to end badly. It continues till the inevitable moment when the native population feels itself threatened, and forces the Government to stop a further influx of Jews. Immigration is consequently futile unless we have the sovereign right to continue such immigration."

.... "Shall we choose Palestine or Argentine? We shall take what is given us, and what is selected by Jewish public opinion. The (Jewish) society will determine both these points."

...... "It might further be said that we ought not to create new distinctions between people; we ought not to raise fresh barriers, we should rather make the old disappear. But men who think in this way are amiable visionaries; and the idea of a native land will still flourish when the dust of their bones will have vanished tracelessly in the winds. Universal brotherhood is not even a beautiful dream. Antagonism is essential to man's greatest efforts."

...... "But the Jews, once settled in their own State, would probably have no more enemies. As for those who remain behind, since prosperity enfeebles and causes them to diminish, they would soon disappear altogether. I think the Jews will always have sufficient enemies, such as every nation has. But once fixed in their own land, it will no longer be possible for them to scatter all over the world."

Theodore Herzl (1896), The Jewish State. Urbana, Illinois: Project Gutenberg. Retrieved
October 16, 2019, from www.gutenberg.org/ebooks/25282.

Efforts for colonizing Palestine were continued by the Zionists in the beginning of the twentieth century, but large-scale settlement activity started after the Second World War, when a large number of Jews were killed by the Nazi regime in Germany. In 1940, Britain, had accepted the plan for creation of a "homeland" for the Jews in Palestine - a declaration of intent known as the Belfour Declaration. Britain continued to help the Jews in this regard. The Jewish colonizers set up, not only the company to buy land from the locals and set up settlements, but also started an organized campaign to terrorize the Palestinians to give up their land and homes, or sell them at cut-throat prices. Britain, then, left Palestine to "itself" and moved its military and civilian administrators out, in coordination with the Zionists. In 1947, the Zionists declared the creation of a Jewish state. The homes, mosques and cemeteries of the Palestinians began to be bulldozed and the local population continued to be driven out of Palestine. This led to a war between the Zionist state and the neighboring Arab states. A ceasefire was reached after the intervention of the newly-established United Nations. No settlement of the newly-created dispute could be reached and the Zionist colony continued to increase its military and economic strength with the help of Britain, France and the United States. Here is what Joe Slovo, of Lithuanian origin and a Jewish leader of the African National Congress (ANC) of South Africa, has to say about it, in his unfinished biography:

"Social theory aside, the dominating doctrine on this kibbutz, as well as on the others, was the biblical injunction that the land of Palestine must be claimed and fought for by every Jew. And if this meant (As it did eventually mean) the uprooting and scattering of millions whose people had occupied this land for 5000 years, more's the pity.

Within a few years, the wars of consolidation and expansion began. Ironically enough, the horrors of the Holocaust became the rationalization for the perpetration by Zionists of acts of genocide against the indigenous people of Palestine. Those of us, who, in the years that were to follow, raised our voices publicly against the violent apartheid of the Israeli state, were vilified by the Zionist press. It is ironic, too, that the Jew-haters in South Africa - those who worked and prayed for a Hitler victory - have been linked in a close embrace with the rulers of Israel in a new axis based on racism." [7]

Helena Dolny, *Slovo: The Unfinished Autobiography*, Ocean Press, Melbourne 1997, p. 47,48

As the Jewish colony in Palestine was expanding and consolidating itself, the process of decolonization, on a global scale, was initiated by the imperial powers under the umbrella of the United Nations - under threat of the Soviet Union. Here is what Dean Rusk, the US Secretary of State during the Kennedy and Lyndon Johnson administrations, has said about creation of Israel, as recorded in conversations with his son:

"Prince Faisal said that with the creation of Israel, the Arabs were being forced to pay for the crimes of Hitler. He held the United States responsible. Those deep-seated feelings have persisted ever since. There likely would never have been a state of Israel had it not been for American support. Relations between Israel and its Arab neighbors have been the most intractable, unyielding problem of the postwar period. I personally bear many scars from it, as have every American president and secretary of state since 1948 and many of my colleagues in the State Department. We were part of a delegation trying to carry out President Truman's instructions, but when we didn't go 1000 percent for the Zionists, they heaped abuse on us all, especially the Middle East desk and its director, Loy Henderson. They claimed Loy and other Middle East hands were hostile to Jewish interests and an insidious element in the bowls of the State Department. Their criticism was unjust; what we wanted was simply a plan that both Jews and Arabs could live with, one that would not produce what has, in fact, happened: an almost permanent alienation between the United States and many Arabs." [8]

In 1956, the Egyptian regime of Jamal Abdul Nassir nationalized the Suez Canal. This led to a joint attack on Egypt, by Britain, France and Israel. The Israelis captured all the territory of the Sinai Peninsula, as a result, but were forced to retreat, as a result of threats by the Soviet Union. In 1967, the Nassir regime of Egypt launched a military campaign against Israel, but was badly defeated by Israel, in "blitzkrieg" attacks by the Israeli military supported by the imperial powers, especially the United States. Israel grabbed not only the Sinai Peninsula this time, but also, the West Bank area and part of the Golan Heights area of Syria.

This was a catastrophic defeat for the Arabs, by the Israeli colonizers and their imperial backers. Arab states have not, till now, recovered from this defeat, despite considerable help from the Soviet Union. Israel continues to occupy territory captured in the war of 1967, in addition to controlling the regimes in Egypt, Jordan and Lebanon, with cooperation of the United States.

> **"Judaism has maintained itself alongside Christianity, not only because it constituted the religious criticism of Christianity and embodied the doubt concerning the religious origins of Christianity, but equally because the practical Jewish spirit – Judaism or commerce – has perpetuated itself in Christian society and has even attained its highest development there. The Jew, who occupies a distinctive place in civil society, only manifests in a distinctive way the Judaism of civil society."**
>
> **Karl Marx, *On the Jewish Question*,**
> **www.marxists.org/archive/marx/works/1844/jewish-question/index.htm, accessed October 16, 2019.**

In 1973, Egypt launched an attack on the Sinai Peninsula, across the Suez Canal, in an attempt to recover the peninsula occupied by Israel. Egyptian forces did manage to cross the canal and establish a bridgehead. The military forces, then, made some progress in capturing part of the territory captured by Israel from Egypt in 1967. However, the Israeli army was able to stage a counter-attack with the help of the United States and other imperial powers, when a gap was noticed by American surveillance satellites, between two groups of Egyptian forces. The counter-attack by the Israeli army was launched through this gap and the army was able to penetrate from the Sinai to the Egyptian side of the canal. Negotiations followed and Israel agreed to withdraw from the former Egyptian territory in the Sinai Peninsula. It however, refused to vacate the Gaza strip which was under Egyptian control before the war in 1967. It also refused to return the territory it had conquered on the West bank of the Jordan River, from Jordan. Further, it kept the Syrian territory of Golan Heights. United States offered to aid the ruling class of Egypt with annual contributions of capital. In return, Egyptian rulers agreed to recognize Israel as a "legitimate" state. This recognition of a *fait accompli* against the Arab people of Palestine as a "legitimate" action, earned Anwar Sadat, the Egyptian dictator, the hostility of the Arabic people everywhere. He was assassinated in 1981. The Palestine Liberation Organization and other Palestinian groups continued to engage in an armed struggle against the Jewish colonization of Palestine.

In 1993, after the disappearance of the Soviet Union, secret negotiations were held between Israel and the Palestine Liberation Organization (PLO) and the "Oslo Accords" were signed. These agreements laid down the basis of a "peace process" that was expected to lead to an agreement to set up two states in Palestine. A Palestinian Authority, with limited self-government functions, was created for this purpose. However, delaying tactics by Israel have succeeded in frustrating all efforts to reach an agreement. The West Bank and Gaza have, basically, been converted into a set of "Ghettoes" and a system of apartheid, like the one that existed in South Africa, has been imposed on Palestine. Also, expansion of the Jewish colonized areas is continuing and more and more Jewish settlements are being created. With the Panamization of most of the oil-rich states of the Middle East by the United States, further expansion of the Jewish colony beyond Palestine and the Golan heights of Syria, is to be expected. Thus, violence and counter-violence between Israel and the Arab nation is likely to continue for a long time. Noam Chomsky, Ilan Pappe and Edward Said have written several books (some are included in the bibliography) about Israel, Palestine and the Middle East.

Indonesia

Indonesia had become independent when the Japanese occupation ended after the Second World War. The Dutch had tried to reoccupy its territory, but stiff resistance was offered by the Indonesian people under the leadership of Sukarno. After independence, Sukarno became the President of Indonesia. Gradually a conflict developed between the Indonesian Communist Party and the Indonesian Army. Indonesian society was a feudal society at that time and, basically, remains so at this time. The Communist Party of Indonesia had become the largest communist party after the communist parties of the Soviet Union and China. It was not an underground political party and, with time, it became close to the Chinese communist party. However, it did not take security precautions in its activities with tragic results. Sukarno tried to keep a balance between the Army and the PKI, but in 1965 a violent upheaval took place. It is said that some units of the Indonesian Army, sympathetic to the PKI, tried to stage a coup but failed and their opponents staged a coup instead. The Indonesian army, and some religious parties aligned with it, went on a rampage of mass murder of the members of the communist party and their families.[9] *Houses of the victims were set on fire. Virtually all the members of the party, their spouses and children were slaughtered and their bodies were thrown into streets, rivers and streams. Eye witnesses, described the ghastly scenes of violence and mass murder, but there was only jubilation in the imperial media.* Their professions of concern for human rights were conveniently forgotten. It has been estimated that between half a million and one million people lost their lives. There are reports that the US Embassy helped the army in this mass murder by providing lists of real and suspected members of the communist party. [4] A military dictatorship was established, headed by General Suharto, who ruled with an iron hand for about forty years. The United States and its corporations quickly developed close economic relations with the regime and exploited Indonesia and its mineral and other natural resources, reaping a high rate of loot and profit.

Chile

Chile had a tradition, of representative governments, that had lasted over a long time. In 1973, a Marxist Political Party, headed by Salvador Allende, came into power by winning a general election. Chile has large copper and other mineral deposits. The new government nationalized the copper mines and carried out other reforms in accordance with its Manifesto. This angered not only the ruling class of Chile but also the United States. The CIA quickly got involved in formulating plans for overthrow of the government. A US corporation, then known as "International Telephone and Telegraph Corporation" (currently named ITT Inc.), became especially active because it was affected by the government's nationalization plans. The US agencies and this corporation encouraged the military to overthrow the legitimate government of the country and launched an economic siege of the country. The President was killed when the President's house was bombed by the Chilean Air Force. General

Pinochet sized power and became the dictatorial ruler of the state. About thirty thousand of the members of Allende's Party and their families were murdered during the first year of Pinochet's rule. Killings of opponents of the Military regime continued for quite a few years. This regime, supported by the United States, surpassed even Hitler's regime in some ways. For example, it trained dogs to rape women, and this form of torture was widely applied to members of Allende's party. [10]

The regime was in power from 1973 till 1990. In 1998, When Pinochet had already relinquished power; he arrived in Britain for medical treatment. A group of his victims had filed a case against him in a Spanish court. The Spanish court indicted him of human rights violations and requested the British government to extradite Pinochet to Spain so that he could be tried for his crimes. Pinochet was arrested in London and held for a year and a half but the British House of Lords refused to extradite him to Spain and Pinochet escaped justice. It is not surprising that the "House of Lords" would do this, since Pinochet, though an especially violent monster, was a capitalist after all and their comrade-in-crime against the people of the world. The "Lords", naturally, wanted to protect him from the people of Spain or Chile. On his return to Chile, Pinochet was indicted by a local court also and was charged with a number of crimes. However, he died in 2006, before his conviction for any of the heinous crimes he and his regime had committed against the people of Chile.

Argentina

Argentina had seen several civil and military dictatorships after becoming independent of Spain in the late nineteenth century. In 1943 a group of military officers, including General Juan Peron, seized power and set up a military regime. Peron managed relations with labor unions and became very popular because he actually cared for the people! He won the 1945 elections by a landslide and was President of Argentina from 1946 to 1955, when he was deposed by a military coup. His followers were called *"Peronists"*,or *"Peronistas"*, and were banned from taking part in politics by the new military regime. Several dictatorships followed but were unable to completely suppress "Peronism". Peron returned to Argentina from Spain in 1973 and again won a landslide victory in an election. He obtained 61% of the popular vote and became President. He was in poor health at that time and died in July 1974. His wife took over the Presidency but could not manage affairs of the state. She was removed by another military coup in 1976. A violent military regime was set up. Several leftist Organizations had emerged in Argentina by that time. Some, like the *Montoneros*, waged urban guerilla warfare against the new regime. The military regime decided to wage an "ideological war" against its opposition and focused on eliminating the social base of insurgency. In practice that meant assassinating many "middle-class" students, intellectuals and labor organizers, most of whom had few ties, if any, to the guerrillas. Human rights groups estimate that over 30,000 persons were "disappeared" (i.e., arrested, tortured, and secretly executed without trial, by means of blows to their heads with hammers or guns) during the 1976–1983 period in which the military regime was in power. Even pregnant women were not spared. They were allowed to give birth to their children and, then, they were murdered. The children were offered to Military officers for adoption. Representative governments have followed that savage military regime, but Argentineans still remember with pain the youth who were "disappeared" by the military murderers. Mass graves of the victims have been discovered and subjected to forensic analysis, providing dramatic and shocking evidence of the savagery.

South Africa

South Africa became a Dutch colony in the middle of the Seventeenth century. As Dutch settlements grew, the inevitable conflict with the native tribes started. Most of these tribes were completely wiped out as the colonization moved northwards. Slaves were brought in from Indonesia by the Dutch East India Company, since the natives were difficult to control. The British grabbed this colony from the Dutch in early nineteenth century. British forces arrived and the level of violence against the native people grew with time. Genocidal wars continued. Slavery was abolished in 1834 and people from India and other British colonies were brought into this colony.

As the conquered territory grew, Botswana, Lesotho were allowed to remain as separate territories while the rest of the territories were organized into the Union of South Africa, in 1910. A policy of racial segregation had been started while South Africa was still a Dutch colony, but the process was intensified after the formation of the Union. Apartheid became official policy of the colonizers, whereby the remaining native South Africans were subjected to severe oppression based on race. Most resources, including 90% of land, were reserved for white colonizers only. The natives started their struggle against apartheid by forming the African National Congress (ANC).

After the Second World War, the Soviet Union started providing military and financial help to the ANC. China also joined in later. The Communist Party of South Africa set up an alliance with the ANC. Both the parties were banned by the racist regime but the struggle against apartheid intensified. The ANC set up its civil and military bases in Angola, Mozambique, Ethiopia, Zambia and Libya. Its military wing, named the *"Spear of the Nation"* was provided military training by the Soviet Union. The US continued its policy of "engagement" with the racist regime and continued to receive "its share" of the loot and plunder of the people of South Africa, through its corporations. As the guerilla war intensified, the racist regime developed and tested nuclear weapons for use against the bordering "frontline" states which were helping the ANC. Israel collaborated with the racist regime in this effort. The racist regime, ultimately, began to collapse. Negotiations were held with Nelson Mandela and the ANC. As a result, in 1994, general elections were held and Nelson Mandela was elected the President of South Africa, after spending twenty-seven years in jails of the racist regime as a "terrorist". The "apartheid" system of racial discrimination came to an end. South Africa has, since then, made steady progress in dealing with the results of the long-term violence and discrimination of its colonial past, against its native Africans. Education and Health care of its population has greatly improved, despite the HIV epidemic that was noticed after ANC came to power. Also, extreme poverty and inequality of income has been decreasing since then, because of government policies. However, the government could not make any substantial progress in reversing the consequences of the forcible occupation of native agricultural lands and eviction of local residents from them, during the colonization of the territory which, now, forms the state. A recently proposed amendment to the constitution would allow for land acquisition, with reasonable compensation, from the current colonial settlers, so that it could be redistributed to the native population.

Ireland

For a long time, Ireland was an imperial possession of Britain. It did not have much natural resources, agriculture was the main industry. Britain exploited the people by means of controlled market subjugation. The people resisted and slowly their resistance grew. Ultimately, Britain agreed to relinquish its imperial control. But, because of the long occupation, some people in Ireland had changed their religion and had converted to the protestant sect of Christianity. Most of the Irish retained the Catholic sect of Christianity as their religion.

So, while leaving, the British demarcated a small part of Northern Ireland to ensure it included all the Protestant majority districts, and retained Northern Ireland as part of Britain. This divided the Irish nation and, naturally, a violent confrontation followed. It continued for a long time. The people in the South started fighting Britain and the North and people in the North started fighting those in the South and the Catholic community amongst them. Both sides were using guerilla tactics. A lot of blood was shed unnecessarily. Britain might still have gained economically, but the gains kept on evaporating due to the military actions it had to take, and the damage that was done to its own economy. In 1999, a "devolution" agreement was reached giving both sides in Northern Ireland some degree of participation in government. Most of the British military bases were also removed and both Protestant and Catholic sides surrendered their weapons. Ireland and Northern Ireland were also brought together because of the British and Irish entry into the European Union. The situation has again become uncertain due to the proposed British exit ("Brexit") from the EU.

Horn of Africa

North-Eastern corner of Africa is sometimes described as the Horn of Africa, because of the shape of its map. During the late seventies, Somalia, Ethiopia and Djibouti were three apparently peaceful states in this region. Ethiopia was ruled by a king, named Haile Selassie. He was also referred to as the "Lion of Juda". Somalia was ruled by a tribal chief, named Saad Barre, who was a former military man. Both these states were former colonies and Djibouti was still under the imperial control of France. Many tribes inhabit this region and this impoverished region has seen famines and large-scale deaths due to starvation. In the late seventies, the "Lion of Juda" was overthrown by a revolutionary organization. Fighting also erupted in Ethiopia's coastal region, because the Muslim tribes in the area wanted to separate from Ethiopia. Somalia attacked Ethiopia to capture the Muslim dominated desert region known as the Ogden desert. The attack failed, but then the Ethiopian regime supplied weapons to Somali tribes and encouraged them to fight the central government. Thus, a civil war started in Somalia also. It completely destroyed the government of the state so that Somalia is now a days referred to as a "failed state". Ethiopia also went through a long period of civil war, but ultimately decided to allow the coastal Muslim tribes to secede. A new state of Eritrea was, thus, created. Peace and development have returned to Ethiopia, but Somalia continues to suffer. The Northern part of Somalia has declared itself as a separate state of "Somaliland" and is now a peaceful state, though it has not been recognized as a sovereign state by any other country. Tribal fighting and terrorist attacks continue unabated in the rest of Somalia.

Sudan

Sudan was a British imperial possession. After independence it became a very large state, as regards area, but it also had a very primitive tribal society. People of Sudan have the same race, but the tribes of the North and the South are very different from each other in religion and language. Tribes of the North were Muslims and those in the South were mostly Christian. The tribes of the South wanted to secede from Sudan and a guerilla war started. It continued for several decades, resulting in bloodshed and killings on a large scale. Ultimately, in 2011, the Southern part became a separate state after a referendum. However, inter-tribal violence has continued after independence. Currently, a new peace agreement is being negotiated between the two main tribes.

Guatemala

In late 1944, the ruling dictatorship in Guatemala was overthrown by a popular uprising. This event is generally referred to as *the "October Revolution"*. It is also referred to as *"Ten Years of Spring"*, highlighting the only years of representative government in Guatemala, in its history till that time. It followed more than one hundred years of oppressive regimes, which focused on coffee production in the country. Significant concessions were granted to the United Fruit Company, which traded in tropical fruit and evicted many indigenous people from their traditional lands. After the "October Revolution", elections were held and a professor, named Juan Arevalo, became the president. He implemented a program of social reform, including a literacy campaign. In 1951, after a land-slide election victory, Jacobo Árbenz became the president. He launched an ambitious land-reform program. The United Fruit Company lost some of its uncultivated land in that program. The company lobbied the US government for the overthrow of President Árbenz, and the Eisenhower administration organized a coup under the pretext that Árbenz was a communist. The US intervention led to a civil war which lasted from 1960 to 1996, and saw the US-backed military commit genocidal massacres against the indigenous Maya people of Guatemala.[11]

Nicaragua

United States entered into conspiracies with the extremists of the ruling class of Nicaragua and occupied this country during 1909 to 1933. This was one of several such US invasions of the small Central American states that the US undertook when it became a global imperial power. The motivation for this occupation of Nicaragua was the possibility of construction of the Nicaragua canal linking the Atlantic and the Pacific oceans. When Panama Canal was constructed, this motivation for another canal disappeared. The US forces, however, stayed till 1933. In 1927 a guerrilla war started against the US and its local allies. It was led by General Augusto César Sandino, who was executed by the "National Guard" organized and equipped by the US. Anastasio Somoza García, an ally of the US, was put in charge of the country and the US marines left Nicaragua. Nicaragua had, thus, become a Bantustan ("Samozaistan"). The Samoza dynasty ruled Nicaragua till 1979.

In 1979, Nicaragua went through an upheaval. Revolutionaries of the Sandino National Liberation Front, commonly known as the Sandinistas, seized power and the country was de-Bantustanized. Sandinistas took over large land holdings and businesses and took steps to empower the people and improve their conditions of life. They also created a large army of about 75,000 men with Soviet and Chinese help.

In Honduras, the members of the "National Guard", who had fled Nicaragua, were organized by the US Central Intelligence Agency into a counter-revolutionary force, generally referred to as "Contras", to oppose the Sandinistas. After some time, this terrorist organization started its activities inside Nicaragua, assassinating Sandinista supporters and destroying schools, bridges and factories. The US, also, directly carried out terrorist attacks on Nicaraguan ships and mined ports of the country. The Sandinista government filed a case in the US-created International Court of Justice against these actions of the US. The ICJ held that the US had violated international law by supporting the Contras in their rebellion against the Nicaraguan government and by mining Nicaragua's harbors. The US, however, paid no attention to this judgment and continued with its terrorist activities, till, under the chaotic and economically stressful conditions for the people of Nicaragua, the Sandinistas held, and lost, new elections and a government acceptable to the US was established.[12]

During the 2006-2016 period, Daniel Ortega, of the Sandinista National Liberation Front (FSLN), was re-elected to his second, third and fourth term. However, after the elections in 2016, he has faced demonstrations by students and other opponents, alleging election irregularities and asking for his resignation.

Iran

Iran has been the center of an old civilization which has gone through a lot of changes over thousands of years. In 1921, after the First World War, a military officer, named Reza Khan, seized power from the ruling dynasty with British encouragement. Oil had been discovered in Iran in 1908. Reza Khan changed his name to Reza Shah Pahlavi and declared himself King of Iran. He managed to avoid complete domination of Iran by Britain and developed relations with European powers including Germany. Because of the German connection, the Soviet Union and Britain invaded Iran at the beginning of the Second World War. The Soviet Union withdrew from Iran after the war and Britain forced the Shah to abdicate in favor of his son, Mohammad Reza Shah Pahlavi. Reza Shah fled to South Africa and died in Johannesburg. In 1951, a Prime Minister of Iran, named Mosaddeq, nationalized the oil industry despite intense pressure from Britain. The Shah fled to South Africa. The Central Intelligence Agency of the US, then, organized a coup by the Iranian army to topple Mosaddeq and restore the Shah back to power. The Shah became closely aligned with the United States. The Iranian government entered into an agreement with an international consortium of foreign companies which ran the Iranian oil facilities for the next 25 years or so. Iran became a member of the Baghdad Pact (later renamed the *Central Treaty Organization, or "CENTO"*). Being a puppet of the United States, the Shah was very unpopular with the Iranian population, although he did a lot to modernize the state. The Shah's secret agency, the SAVAK, was dreaded by the population because of its savage atrocities against those who fell afoul of the Shah and his regime. In 1979, the people of Iran rose against the Shah's tyrannical regime and the Shah was overthrown. He fled first to South Africa and then to Panama. Iran became an Islamic Republic under the leadership of Ayatollah Khomeini and a theocratic regime was set up after a lot of bloodshed. Then, with US encouragement, Iraq invaded Iran and a war, lasting several years, started. The fighting was then encouraged, on both sides, by the United States to weaken both Iran and Iraq, in a repeat of its strategy during the Second World War – when it desired that both the Soviet Union and the Hitler regime suffer maximum casualties. About a million Iranians and Iraqis died in this war. Because of the new religious, and naturally intolerant, regime the whole middle class of Iran fled to Europe and North America. All opposition parties were crushed by the theocratic regime, by mass murders of about 30,000 of their members.

Afghanistan

A military take-over occurred in Afghanistan in April, 1978. Afghanistan was, and still is, a very primitive feudal state. It is a multi-national state with basically five nations - Pashtuns, Tajiks, Uzbeks, Hazaras and Turkmens. All nations are at the stage of primitive feudalism. The Tajiks are somewhat advanced since the ruling class of Afghanistan was mainly composed of Tajiks, although the ruling king and his family were Pashtuns. About half the population is Pashtun, 25% are Tajiks and 20% or so are Uzbek. It, mostly, has a dry hilly terrain. Pashtuns and Turkmens live in parched, dry areas. In the Pashtun area, the Eastern Provinces of Nangarhar (with its capital of Jalalabad) and Qandahar have relatively large areas where agricultural production takes place. The Kabul and Helmand rivers flow through these areas. In the North of Afghanistan, the cities of Mazar e Sharif in the Uzbek area and Kunduz in the Tajik area, have green agricultural areas around them. The Communist Party of Afghanistan was a tiny party and, because of the Sino-Soviet split, had splintered into two factions known as the Khalq Party (or "People's Party") and Parcham Party (or the "Flag Party"). The Parcham Party was based in urban areas and was pro-Soviet in its outlook. The Khalq Party had a following both in urban and rural areas, including among intellectuals in educational institutions and civil and military bureaucracy. The total membership of the two parties combined was perhaps about 10,000. Three years, or so, before the military take-over, the King had already been overthrown by his brother-in-law, Sardar Daoud. Sardar Daoud was overthrown and killed as a result of the coup by the military's Khalq-related officers. This "Revolution" was not really a revolution, it was a takeover of the state by a small political party of self-styled Marxists, who, as it soon became clear, were highly tribal and feudal and had very little to do with Marxism.

The new regime started a process of land reform. Large land holdings of tribal leaders and religious institutions were taken over and redistribution of the land to poor peasants and share-croppers was started. New state-owned schools were set up as part of a new education system. Drugs and sale of women (an ancient and widespread practice in Afghanistan, even at this time.) were banned. Afghanistan did not have much industry, but whatever was there, was taken over by the state, immediately. The training and expansion of the small, rag-tag 40,000-man ill-equipped army was taken in hand with Soviet help. These measures, at once, created intense hostility for the regime among the tribal leaders, land owners, religious leaders and the primitive illiterate population of Afghanistan.

By 1975, the ruling classes of imperial powers had been forced out of virtually all imperial possessions in Africa, Asia and Latin America. Only the racist regime in South Africa and the Jewish colony in Palestine continued to abuse the people controlled by them. The Soviet Union, however, continued to push the newly-independent states towards greater state control of their economies. It also sponsored and supported liberation movements wherever they developed. It was clear at, this time, that the desire of the under-developed countries for independence from imperial powers had been mostly met. Their societies were primitive and systems of modern government were unknown to them. A resistance towards further socialization and creation of state-controlled economies started developing as the local ruling classes began to succeed in convincing their people that "complete freedom" had already been won by them and that state-control of an economy was against their traditions. Thus, the Soviet Union had reached a point where further liberation was likely to be seen as its domination, by the newly-independent countries. But the Soviet Union did not change its policies.

In this atmosphere, the US decided to try to reverse the "gains" made by the Soviet Union by using terrorist tactics of its own. Zbigniew Brzeziński was the national security advisor to President Carter, at that time. Terrorism is not the kind of activity that President Carter would have willingly agreed to. It is not clear how Mr. Brzeziński managed to get the President to approve it. Some retired CIA officials were sent to Pakistan to assess the situation in Pakistan and Afghanistan. Based on their assessment that the situation was "excellent" for the project proposed by Brzeziński, the US approached the newly-established dictatorship of General Zia ul-Haq in Pakistan. Zia was an extreme religious fanatic and was more than willing to join the US in its terrorist plans. He had overthrown the first elected constitutional government of Pakistan and his planned murder of the over-thrown Prime Minister Zulfiqar Ali Bhutto, through his kangaroo courts, had reached an advanced stage. Zia was in urgent need of financial help in consolidating his usurpation of political power in Pakistan. Brzeziński had, thus, found the ideal general to organize and execute his project. Large-scale military and financial help was provided by the US to the Pakistani army and its intelligence agency, the Inter-Services-Intelligence (ISI). Many terrorist training camps were established in Pakistan, near the Pakistan-Afghanistan border, starting in late 1978. However, no Afghan recruits were available to launch a terrorist war against Afghanistan. Instead, men from the tribal areas of Pakistan were recruited to serve as "Afghan Mujahidin" ("Afghan holy warriors"). They were trained in the usual terrorist tactics of attacking road transport vehicles, blowing up road bridges and schools, poisoning water supplies and killing doctors, nurses, school teachers and officials of the Afghan government. The US supplied military equipment, initially through Egypt, including Kalashnikov rifles and rocket-propelled grenades for those camps. Fully-trained terrorists were, then, sent into Afghanistan to start the "holy war".

The Afghan government compounded its blunders by reacting to violence with violence. As the fighting between the Afghan army and the terrorists increased, the civilian population of the affected towns and villages began to migrate towards Pakistan, where they were welcomed as refugees and became a desperately-sought source of new genuinely-Afghan recruits for the "holy war" against the "Godless" rulers of Afghanistan. General Zia himself began to direct the war against the people of Afghanistan and CIA officials and Afghan warlords began to attend the meetings of his "holy war cabinet". Zia's regime also approached many organizations of religious fanatics all over the Islamic world for help in men and material to wage this holy war. Men from many countries arrived in Pakistan to wage war. *Included in them, was one Osama Bin Laden from Saudi Arabia.* At one time, even Israel supplied weapons for this war. Thus a "holy war", or "Jihad", in the service of Israel and the US, had been started by General Zia of the "Islamic Republic" of Pakistan![13]

Figure 8-03: President Reagan with Representatives of the Jihadists ("holy warriors") of Afghanistan in the White House, in 1983

Photographer/Author : Michael Evans, of the US federal government.

Within a year of this insurgency, it was quite clear to the Afghan government that it could not suppress it on its own and it approached the Soviet Union for military help, since the Afghan army could not be expanded and trained in time. Finally, the Soviet Union responded on the 25th of December, 1979, when the Soviet forces were sent to Afghanistan. The Soviet army units set up bases mainly in Afghan cities and took over the defense of the cities, freeing up Afghan forces for combat in the countryside. The introduction of Soviet forces into Afghanistan was a big blunder of the Soviet leadership. This was exactly what Brzeziński had hoped for. The primitive, feudal society of Afghanistan could not be modernized in a few years to support a modern, let alone "socialist" state. Now, the war began to be described, and is still described, as a holy war against the Soviet occupation of Afghanistan.

People soon forgot that the Soviet Union had not invaded Afghanistan and had only responded to Afghan requests for help against the US-sponsored war against the government and people of Afghanistan. Soviet forces had been welcomed and received with warmth by the Afghan military and government. There was no government opposition to the arrival of Soviet forces. About one and a half million Afghans were killed in this war and 13,000 Soviet soldiers also died in it.

The "holy warriors" had gone berserk, poisoning school children, engaging in random murders of men and women and raping women whenever they captured a town. It had become a new kind of Jihad, where even gang rapes of Afghan girls and mass murders of children could be considered *"holy Islamic deeds"!* [14,15] President Reagan had greatly increased financial and military support for the Islamic terrorists, starting in 1980. He invited their leaders to the White House in 1983 and described them as "freedom fighters". At this time, YouTube is hosting a video of this meeting and another one which covers Brzeziński's speeches to the "holy warriors", in early 1980.

Chapter - 9

Economics of Benevolent Capitalism

As described in the previous chapter, starting in the middle of the Second World War, the US imposed the following military, economic and political changes on the global community of states. It adopted the following policies to create the "new world order" its ruling class desired, i.e., its new extended empire:

- Creation of the United Nations to manage political and military problems in the future.

- Creation of IMF to manage global inter-state trade. Creation of the World Bank and, later, the Asian Development Bank, to manage the economic containment of the Soviet Union and its allies.

- Creation of NATO, CENTO and SEATO for military containment of the Soviet Union and its allies. Further, it embarked on creation of military bases in Asia and Europe and extension of the US naval presence into all oceans of the world.

- De-colonization of all imperial possessions of other imperial powers, except the Jewish colony in Palestine.

- Panamization of Japan, South Korea and Taiwan and establishment of military pacts with these states.

- Creation of European Economic Community and Panamization of all of Western Europe, Canada and Australia.

- Take-over of all British-created Bantustans in the Middle East.

These were major changes and, as a result, the economic and political map of the world was totally transformed in about thirty years. In the following sections, we would examine the economic system of benevolent capitalism that emerged around 1985, when all the objectives of the US, as outlined above, had been achieved. The political and military changes during this period have already been described in the last chapter.

Production & Service Centers under Benevolent Capitalism

The organization of various production centers remains the same as intensive capitalism evolves into benevolent capitalism. However, the *rate of exploitation* (ROE), does go down. Also, because of the increased need for skilled labor, creation of human capital is increased.

Schools, universities and training centers are set up for this purpose. Also, hospitals and day-care centers are set up to make healthy men and women available for production and service jobs. The organization of universities and hospitals, etc., remains the same as under intensive capitalism, but their numbers and sizes increase and these institutions are made available to the people in addition to individuals of the "elite" class.

Capital Accumulation under Benevolent Capitalism

As the economic system becomes benevolent under the pressures of labor shortages, threats of labor unrest, etc., the conditions of the workers improve. The result may be that the workers are no longer working as wage-slaves, but are able to have a living above

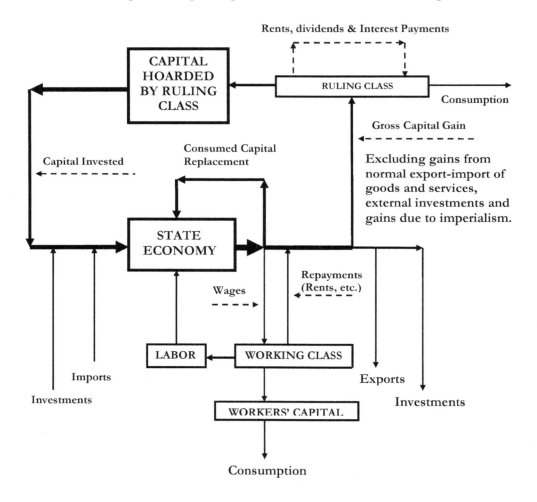

Figure 9-01: Capital Accumulation under Benevolent Capitalism

subsistence. This happened in Europe, the United States, Australia, Japan, South Korea and Taiwan under the threat to the ruling classes posed by the Soviet Union.

Under the new stage of evolution of the social system, the workers are able to save part of their wages to meet contingencies that may arise in the life of a family. This happens mainly due to the changed labor market. Wages increase because, for high technology industries, highly educated and trained professionals are required. Children of working people are able to get education, they may be allowed to have access to some level of health-care and they may be able to own their vehicles, etc. Thus, the only change from the conditions of intensive capitalism is the small amount of capital in the hands of the working people and the higher skills they have had to acquire to perform the labor that high technology demands. Thus, the *human capital* of society grows with technological development. The model of capital accumulation within a state, then, becomes as shown in Figure 9-01. Total capital accumulation does not slow down, as the social system develops further in terms of specialization and productivity of labor.

As the productivity of labor increases, as a result of mechanization and automation of production processes and the movement towards higher technology and higher wages, the rate of profit tends to fall, because of greater competition between owners of various sectors of the economy. We refer to the rate of profit, in the domestic economy of a state under benevolent capitalism, as the *rate of benevolent profit* (RBP). Another aspect of this technological change is that the owners of the production centers, in the highly automated sectors of the economy, tend to lose interest in further increases in automation, because the resulting increases in productivity cannot keep pace with increases in wages. They tend to move their investments into other sectors of the economy. If no such opportunities are available with substantial increases in capital accumulation, then the ruling class tends to move its investments to foreign states. The ruling class, thus, tends to drive its government towards imperial adventures, if possible, as under intensive capitalism. This happened to the US during the period between 1945 and 1980, along with improvement of the standard of living of American people. However, confidence returned to the ruling class, when the Soviet Union seemed to be facing economic difficulties - with the resulting decreases in its possible economic and military reach. Thus, since the beginning of 1980s, the ruling class of the US has reversed its policies - and economic disparities between the two main classes, i.e., the ruling class and the people, have been growing ever since. Incomes of most sub-classes of the working people have been declining since then. The drive towards imperial domination over states has become very difficult for most of the recent imperial powers, who find it very difficult to hold onto their surviving imperial relationships, let alone expand them independently. The only exception is the United States, which has continued to expand the semi-global empire dominated by its ruling class.

Effects of Panamization

As has been discussed before in Chapter-6, the Panamized states are subjected to much higher levels of economic exploitation by their imperial masters, as compared to independent states which are subjected to exploitation by the "normal" methods of inter-state trade and controlled market subjugation. Thus, for example, Japan and South Korea may lose a quarter, or one-third, of what they have earned by inter-state trading. Iraq may lose even more than that. It all depends on the imperial arrangements that the US has made for "sharing" of accumulated capital between its own ruling class and the ruling classes of the subjugated states, through the mechanisms that we have described in Chapter-6. However, as mentioned before, Panamization is less abusive politically and economically on a long-term basis, than Bantustanization, because of an inherent contradiction – The US has to pretend that it is setting up a democracy in the subjugated state. The resulting pseudo-democracy, thus, has to provide some space for democratic organizations to develop – thus giving rise to the possibility of development of resistance to the imposed regime.

Role of the World Bank

As discussed in the last chapter, the planners of the Bretton Woods system had created the IMF and the World Bank. Voting power in these institutions was concentrated in the US and its sub-imperial powers. Initially, both these institutions focused on Europe and East Asia. Later on, they mainly focused on development of trade relations and economic development of states bordering the, so-called, "Socialist Block", as part of the containment policies of the US. Political strings were always attached to the loans extended by the World Bank. For example, urbanization in developing states was considered a threat to "stability" in those states. So, loans were issued for dams for irrigation, so that urbanization would be slowed down. Its *"structural adjustment"* policies were described as "designed to streamline the economies of developing states", but resulted in eliminating social programs for the people. Recently, as a result of large-scale protests, some social objectives have been added to the bank's priorities but, basically, the main policies of the World Bank and the IMF are what are referred to as the *Washington Consensus - deregulation, liberalization of markets, privatization and the downscaling of government, including elimination of state-owned enterprises – thus encouraging or forcing developing countries towards greater ownership by the local ruling classes, making them vulnerable to encroachment by trans-state corporations and imperial institutions of political control.*

Role of the Asian Development Bank

The Asian Development Bank (ADB) was established in 1966 to give special treatment to Japanese preferences in development of the Asian economy. Its activities are very similar to the World Bank. The US and Japan are the major share-holders and these two states control the policies of the bank. The activities of the bank are focused on the states constituting Japan's former *"Greater East Asia Co-Prosperity Sphere"* - Korea, Taiwan, Philippines, Thailand, Indonesia, Malaysia, etc.

The Business Cycle

When intensive capitalism evolved into benevolent capitalism in Europe, the following changes were introduced in the social systems of the United States, Canada, Europe, Australia and Japan:

- Wages of all workers were allowed to increase above subsistence levels. Laws were introduced to guaranty minimum wages.

- Workers were given access to free education for their children up to high school level.

- Unemployment insurance was extended to all workers. However, the level of insurance has been different in these states.

- Systems of universal and comprehensive healthcare were introduced, except in the United States. In the United States, the system was not universal.

- Social Security systems were introduced to provide for medical care and retirement income for retirees.

- Laws were enacted to ensure environmental protection and worker safety.

The ruling classes of these states allowed these changes to happen because of two developments. A shortage of labor had developed due to the higher death rates of workers who had not been given adequate income and medical care during *intensive capitalism*. Large numbers of young men and women had also died in the two world wars, causing further shortages of working people. The emergence of the Soviet Union caused the ruling classes of these states to feel threatened by their own working populations. Thus, concessions had to be made, and were made, to coopt the communist parties clamoring for rights of the working populations in those states.

As a result of the social and economic changes brought about by the shortages of labor and demands of working classes of European and North American states and Australia, etc., the processes involved in the business cycle have also been changed. The Keynesian system of management of the business cycle has been adopted. When the economy of a state is in a deep recession under the system of benevolent capitalism and is no longer contracting substantially, conditions are somewhat different than those when this happens under intensive capitalism. Unemployment, generally, does not become very high and wages do not fall below subsistence level. Many families in the working class still own their homes, although many would have lost most of their equity due to the need for refinancing of their mortgages. Many would have lost their homes due to foreclosures. Income of most working families would have been reduced drastically due to unemployment and reductions in their savings if any. Thus, the buying power of the people has fallen to a very low level, but is cushioned by social benefits like free high-school education for children, health care and food stamps for poor families, unemployment benefits, etc. These social benefits are called "automatic fiscal stabilizers", because they increase automatically during a recession and help to reduce its intensity. The buying power of the small ruling class would stay low at this stage. Thus the "aggregate demand" for goods and services would be very low at this stage. Investment by the ruling class would have virtually stopped. Only producers in the basic consumer industries would still be able to sell their products, since essential consumer items like food, clothing, medicines, etc., would still be needed. However, the demand for these goods, and services like transport, would also have decreased. In short, at this stage, the aggregate demand would have reached a minimum, unemployment would be at a maximum and investment would have come to a virtual halt. As the recession starts, the government

225

increases the money supply gradually and reduces interest rates. At the same time, it would reduce taxes to make money available in the hands of the population. Its expenses would go up due to unemployment insurance and other social spending. It may embark on stimulus spending by making payments to its citizens to stimulate demand. These measures, ultimately, increase economic activity and the recession, or the "bust" period, begins to end.

When the growth, or 'boom" is well underway, the government raises taxes and reduces expenses to reduce aggregate demand. It, also, gradually raises interest rates, with the same objective, if inflation develops. Under benevolent capitalism, increases in wages are considered "inflation", because when an economy reaches close to its optimum performance, the bargaining power of working people increases – leading to wage increases. The employers try to pass these wage increases, or "costs", to consumers by increasing prices of their products. As the price-increases become unsustainable, because of over production, and inventories begin to build up, the workers are laid off. Thus, wage increases are considered to be the source of higher prices of products. *Thus, the main objective of the Central bank in the economy becomes the maintenance of "sufficient" unemployment.* The raising of interest rates tends to reduce investment and encourages layoffs for this purpose. Governments also tend to encourage immigration to make it easy for the central banks to maintain "sufficient" unemployment during boom periods of an economy. These measures, by governments and their central banks, reduce the rate of growth of the economy during the "boom" periods and prepare it for a "soft landing" when the unavoidable recession arrives. In 1997, as chief of the Federal Reserve, Alan Greenspan, told the Senate Budget Committee : *"As I see it, heightened job insecurity ... explains a significant part of the restraint on (wages), and the consequent muted price inflation".* However, even under the conditions of *benevolent capitalism*, this phenomenon of the business cycle does not come to an end, since greed still remains in control, although the intensity of its control over an economy is reduced, to some extent.

This system of management has also introduced new problems. One of the problems is the depreciation of money, because of the easy money policies that have to be adopted during a recession. The creation of money out of proportion to increases in production and the stimulus payments to the people to increase the aggregate demand, have the effect of reducing the buying power of savings. These processes also cause the prices of real estate, shares of companies listed on stock exchanges and "futures" contracts, on mercantile exchanges, for supply of commodities, etc. to increase greatly, due to speculation. Thus, real estate bubbles and stock market "bubbles" form to create a false impression of economic growth as compared to growth of the real economy. These 'bubbles" burst ultimately, when reality hits the speculators, causing severe economic dislocations. Specially, stock market crashes cause life savings of many people to be wiped out. Real estate bubbles cause many home owners to lose their homes.

Keynesian Scheme & Reactions of the Ruling Classes

The ruling classes found this very difficult to accept the Keynesian scheme of management of a business cycle, because it meant that governments would increase social spending when a recession arrived and would, thus, help the people and not the members of the ruling classes. Search continued for alternative schemes, which would help the ruling classes in a recession.

Monetarists stressed the need for a stable rate of growth of money stock. Their claim was that this action would alone reduce the severity of business cycles, if not put an end to them. The aim, of course, was to avoid putting money in the hands of the people and to rely on money creation through the banking system, so that it would end up in the hands of business owners. They did not wish to reduce the sufferings of the people, even in a recession. Instead, they continued to worship the capitalist class and money. Monetarist can be better described as *"Money Worshipers"*, because that is what those Shylocks were, and are. They are blind, otherwise. The fact is that it is very difficult to decide which measure of money stock should be controlled. Also, it is not really clear what is and what is not money. In addition, the central bank of a state can control the reserve ratio, and the discount rate for commercial banks, but there are other factors that influence the creation of money. Those factors cannot be easily controlled by a state and its central bank. The results of the use of the monetarist strategy for dealing with a recession can be catastrophic for the people, as was seen during the adoption of such policies by the Pinochet regime in Chile, during the seventies.

Supply-siders wanted to go back to the days of Adam Smith. When taxes are reduced, they wanted to reduce the taxes on the ruling-class with the claim that this would also increase the aggregate demand. In the United States, this claim was given wide publicity, during the Reagan administration. The Supply-siders, may be better described as *"Capitalist Siders"*, as the real desire behind their "technical" reasons was the desire to benefit the ruling class and not the people even in a recession. The tax-breaks to the rich do not increase consumption, or investment, as they claimed. There is no "trickle down" effect, as was claimed. The ruling class simply hoards the additional capital it gains because of tax breaks and does not invest it because of the recession of the time and the prevailing pessimism about profits. The additional money in the hands of the ruling class is generally used in buying back shares of their own corporations.

As time passed and the Keynesian Scheme of management was used with success by many governments, it was accepted everywhere and is now universally used by all states for management of the business cycles. However, the ruling classes have continued to develop strategies to enrich themselves and protect "their" capital even at times of severe recession in their states; and to pass the cost of management of the recession to the people.

Characteristics of Benevolent Capitalism
Trans-State Corporations

During the era of benevolent capitalism, the existing trans-state corporations began to grow much faster than before. Their merger and acquisition activities were mainly concentrated in the areas Panamized by the US, i.e., in Canada, Western Europe, Japan, South Korea and Australia. European trans-state corporations also made much headway in these activities. Further, South Africa and South America also experienced expanded activities of these US and European trans-state corporations. During the last two hundred years, or so, these corporations have grown in size and economic influence. Their operations are spread over many states. For example, General Electric, General Motors, Boeing, CITI Bank, Bank of America, American International Group (AIG), Exxon Mobil, etc., are huge, US corporations that have global operations. Similarly, Toyota and Mitsubishi are Japanese trans-state corporations. Rio Tinto, Vale and BHP Billiton are trans-state mining corporations. These corporations dwarf many states in terms of capital and have factories and mines located all over the world, making states helpless in controlling their activities. Most of these corporations started to become trans-state corporations during the era of intensive capitalism.

As mentioned before, a corporation is purely an economic social group and its only purpose is to maximize capital accumulation for its owners or investors, most of whom tend to be in its "home" state. Corporations are, basically, unaccountable private tyrannies – workers, generally, have no say in how they are run. The share-holders of a corporation select, or "elect", a board of directors for a corporation. The board generally deals with policy matters. The board of directors and the share-holders, then, select the executive mangers who actually run the corporation. The executive managers can hire and fire workers at will. Corporations gobble up other corporations and keep growing in size and influence not only on their home states but all states in which they have operations. Some corporations can exercise control over their "host" states. The economic power of trans-state corporations increased dramatically during 1801 - 1985, i.e., during the age of *intensive capitalism and the era of benevolent capitalism*. These corporations also increased their direct operations in foreign states and their foreign direct investment *(FDI)* became another major factor in their global exploitation of mankind. In order to consolidate the hold of the ruling class of the US on its government, the power of corporations was further increased by declaring them as "persons", who could voice their views by using media owned by their owners and could make contributions to campaigns of political parties and politicians![1]Thus, issues of inter-state trade and foreign direct investment and political processes became very closely related – especially in the US.

Imperialism & Colonialism

All the modern empires had to be dismantled as imperial powers faced the Soviet Union after the Second World War. All imperial powers staged a retreat from their imperial possessions. However, the desire, driven by insatiable greed for imperial possessions and colonies, remained. After the Second World War and the disappearance of the Soviet Union, the US became the dominant imperial power of the world. It continues to expand its virtually global empire, with the active participation of its Panamized satellite sub-imperial powers. The desire for the global empire is driven by the same *insatiable greed of the ruling classes* of these imperial powers, as had resulted in the empires before the Second World War. This tendency has in fact increased, since the domestic economy has developed a shortage of labor and, as a result, its rate of benevolent profit (RBP) has decreased, in the domestic economies. However, the ability of the former colonies and imperial possessions to resist conquest has increased greatly. Thus, it is virtually impossible to re-create the empires of the past, and obtain the same level of "tribute". Only, the United States is in a position to extend its empire and it is doing so in a systematic way.

Capital Flight

The members of the ruling classes may move from the relatively under-developed states to the states in Europe and North America, or they may move their accumulated financial wealth to these destinations. This capital flight is a very heavy additional burden on the under-developed states and societies, because of their inter-state economic relations.

Social Inequality

As concessions were made to the working classes in Europe, North America, Australia and Japan, the standards of living of the people began to improve. Free high school education for children, programs of health care for some segments of the population including some in the working class, the Social Security system including protections for the disabled and healthcare for the elderly, helped the poor population to feel secure and improve their education level. Thus, initially, there was some reduction in the social inequality of the population. But later on, inequality began to grow again as the ruling classes of the imperial powers made huge profits from investments abroad, especially from the newly Panamized states. After 1980, inequality began to grow much faster in all the above-mentioned states.

Population Changes

By the end of the Second World War, population growth had slowed down considerably in Europe, the US, Canada, Australia and Japan, due to losses of millions of human lives in the war and the very high death rates and low birth rates of the working population. Due to the concessions made by the ruling classes to the people, population growth did stabilize to some extent, but as the European, Japanese and South Korean economies recovered, the population growth still remained below the demand for labor, because of continuing growth of business and industry.

The improvements in education and increase in the knowledge of the people had another unexpected effect also. It increased the availability of information about the cultures of the rest of mankind, increasing the desire for tourism and education in foreign institutions. The increased availability of consumer goods and housing on mortgage basis, developed the desire to own homes and to own cars, boats, motor bikes, etc. The transformation of family relations into "money relations" as Karl Marx had described it, also became much more acute due to these developments. Families were, more and more, pulled into travel and ownership of homes and other conveniences rather than having children who were no longer a "good investment" for the future, because as they grew up, they did not care for their parents, as they used to in the previous generations. This tore up the foundation of natural love and care, on which family relations are based. The result was that more and more families decided to postpone having children. Some, even, decided to have none. They began to focus on other pleasures instead. This process considerably slowed down population growth.

By 1980, the fear of the Soviet Union, that had prompted the ruling classes to make concessions to the people, began to decrease as the Soviet Union began to face internal economic difficulties. Since that time, starting with the Reagan administration in the US, the standard of living of the people began a continuous slide downward. It did tend to slow down during the administrations of the "Democratic Party", but the downward trend has continued till now. As a result, the population growth of the native populations, in the states enumerated above, slowed down even further.

To meet the increased demand for labor in the atmosphere of low population growth, all imperial powers had changed their immigration policies. Then, starting in mid-sixties, immigrants from Asian and African states were also welcomed to meet the labor shortages and to maintain *sufficient unemployment* for management of business cycles. The immigration policies were designed to encourage educated individuals and technical professionals and experts to move to these states from the under-developed regions of the world. This *"brain drain" was, and continues to be, another big burden on the under-developed societies and states.*

Chapter-10

States of Managed Transformation

When a revolution occurred in China it basically had an advanced feudal society, in which intensive capitalism was developing. As a result, its ruling class was overthrown and replaced by its working class. It, now, has a mixed economy and a restricted economic market. It has been ruled by its communist party since 1949. It has embarked on a path of "socialist" construction and management which is unique. Some smaller states, ruled by their communist parties, have a similar status, because of their mixed economies. They include Vietnam, North Korea, Laos and Cuba, etc. Their communist parties are managing their social, economic and political transformation into modern states. Because of their unique political and economic management, these states do not fit into any other category, although they have been and are going through evolutionary socio-economic processes that are similar to the four stages of currently existing human social systems on our planet. They are going through a *managed transformation after revolutionary changes*. These transformations have, basically, speeded up their evolutionary processes.

Emergence of China

Although China had been the biggest economic power in the world for several centuries, it faced attacks by foreign imperial powers in the nineteenth century and the first half of the twentieth century. People of different parts of China were, thus, subjugated by Japan, Britain, the US and Portugal. China did, however, manage to free itself from foreign control after a long war against Japan and a civil war that ultimately resulted in a revolutionary change. The struggle for independence had cost China more than thirty million lives. A one-party state was established in 1949 under the control of the Communist Party of China (CPC) and was named the *People's Republic of China*. The US, however, after its failure to establish Panamized states in Siberia and mainland China, established a Panamized state in Taiwan, using Chiang Kai-Shek and his supporters. Thus, mainland China became a state basically somewhat on the model of the Soviet Union; however, it did not develop the kind of bureaucratic system that the Soviet Union had. Its agricultural system was based on *People's Communes*, which were also different from the state farms of the Soviet Union. Its military was organized differently; and was based on the defense of its territory by means of *guerilla warfare*, if invaded. China's society was at a lower stage of advanced feudalism, as compared to Soviet society, when the revolution occurred. However, it focused on educating its population and on developing the skills of those of working-age. As compared to the Soviet Union, the Communist Party of China maintained the support of its people throughout its struggle against foreign powers and its domestic opponents supported by the US. After several political and economic mistakes, resulting in economic stresses and hardships for its people, the CPC managed to set up a modern system of government and economic management in 1979.

China re-introduced the economic market in most sectors of its economy, starting in 1979. Also, a one-child policy was adopted for the urban population, while couples in rural areas were, somewhat later, allowed to have a second child if the first child was a girl. These rules were not applied to minorities. Economic changes were introduced carefully and the market has been closely monitored and regulated. The one-child policy continued till the end of 2015. Beginning in January, 2016, the policy has been relaxed to a two-child policy for most couples. The one-child policy for Chinese couples has prevented birth of about 500 million children and has acted as a brake on growth of China's population, which is currently about 1.4 billion and is growing at about 0.4 percent per year. The growth rate is, however, decreasing.

"These measures will of course be different in different countries. Nevertheless, in the most advanced countries, the following will be pretty generally applicable:

1. **Abolition of property in land and application of all rents of land to public purposes.**
2. **A heavy progressive or graduated income tax.**
3. **Abolition of all right of inheritance.**
4. **Confiscation of the property of all emigrants and rebels.**
5. **Centralization of credit in the hands of the State, by means of a [state] bank with State capital and an exclusive monopoly.**
6. **Centralization of the means of communication and transport in the hands of the state.**
7. **Extension of factories and instruments of production owned by the State; the bringing into cultivation of waste-lands, and the improvement of the soil generally in accordance with a common plan.**
8. **Equal liability of all to labor. Establishment of industrial armies, especially for agriculture.**
9. **Combination of agriculture with manufacturing industries; gradual abolition of the distinction between town and country, by a more equable distribution of the population over the country.**
10. **Free education for all children in public schools. Abolition of children's factory labor in its present form. Combination of education with industrial production,**
   &c., &c."

Figure 10-01: Measures for Socialism in Advanced European States

Karl Marx & Friedrich Engels. (1848). *The Communist Manifesto*. Urbana, Illinois: Project Gutenberg. Retrieved October 15, 2019, from www.gutenberg.org/ebooks/61.

It is expected that the population would peak at about 1.5 billion in 1945 and would, then, begin to decrease. The working age population is already decreasing and the elderly population has begun to increase. These trends are likely to continue for the next four or five decades. It is expected that by the year 2100 the population of China would stabilize at about one billion. By that time, it may consist of six to seven hundred million well-educated and highly-trained working age adults.

Chinese economy has been developing fast as a result of the reforms that were implemented after 1979. The emerging social system is officially referred to as *"Socialism with Chinese characteristics"*. To see what it implies, let us look at what reforms have been implemented and compare them to what was recommended for the initial stages of socialism in the Communist Manifesto, as indicated in Figure 10-01:

1. Land Ownership and Management

All land ownership had been taken over by the state before 1979. Collective Farms, known as "People's Communes", had been set up in the countryside and villages. After 1979, People's Communes were abolished and families were allowed to rent land on long-term lease basis and set up private farms on it. Later on, agrobusinesses were allowed to be set up by local governments of villages and counties (or "prefectures") involving joint government and private investment.

2. Progressive income taxes

Progressive income taxes, value-added tax (VAT) and import/export duties have been imposed. A number of free-trade zones were set up along the coast, where foreign firms could set up industries. Currently, China's tax to GDP ratio is about 21 percent.

3. Abolition of all Rights of Inheritance.

All rights of inheritance have not been abolished. However, some inheritance taxes have been proposed, but not yet implemented.

4. Confiscation of all Property of Emigrants and Rebels.

This did happen after the People's Republic was founded in 1949, as the elements of the KMT retreated to Taiwan, under US military protection.

5. Centralization of Credit in the Hands of the State, by Means of a [State] Bank.

China's banking system is basically state owned. There is a central state bank, named the People's Bank of China, on the lines of the Federal Reserve Bank of the US. Its functions are also the same, in terms of control of all other banking transactions, reserve ratio, prime interest rate and control of money supply.

Four big state-owned banks deal with different sectors of the economy and are named the "Bank of China" (BOC), the "China Construction Bank" (CCB), the "Agricultural Bank of China" (ABC) and the "Industrial and Commercial Bank of China" (ICBC). BOC specializes in foreign-exchange transactions and trade finance.

Three state-owned "policy banks" have also been set up. These are named the "Agricultural Development Bank of China" (ADBC), "China Development Bank" (CDB) and the "Export-Import Bank of China" (CHEXIM).These banks are mainly responsible for financing economic and trade development of state investment projects.

In addition to the above seven main state-owned banks, numerous, "county banks", "city banks" and "village and township banks" (VTBs) have also been set up, primarily to engage in real-estate lending and rural development in general. Some of these banks do allow minority participation of private capital. Recently, some foreign country sponsored banks were allowed to set up retail commercial operations in joint ventures authorized and controlled by the People's Bank of China (PBC). China's banking system, thus, remains, almost wholly state-owned.

6. Centralization of the Means of Communication and Transport in the Hands of the State.

Road, rail, air and sea transport systems of China are basically state owned. However, privately owned cars, buses and taxies are allowed. Bus transport systems are available in all cities and bike-sharing companies are operating in many cities. China has the biggest high-speed railway system and the biggest expressway system in the world. It is planning to complete under-ground metro systems and airports in all its major cities by 2022. It has several mostly state-owned airlines with large passenger aircraft that have been acquired from abroad. However, it has built a short-range airplane which is already in service with Chengdu airlines. Another prototype mid-size passenger airplane, the C919, is being tested extensively to obtain recognition of its global air-worthiness. China is planning to mass-produce this aircraft in the near future and has embarked on joint development of a larger aircraft, the CR929, in cooperation with Russia. It is building sea-ports and civilian transport ships and submarines. Its aircraft manufacture, railway construction, railway engine and train-bogey production enterprises are all owned by the state.

China has a diversified communications system that links all parts of the country by Internet, telephone, telegraph, radio, television and other means of digital communication. The country is served by an extensive system of automatic telephone exchanges connected by modern networks of fiber-optic cable, coaxial cable, microwave relay, and a domestic satellite system; cellular telephone service is widely available and telecom operators in China are exclusively state owned: two fixed-line operators with nationwide licenses - China Telecom and China Unicom and three mobile carriers - China Telecom , China Mobile and China Unicom. The State has control and majority ownership of all of them. However, consumer equipment, i.e., cell phones, radios and television sets, etc., are manufactured mostly by privately owned companies. Thus, China's means of communication and transport are mostly in the hands of the state.

7. Extension of Factories and Instruments of Production Owned by the State, Cultivation of Waste-lands.

Most basic and essential industries of China are in state control. There are about a hundred state-owned enterprises (SOEs) controlled by the central government, about a thousand are controlled by provinces and more than a hundred thousand are controlled by county and local governments. The SOEs include companies engaged in oil, coal, aluminum, steel and electricity production and oil, gas, water and electricity distribution. Many dams for hydroelectric power generation and water management have been completed, including the three-gorges dam – the world's biggest. Many others are under construction, or in the process of planning and investigation. Companies constructing nuclear power reactors are also in the state sector and are constructing nuclear power plants in Pakistan, Britain and Argentina. Thus, the "commanding heights" of China's economy are under state control. Mainly, consumer- oriented industries are privately owned.

8. Equal Liability of all to Labor. Establishment of Industrial Armies.

There were large-scale agricultural and industrial campaigns in the early years of the establishment of the People's Republic. Now, job-related training and kind and duration of work are decided by the workers themselves.

9. Combination of Agriculture with Manufacturing Industries; Gradual Abolition of the Distinction between Town and Country.

Policies have been formulated to encourage people to move from rural areas to cities and the urban population has grown rapidly, as a result. The rate of urbanization is about 3% per year. At this time, more than sixty percent of China's population has moved to urban areas. Agrobusinesses are growing in number and size and the mechanization and productivity of agriculture has been increasing. Thus, living standards of the rural population have been rising steadily.

10. Free Education for all Children in Public Schools. Abolition of Children's Factory Labor in its Present Form. Combination of Education with Industrial Production.

China has a state-run system of public education and its spending on education is estimated to be about 5% of its GDP. All children are required to attend school for at least nine years and factory labor is not allowed for them. Its literacy rate is about eighty percent and the school life expectancy is 14 years. In 1985, the government abolished fully tax-funded higher education, requiring university applicants to compete for scholarships based on academic ability. However, higher education still remains highly state-subsidized. In the early 1980s the government allowed the establishment of private schools as joint ventures with foreign institutions. Foreign participation can be up to 20%. Several US and European universities have set up campuses in the country. Education is geared to the technical requirements of its economy. Several Chinese universities are now in the top 100 universities in the world. Some have made it into the top 20 list. In 2016, China produced 7.5 million graduates. This number increased to eight million in 2017.

In addition to meeting the requirements of a socialist state, i.e., a state in socio-economic transition, as described in the Communist Manifesto, China has also introduced the following programs:

1. A good health care system has been developed for its people and universal health care is planned to be achieved by 2022. The quality of healthcare is, however, much below European standards, but is steadily improving.

2. Under its poverty alleviation program, more than 800 million people have been pulled out of poverty since 1979. Currently, less than one percent of the population is living below the poverty line. This number is planned to be reduced to zero percent by 2022, when China is expected to achieve the middle-income status.

3. Creation of a comprehensive social security system has been started. Workers in the cities, working in various enterprises, state owned or private, and those working in offices of the state, are already covered by this system, however those in rural areas, working as farmers, are not, although even they are covered by an old-age support system and the poverty alleviation program.

As mentioned before, a one-child policy was introduced in early 1980s to control China's population. It has, now, been revised to a two-child policy. China's birth control program has prevented birth of about half a billion children and, thus, *deaths due to starvation, extreme poverty and lack of medical care, a characteristic of intensive capitalism, have been avoided under its socialist system.*

In addition to the system of universal free education for all children, a modern comprehensive higher education system has been developed, covering all areas of modern science and technology. A comprehensive system of research and development, including technology development centers, has also been set up. Higher education is not free but it is highly subsidized.

China's defense production has also grown in quality and quantity. It is meeting virtually all its defense needs from its domestic production. Its modern conventional and nuclear defense is, however, based on the principals of sufficiency and minimum deterrence. Looking at its achievements over the last 40 years or so, it is quite clear that China has established a transitional system that is gradually moving its society toward economic equality and freedom under the constraints of the evolutionary status of its society and its economic and political relations with the rest of the world, i.e., *"socialism with Chinese characteristics"*, although it was not like one of the *"most advanced countries"* of Europe of 1850, when it started on this road in 1979.

Within the structure of its economy as adopted after 1979, China has created free trade zones and encouraged establishment of joint-venture companies, to increase exports and ensure technology transfer to local enterprises. It has introduced changes slowly after evaluating their effects under pilot projects. In addition to manufacturing and education, foreign investment is being allowed in other service industries also, e.g., in financial services. These measures have led to a great boom in production of consumer goods and development of new technologies.

China has been engaged in a huge infrastructure program throughout its territory. Its networks of freeways and high-speed railways have become the largest in the world. The number of vehicles on roads have increased dramatically. It has become the largest car market in the world and the number of cars, including electric cars, on its roads have been growing exponentially. In 2013, China announced a huge infrastructure program for East and Central Asia and Africa, to link it to Europe and Africa by land and sea. Most countries of Asia, Africa and Europe have become involved in this program and under this program development projects have been planned in more than 100 countries. The program has been named the "Belt and Road Initiative (BRI)". The "belt" in its name refers to a geographic belt of countries engaged in economic cooperation and the "road" refers to road and rail links and seaports. China has also become engaged in additional huge investment projects in many countries as part of its commitment to a "shared future for mankind".

Chinese cities have been categorized as primary, secondary and tertiary cities for urbanization and movement of population to urban areas. Limits have been set for maximum population in primary cities. People in rural areas are encouraged to move to secondary and tertiary cities instead. The level of urbanization has reached almost 60 percent. The government has started to shift China's economic growth from one based on exports to that based on consumption. Also, control of environment, including reduction of air and water pollution and CO_2 emissions, has been given a high priority in its economic development plans. It has started implementation of a huge plan for development of renewable energy resources and improvement of energy efficiency in its economy.

In the political field, China has introduced direct elections at the local government level. Governments at the "prefecture" level, i.e., county level, and also those at the provincial and central government level are elected by indirect elections. The performance of each government is periodically evaluated by the next higher level of government, taking into account the administrative performance of office holders and their performance as evaluated by their citizens in opinion surveys. The Central Committee of the communist party, i.e., the "house of representatives", elects the President and his cabinet. A system has been set up for investigation and punishment for corruption related offences, since government officials are vulnerable to bribery by the rich owners of private firms, who have no political power otherwise. Thus, an effective government of the people, by the people and for the people has been set up. In a recent address in the World Economic Forum, at Davos, President Xi Jinping stated the basic fact that China is developing "an economy of the people, by the people, for the people".

In foreign affairs, government of China has announced its policy of a "shared future for mankind" and it seeks win-win cooperation with other states in trade and economic development activities. It has also announced rules of its foreign policy – that it would respect the sovereignty and territorial integrity of other states, it would not engage in attacks on other states, it would not seek hegemony over other states and, also, would not interfere in internal matters of other states. Further, China seeks win-win cooperation with all other states for a "shared future".

These policies would result in creation of increasing numbers of independent and sovereign states and promotion of equality throughout the world. Each state would, thus, progress under the control of its own ruling class. In today's world this amounts to promotion of economic markets, or capitalism, in countries with societies at the stage of primitive feudalism, or advanced feudalism – mainly in Asia and Africa. This policy is meant to allow natural evolution of the society of each state – as it marches towards benevolent capitalism and beyond.

In recent years, China pushed for an increase in its voting rights in the IMF. Its voting rights at that time were about the same as those of Belgium! After much deliberations by major powers, an unreasonable and insufficient increase was agreed to. What was needed to be accepted is that China is the biggest trading state of the world. Even the US has fallen behind it in terms of inter-state trade. A reasonable increase in its voting rights was not agreed to by major powers. As a result, China set up the Asia Infrastructure Investment Bank (AIIB), to promote construction of infrastructure projects, and the New Development Bank (NDB) to promote infrastructure development geared to the export-related activities of the BRICS (Brazil, Russia India, China and South-Africa) countries. Both these banks have become quite active in issuing loans to their members, to ensure their sustainable development. China has also reached agreements with Russia, Iran, Pakistan and some other countries to use the Chinese Yuan for their inter-state trade. Recently, the commodity exchange of China in Dalian has started issuing "Futures" contracts for crude oil. New reforms of the financial sector of China have also been announced to further integrate the economy of China into the global economy. The main objective seems to be to dilute the role of the US Dollar as the reserve currency of most countries of the world. Ultimately, this may make the Yuan into another reserve currency for inter-state trade.

Other States of Managed Transformation

The dramatic failure of the political system of the Soviet Union, deeply affected other states ruled by their communist parties. All communist parties studied the history, development and the causes of failure of the Soviet state and made changes to their own organization and policies. The state structure and policies of their governments were also changed by the communist parties of Vietnam, Laos and Cuba. Only North Korea has not made any substantial changes, perhaps, because of the unique military threats it has faced from the US forces in South Korea and Japan. All these states, like China, have developed fast developing economies, especially Vietnam and Laos. Exports have grown dramatically and the high level of poverty in 1991 has decreased in both Vietnam and Laos and foreign trade and investment has grown. North Korea has developed a nuclear deterrent after an effort lasting more than fifty years after the Korean war. Now, its government seems to feel that the country is militarily secure. It may, now, shift its focus towards faster social and economic development.

Several European states have made substantial progress in inequality reduction and extension of social security systems to cover all their populations. Norway, Sweden, Finland and Denmark are outstanding examples of states where working people have achieved substantial control over their governments and their political systems. People of these and several other European states have moved their social systems beyond benevolent capitalism. Their achievements underscore an evolutionary path toward higher levels of socialism and equality.

Chapter - 11

Globalization

By the late 1970s, it was clear that the global class war, unleashed by the Soviet Union, had basically succeeded in its first objective of driving the ruling classes of the imperial powers out of their empires. The second Soviet objective was to support the local working classes in achieving revolutionary changes in their states. Major social and political changes did occur in a few states, like Cuba, Vietnam, Ethiopia, Angola and Mozambique, but other societies were not ready for such changes. Thus, a stage had been reached such that further radical changes had to wait for the local societies to develop. At this stage only the Jewish colony in Palestine was continuing to try to expand and consolidate itself and South Africa was trying to resist change, while the ANC was waging an increasingly successful guerilla war. The Soviet Union had played a great role in the liberation of mankind from imperial exploitation. But all was not well with the Soviet Union itself.

The causes of the immense stresses that the Soviet Union faced in around 1980 were traceable to the very creation of the state. The Soviet Union was organized as a "dictatorship of the proletariat", as a means to achieve a smooth transition to Socialism. By the *"dictatorship of the proletariat"*, Marx had meant a state of affairs in which the working people would have an overwhelming voice in the political management of the state. Those who wanted to continue the system of exploitation would not be able to have the political power to restore the previous system of *intensive capitalism*. What he visualized was a democratic system on the lines of the Paris Commune. What was actually set up to replace the Tsarist Empire was literally a dictatorship. As discussed in Chapter-6, it was, initially, not the dictatorship of an individual, but the dictatorship of the party of the reds. Marx's advice had been ignored.

Initially, the Soviet Union had made big progress in setting up new industries and educating its population and providing it with health protection. However, a flawed understanding of agricultural production and the traditional rural population engaged in it, led to what was called "collectivization" of agricultural. As mentioned before, this process was resisted by the rural population but was forcibly carried out during Stalin's ruthless dictatorship. At the same time, freedom of independent decision-making by governments of Eastern European states was severely restrained by Stalin, under his paranoid fears about what they might do. As a result, those parties lost their popularity with time. From the beginning, the stress was on heavy industry for production of armaments. This priority did not change even after seventy years. The Soviet Union was continuously engaged in a military confrontation, due to the policy of containment adopted by the United States. With time, the command economy began to stagnate, because of the unnecessary total suppression of the economic market and micro-management of state enterprises, the stress on heavy industries and the burden of maintaining a huge military profile globally.

After the Second World War, the Soviet Union had been behind the US in development of nuclear technology. It tested its atomic, or fission weapons in 1949 - about four years after the United States. It had to keep its armies at full strength during the war and even afterwards, because of the danger of a nuclear attack by the United States. It tested its first hydrogen, or fusion, bomb about a year after the United States. It was, thus, catching up with the United States in the nuclear arms race. As the US developed more advanced weapons and their delivery systems, the Soviet Union did the same. When Britain and France joined the nuclear club, the Soviet Union decided to equal the total nuclear force of *the US and its imperial allies combined.* All these military developments placed a heavy burden on the Soviet economy which was much smaller than that of even the United States alone.

The Soviet Union actively supported wars of liberation in Asia, Africa and Latin America and, at the same time, continued to engage in conventional and nuclear-armament races with the United States and its allies. Sharp differences, between the Soviet Union and China, arose in the sixties. Fundamentally, the conflict had arisen because of the desire of the Chinese communist party to make its sovereign decisions independently. Since the days of Stalin, the Soviet communist party had controlled the decision making of other communist parties globally and it had refused to allow independent actions by these parties. It had reacted violently to any such demonstrations of independence. In this spirit, when the Chinese "problem" arose, the Soviet communist party reacted as it always had. When the Chinese refused to listen to diktat, Soviet forces were deployed along the Soviet-China border and economic help was stopped to the Chinese military and China's economic projects. These actions placed a further burden on the Soviet economy, in addition to causing dislocations in the Chinese economy. Further, the Soviet Union had become such a bureaucratic state that its decision-making had become a hostage to jockeying by members of its communist party for power for its own sake, resulting in decisions that helped its members in advancing themselves rather than the interests of the Soviet people. Ultimately, all this spelled disaster for the Soviet Union.

The End of the Soviet Union

It was clear to some members of the Soviet communist party, like Mikhail Gorbachev, that something had to be done to reform the Soviet Union. Previous efforts at gradual reform during the days of Khrushchev, Kosygin and Podgorny had failed. *It was, therefore, decided to initiate reforms in the economic, military and political fields at the same time and to move quickly in implementation of these reforms, so that the process would not bog down in bureaucratic wrangling as before.*

Mikhail Gorbachev became the Soviet Premier in March, 1985. He and his supporters in the Soviet communist party, then, initiated a program of reforms they had planned. Gorbachev referred to his program as *"Perestroika and Glasnost"* in the Russian language. "Perestroika" means restructuring and "glasnost" means openness. The commitment to openness was quite obvious and became quite clear, when he started visiting communist party offices, industrial units and agriculture farms and discussed their problems with the managers and workers of these establishments in an open atmosphere, encouraging discussion and criticism with live television coverage. Among the first steps he took, was to drastically increase the prices of vodka, wine and bear. This was noticed by Soviet consumers at once, many of whom had become alcoholics because of their boring lives of non-involvement in political and social activities. Gorbachev discussed alcoholism openly and in good humor. His unassuming nature was quickly noticed by the people and they began to take part in discussions without fear. The reorganization of bureaucracy was started with mergers of unnecessary departments and removal of excess staff from economic and political organizations and government departments. Many opponents of Stalin were rehabilitated.

Being aware of the great threat posed by nuclear weapons in possession of the two super powers, Gorbachev announced, in the beginning of 1985, the suspension of the deployment of SS-20 medium range missiles in Europe as a move towards resolving intermediate-range nuclear weapon issues with the United States. He also announced a unilateral moratorium on nuclear tests a few months later. In late 1985, Gorbachev proposed that the Soviets and Americans both cut their nuclear arsenals in half and proposed a summit meeting with President Reagan of the United States. He went to France on his first trip abroad as Soviet leader. As a result, a summit meeting between Gorbachev and Ronald Reagan was held in Geneva. Though no concrete agreement was finalized, Gorbachev and Reagan struck a personal relationship and decided to hold further meetings. In the beginning of 1986, the 27th Congress of the Communist Party of the Soviet Union (CPSU) discussed the global issues of security in detail and proposed an all-embracing system of international security, under Gorbachev's leadership.

In early 1986, the moratorium on nuclear testing, announced by Gorbachev earlier, was extended till the beginning of 1987. The US, however, did not reciprocate immediately. Gorbachev made his boldest international move then, when he announced his proposal for the elimination of intermediate-range nuclear weapons in Europe and his strategy for eliminating all nuclear weapons by the year 2000. The people of the Soviet Union were, now, fully involved in discussions regarding domestic problems and the threat posed by nuclear weapons.

Gorbachev began discussing other military issues openly. His basic thesis was that the Soviet Union had overextended itself by focusing on development of weapon systems and by deploying its military forces outside its borders. He discussed the Soviet troop deployments on the Chinese border, in Eastern Europe, Afghanistan, Cuba, Ethiopia and Vietnam and deployments of the Soviet naval forces in the Atlantic and Pacific Oceans. According to him, all this was unnecessary and the *Soviet Union needed military forces on a "reasonable sufficiency" basis only*. He travelled to Eastern Europe and discussed the status of Soviet troops based there with the governments of those states. Proposals were made to China regarding reduction of tensions between the two states. China demanded that, first, the Soviet Union should fulfill three pre-conditions - removal of Soviet troops from Afghanistan and the Chinese border and the withdrawal of Vietnamese troops from Cambodia. Small scale withdrawals of troops were started in Afghanistan and Mongolia, while debates and diplomatic discussions continued. The Soviet Union, also, requested Cuba to withdraw its medical and military contingents from Ethiopia and Angola. In Angola, the Cuban troops were involved in defending the state against frequent attacks by the South African racist regime. Similar suggestions were made to the Vietnamese government regarding its troops in Cambodia.

In late 1986, Gorbachev and Reagan met in Reykjavik, Iceland, to discuss reducing intermediate-range nuclear weapons in Europe. There was an agreement in principle to remove all intermediate range missiles from Europe and a global limit was set, of 100 such missile warheads for each side. Reagan had been making propaganda speeches about global elimination of all nuclear weapons, the so called "zero option". In response, Gorbachev actually proposed the elimination of all nuclear weapons within ten years. His proposal was for elimination of half of all types of weapons on each side, within five years. He proposed to bring China, Britain and France into negotiations within that period and eliminate all nuclear weapons within the next five years. This was a big surprise for Reagan and his advisors, who had been making their propaganda speeches on the assumption that the Soviet Union would never agree to elimination of all nuclear weapons. Faced with this bold proposal against their expectations, they had to make an embarrassing retreat. Their modified position was that *they had only meant elimination of all missiles on both sides and not all nuclear weapons!* They, thus, proposed a modified disarmament plan, proposing to eliminate all missiles on both sides, if the Soviet Union agreed to continuation of Reagan's anti-missile defense program, the so-called Strategic Defense Initiative (SDI). Gorbachev had a question to ask after the new US proposal – *If all nuclear missiles on both sides were to be eliminated, why did the US need an anti-missile program?* The US had no reasonable answer to this, but refused to agree to the Soviet proposals. Gorbachev was disappointed by this refusal to his proposal. However, the United States and the Soviet Union signed the Intermediate-Range Nuclear Forces (INF) Treaty, eliminating all nuclear ground-launched ballistic and cruise missiles of intermediate range in Europe.

By1988, the press in the Soviet Union became far less controlled, and thousands of political prisoners and many dissidents were released. Freedom of speech and expression became legally protected. A Law on Cooperatives was enacted. It was a radical measure permitting private ownership of businesses in the service, manufacturing, and foreign-trade sectors. As a result, cooperative restaurants, shops, and manufacturers became part of the Soviet scene. Under the new law, the restructuring of large industrial organizations also began. Aeroflot was split up, eventually becoming several independent airlines. These newly autonomous business organizations were encouraged to seek foreign investment. Also, radical reforms were launched in the political field, meant to reduce party control of the government apparatus. A presidential system was approved and a new legislature, called the *Congress of People's Deputies* was elected in free elections. Gorbachev was elected as the first *Executive President* of the Soviet Union. These were multi-candidate elections. Non-Party members began to be appointed to government positions. The Congress, then, met to elect representatives for the *Supreme Soviet (House of Representatives)* of the Soviet Union. Its sessions were televised. Soviet citizens began to see live discussions and disagreements on critical issues affecting them. The level of public interest was such that, during the sessions, work essentially stopped in most industrial centers and government offices. Gorbachev had, already, announced the full withdrawal of Soviet forces from Afghanistan. The withdrawal was completed in 1989, although the US-sponsored terrorism continued with mass murders of the civilian population by the terrorists trained by the Pakistani army, under a US program. In Pakistan, the cooperation of the Zia dictatorship had been bought by the US for this purpose, by offers of several billions of dollars. The plan was to overthrow the Afghan government. About 13,000 Soviet soldiers had died in this fighting, while rapes, water-poisonings of school children and mass murders mainly by the "Islamic holy warriors" resulted in about *one and a half million civilian deaths*. Mostly, members of the Pashtun nation in Afghanistan were killed in this effort to overthrow the pro-Soviet government of the Khalq Party ("People's Party") of Afghanistan.

The Eastern European Upheaval

During 1988, Gorbachev clearly indicated that the Soviet Union would abandon the practice of defending the Stalin-installed communist party dominated regimes in Eastern Europe and would allow those states to freely determine their own internal affairs. Jokingly dubbed the *"Sinatra Doctrine"* by Gorbachev's Foreign Ministry spokesman Gennady Gerasimov, this policy of non-intervention in the affairs of the other Warsaw Pact states proved to be the most momentous of Gorbachev's foreign policy reforms. In his 6th of July, 1989 speech arguing for a *"common European home"* before the Council of Europe in Strasbourg, France, Gorbachev declared: "The social and political order in some countries changed in the past, and it can change in the future too, but this is entirely a matter for each people to decide. Any interference in the internal affairs, or any attempt to limit the sovereignty of another state, friend, ally, or another, would be inadmissible." This change in Soviet policy, regarding confrontation with the US and its "allies", allowed the East European states to change their systems of government. Imprisoned politicians were released. Demonstrations and the resulting political changes followed. There were political upheavals in virtually all East European states and, in time, the communist parties, more or less, lost power in all of these states. With the exception of Romania, the popular upheavals against the pro-Soviet regimes were all peaceful ones. People in East Germany were suddenly allowed to cross through the Berlin Wall into West Berlin, following a peaceful protest against the country's dictatorial administration. Unlike the previous riots which had been suppressed by force, Gorbachev, who came to be lovingly called "Gorby" in West Germany, now decided not to interfere with the process of reunification of the two German states. He stated that *German reunification was an internal German matter*. The loosening of Soviet control over Eastern Europe and the military withdrawal from other areas of the world led to a welcome break in the global class war.

Despite international détente reaching unprecedented levels, with the Soviet withdrawal from Afghanistan, completed in early 1989, and US-Soviet talks continuing between President Gorbachev and President H. W. Bush, the new US president, domestic reforms were suffering from increasing divergence between reformists, who criticized the pace of change, and conservatives, who criticized the extent of change. Gorbachev tried to find a middle ground between the two groups. Also, the old nationalist feelings were re-awakened by the openness provided by the changes in the system of government. Calls for greater independence from Moscow grew louder, especially in the Baltic republics of Lithuania, Latvia, and Estonia. These republics, which had been part of the Tsarist Empire, had been conquered by Germany after the First World War when large parts of Belorussia and Ukraine had also been occupied by it. The Soviet Union had reasserted its control over these territories before the start of the Second World War and Hitler had failed to recapture them during the war. Similarly, nationalist feeling also took hold in Georgia, Ukraine, Armenia and

243

Azerbaijan. Violence erupted between Armenia and Azerbaijan over the Nagorno-Karabakh enclave of Armenia within Azerbaijan.

In the beginning of 1989, elections to the *Congress of People's Deputies* took place throughout the Soviet Union. This returned many pro-independence politicians in the Soviet republics, as many candidates of the Communist Party of the Soviet Union (CPSU) were rejected. The televised Congress debates allowed the dissemination of pro-independence propositions. The CPSU which had already lost much of its control began to lose even more power as Gorbachev moved the political reform process forward. Local elections returned a large number of pro-independence candidates. The Congress of People's Deputies then amended the Soviet Constitution, removing Article 6, which guaranteed the monopoly of the communist party. The process of political reform was, thus, coming from above and below, and was gaining a momentum that would encourage nationalism in the Soviet republics. Soon after the constitutional amendment, Lithuania and other Baltic republics declared independence. Gorbachev was elected the President of the Soviet Union by the Congress of People's Deputies and chose a Presidential Council of fifteen party members. Gorbachev was essentially trying to create his own political support base independent of the main communist party. The new Executive was designed to be a powerful position to guide the already spiraling reform process, and the *Supreme Soviet of the Soviet Union* and *Congress of People's Deputies* had already given Gorbachev increased presidential powers. Despite the apparent increase in Gorbachev's power, he was unable to stop the process of increase in the power of nationalist leaders. Further, ambitious Boris Yeltsin was elected Chairman of the Russian Parliament, and began to grab powers from Gorbachev, engaging in conspiracies with other nationalists and the Soviet military on Russian soil.

After the 28th Congress of the CPSU, Gorbachev further reduced Party power when he issued a decree abolishing Party control of all areas of the media and broadcasting. At the same time, Gorbachev was working to consolidate his presidential position, culminating in the Supreme Soviet granting him special powers to rule by decree in order to pass a much-needed plan for transition to a market economy, which however, could not be finalized. *Gorbachev pressed on with political reform, his proposal for setting up a new Soviet government, with a Soviet of the Federation consisting of representatives from all 15 republics, was passed through the Supreme Soviet.* Meanwhile, Gorbachev continued to lose ground to the nationalists. A draft of a new *union treaty* was published, which envisioned a *Union of Sovereign Soviet Republics.* This would have created a truly voluntary federation in an increasingly democratized Soviet Union. The new treaty was strongly supported by the Central Asian republics, which needed the economic power and markets of the Soviet Union to prosper. However, Boris Yeltsin and his gang were working for absolute power in Russia and did not care if their plans led to the disintegration of the Soviet Union. Nevertheless, a *referendum* on the future of the Soviet Union was held, which returned a favorite "yes" vote for the creation of the new union. *There was an average of about 76% "yes" vote in the nine republics where it was taken, with a turn-out of about 80% of the adult population.* Six of the smaller Soviet republics did not participate.

The Conservative Coup

In contrast to the reformers' moderate approach to the new treaty, the conservatives in the Communist Party were completely opposed to anything which might lead to the break-up of the Soviet Union. On the eve of the treaty's signing, they launched a coup in August, 1991 in an attempt to remove Gorbachev from power and prevent the signing of the new union treaty. They called themselves *"The State Emergency Committee"* Gorbachev was placed under house arrest in a "dacha" in the Crimea. However, the coup failed when the armed forces refused to obey the orders of the coup leaders. Gorbachev was, then, freed and restored to power after remaining under house arrest for only a few days. However, Yeltsin took advantage of his absence and stood up to the committee till the collapse of the coup and grabbed control of most state institutions, in collusion with the Russian military commanders. On his return, Gorbachev advised the Central Committee to dissolve itself, resigned as General Secretary of the Communist Party and disbanded all party units within the government. Shortly afterwards, the Supreme Soviet suspended all Party activities on Soviet territory. Congress of People's Deputies, also, dissolved itself in September. Though Gorbachev and the representatives of eight republics (excluding the other small seven republics) signed an agreement on forming a new economic community, events were overtaking Gorbachev, as one republic after another declared independence.

Yeltsin's Reactionary Coup

Yeltsin suspended all CPSU activities on Russian territory and began taking over what remained of the Soviet government. He and the presidents of Ukraine and Belarus met in the historical town of Brest in Belarus, founding the Commonwealth of Independent States and declaring the end of the Soviet Union. His coup against Gorbachev was now virtually complete. Gorbachev had to resign *on 25 December, 1991,* when even the communication links of his office with the rest of the country were cut off by Yeltsin's gang, and the Soviet Union was formally dissolved the following day when Yeltsin occupied Gorbachev's office in the Kremlin. Gorbachev had intended to maintain the Soviet communist party as a united party but reform it into a Social Democratic Party. He also wanted to create a new union of the successor states of the Soviet Union, but faced with Yeltsin's coup, he had to give up. Yeltsin's coup was a disaster for the successor states of the Soviet Union, especially Russia. As Gorbachev put it - *"Behind our backs there was treachery. Behind my back, they were burning down the whole house just to light a cigarette. Just to get power. They couldn't get it through democratic means. So, they committed a crime. It was a coup."*

The Soviet Union survived for about three quarters of a century. In its brief life, it brought about dramatic changes on the world scene. Colonial and imperial powers were forced to end their direct rule over Asian and African states. Their control over Central and South American states was also reduced considerably. China, North Korea, Vietnam, Laos and Cuba went through revolutionary changes. Also, Soviet industry became highly developed, especially the industry geared towards development of military equipment. Population's education and healthcare was greatly improved. However, the Soviet people paid a heavy price for their achievements. Millions lost their lives in the German invasion of their territory during the Second World War and the intervention of imperial powers during the civil war.

The forced collectivization of agriculture, the unnecessary stress on heavy industry, development of an unnecessarily huge pile of nuclear weapons and the *Great Terror*, also, led to a heavy loss of their blood and sweat. Human beings all over the world owe an immense gratitude to them for their sufferings and sacrifices in the service of humanity.

Effects of Soviet Dissolution

The Chinese leadership had realized the defects in the original political system of the Soviet Union, much before the Soviet leadership. The Chinese communist party had thus introduced major changes in the system it had adopted after 1949. The new approach is described as *"Socialism with Chinese characteristics"*. In agriculture, the system of peoples' communes, or collective farms, was abolished after 1979. However, state ownership of all land was maintained. A system was introduced instead, which allowed families and companies to lease land from the state to set up farms, build houses, or to set up factories and offices. Also, individuals were allowed to set up businesses and to hire workers. Initially, only consumer goods industries were allowed to be taken over by individuals and companies, both domestic and foreign. Later on, more sectors of the economy were opened to foreign investment. Only, major industries of critical importance to the Chinese economy were kept in the state sector. Thus, all sectors of the Chinese economy, except its *"commanding heights "*, were gradually merged with the global economy. Thus, an economic market re-emerged in the Chinese economy. The process of reforms was, however, gradual and systematic - in coordination with the changes in China's external political and trade relations. A population control program was also introduced. It was based on the controversial one-child per family policy. Such a population control program had never been adopted by any state in the past and it was met with great skepticism in the beginning. Ultimately it came to be accepted by the Chinese people and other states as a reasonable solution to China's population problem. China, also, created several free-trade zones to attract foreign investment. These reforms were underway in China when the changes began to occur in the Soviet Union. Thus, as a result, China did not have to make major changes in its policies. However, its relations with Russia and the other successor states of the Soviet Union had to be reorganized.

The communist parties in most "socialist" states had to reconsider their ideas about state organization, in light of what happened in the Soviet Union. Thus, the systems of government in Vietnam, Laos and Cuba went through a period of major changes. In South Africa, ultimately, the racist regime had to give in and negotiations were started with the ANC for dismantlement of apartheid. Nelson Mandela became the President of South Africa after general elections were held and the process of dismantlement of apartheid was taken in hand. Britain finally started negotiating with the Irish to end, or decrease, its control over them. An agreement was signed with the political leadership of the Irish Republican Army and implemented gradually. Britain, however, continues to seek to maintain the status quo, to the extent possible.

In Sudan, an organization representing the tribes of the South finally reached an agreement with the Central Government. According to the agreement a referendum was held in January, 2011. The population of the South was to be allowed to secede if the referendum indicated that it is the will of the people to secede. It was hoped that, then, the mass killings would stop. The referendum was held and the population of the South overwhelmingly voted to secede and the new state of South Sudan came into being. The process is not complete though, as the will of the people in a small region between the regions of South and North Sudan still remains undetermined, leading to more violence between the two states. A few years ago, violence had erupted in the Western part of Sudan also. This region is known as the Darfur region. The reason for this new source of violence was that oil had been discovered there and the tribes of the region were demanding some benefit from this discovery. The central government of Sudan, however, responded, as usual, with violence. The United Nations and the African Union tried to restore peace in the region by sending troops of the African Union. The level of violence did decrease. However, tribal rivalries within the South Sudanese leadership led to renewed violence.

A new conflict developed between the US and Iran, when it was discovered that Iran had set up Uranium enrichment facilities. Iran, a signatory of the US-Sponsored Non-proliferation Treaty during Shah's time, is permitted by the Treaty to engage in this effort, however, the US sees this as an attempt to develop nuclear weapons, which goes against the US desire to have only Israel as a nuclear power in the Middle East. Iran has vast reserves of natural gas and large reserves of oil, so that it does not, at this time, really need this Uranium enrichment program for electric power generation, as it claims. An agreement was reached between Iran and the five permanent members of the Security Council of the UN, plus Germany and the European Union, in 2015. Under this agreement, generally referred to as "the Iran nuclear deal", Iran agreed to curtail its nuclear development and uranium enrichment programs for about 15 years, in return for removal of sanctions. It seems the US desired this freeze of the Iranian nuclear program to use the period of 15 years to, first, bring Libya, Iraq, Syria, Egypt and Yemen firmly under its imperial control, before it moved forward with its project of Panamization of Iran. This change in strategy was necessitated by the appearance of the "Islamic State" in Iraq, Syria, Libya and Afghanistan and the "Arab Spring", leading to loss of control by the US and the resulting chaos in those states. *However, its possible development of nuclear weapons could be used by the US, in the future, to invade and conquer Iran, in order to Panamize it. That, seemed to be the ultimate US objective at that time.*

As the systems of government in the states of Eastern Europe were changed and new political parties came to power in those states, nationalism as a political force reappeared in this region. In Yugoslavia a civil war started as some nations decided to secede. Since, the boundaries of the national territories in Yugoslavia were not clearly demarcated, violence naturally emerged as the method of resolution of disputes. After a lot of fighting and bloodshed, Yugoslavia split up, effectively, into seven states - Slovenia, Croatia, Bosnia, Serbia, Monte Negro, Macedonia and Kosovo. The secession of the Slovak nation from Czechoslovakia was not violent, however, and Czechoslovakia split up into the Czech Republic (referred to as "Czechia" now) and Slovakia. Earlier, Germany had reunited into one state. Similarly, in Arabia, South and North Yemen united into one state - now simply called "Yemen".

As soon as Yeltsin had seized the Presidency of Russia, with no limits on his power, and had gathered a coterie of his family members and his gang of "reformers" and American advisors, this once-primary-school-drop-out went berserk "reforming" the Russian economy and defense forces. His understanding was that the economy and "everything" else in the Russian Federation would be fixed by "July". He made this emphatic claim repeatedly. By the time July arrived, he was dead-drunk with his cronies and had completely forgotten July of what year, or what century, he had promised to be the dead-line for fixing "everything" in Russia! So, he naturally promised to fix everything by the next July. The members of his cabinet were always seated in a fixed order, around a table in the Kremlin. Since, they and Yeltsin were constantly drinking, Russians referred to them as "bottles". Thus, a cabinet meeting was a gathering of "bottles". Everyone in Russia knew what had been said by the third bottle on Yeltsin's right and what the first bottle on his left was doing at that time! So, Yeltsin continued promising to fix everything by "July", in all his meetings with his "bottles" and in his public pronouncements. Meanwhile, members of his gang were busy doing what they liked most – grabbing state property and, by means of fraudulent "transactions", converting it into their "private" property.

To restructure the Soviet administrative command system and implement a transition to a market-based economy, Yeltsin initiated a *"shock therapy* program" on the advice of his American advisors, shortly after seizing power. The subsidies to money-losing farms and industries were cut blindly, price controls abolished, and the ruble given a powerful kick towards convertibility, causing the Ruble to nose-dive like a piece of paper. New opportunities for Yeltsin's family and other gang members to seize former state property were created, thus "restructuring" the old state-owned economy within a few years. After grabbing power, the vast majority of "idealistic reformers" gained huge possessions of state property using their positions in the government and became "self-made" business tycoons, with the active help of what the US media always described as the "great democrat" in Moscow. Existing institutions were conspicuously abandoned prior to the establishment of new legal structures of the market economy such as those governing private property, overseeing financial markets, and enforcing taxation. Yeltsin's advisors had told him that the dismantling of the administrative command system in Russia would raise Russian GDP and living standards by allocating resources more efficiently. They also thought the collapse of the existing system would create new production possibilities by eliminating central planning, substituting a decentralized market system, eliminating huge macroeconomic and structural distortions through liberalization, and providing incentives through "privatization", i.e., transfer by hook and crook of state property to his gang of looters and criminals. *"Looterization"* did occur very fast under these conditions and a new class of super-looteras emerged, referred to, now-a-days, as Russian "oligarchs". According to Gorbachev - "Bureaucrats stole the nation's riches and began to create corporations".

As the class of super-oligarchs, came into being, the people of Russia were deprived of what was called "their share" of the relevant enterprises, by deception schemes hatched by Yeltsin's gang. The rising inflation caused even the pensions of ordinary retirees to evaporate, causing hunger and deprivation. Most retirees were not left with enough to even buy their medicines. The health-care system, one of the best in the world, collapsed completely and diseases began to spread with alarming speed. There was wide-spread dislocation and unemployment, as enterprise after enterprise was sold off at a fraction of its value to the domestic and global capitalists. Production declined dramatically and the Russian economy went into a *more severe depression* than the *Great Depression*. More than a *quarter* of the Russian population was pushed under the poverty line, life expectancy fell dramatically as the death-rate increased and the birth-rate declined. *The GDP of the Russian Federation fell by more than 40%.* Yeltsin, however, continued to travel to his favorite destinations in the US and Europe, having great "intellectual discussions" about "freedom and democracy" and the market economy, while half-drunk all the time. On one of his foreign tours, when he arrived in the Irish capital of Dublin, he was so drunk that even his care-takers could not bring him out to attend the red-carpet reception ceremony elaborately prepared by the Irish government! On his tours, especially to the United States, Yeltsin made many "large hearted" giveaways and voluntary concessions regarding military matters. His nuclear concessions to the United States were especially dramatic, prompting Gorbachev to refer to him, later, as a *traitor*! As popular discontent and protests spread throughout Russia, and *"July"* did not arrive, Yeltsin had to resign and hand over all power to Putin. It is not clear whose advice he followed in this respect because it did turn out to be one constructive decision for Russia's future. It is said that his hope was that Putin would protect him from any legal action for his *"achievements"* - destruction of the economy of Russia*; Intensive Death Rate* of its people by starvation and denial of health care (maybe 4-7 million souls), especially, the sick and the retired; his creation of a group of mega-looteras, or "oligarchs", including his family members; mass murder of the people of Chechnya and destruction of most of its population centers including the capital, Grozny; damage to the security of Russia and its nuclear forces, by treasonous secret verbal agreements with the rulers of foreign powers and creation of a corrupt and dictatorial regime, in the name of "democracy", by letting loose tank-fire on the Russian parliament, the very parliament that had previously elected him President of the Russian Republic of the Soviet Union!

Following a failed run for the presidency in 1996, Gorbachev established the Social Democratic Party of Russia, which was banned because of laws adopted by the ruling party which do not allow parties which initially do not have a specified number of members in most regions of Russia. The purpose of this law is to prevent creation of new political parties. Gorbachev, however, did succeed in founding a new political party, called the Independent Democratic Party of Russia. Putin did bring some economic and political stability to Russia. Slowly, the economy of Russia began to improve, due to local and foreign investment and business development and the high prices of natural resources in the global market. Due to the collapse of the health-care system and the virtual destruction of the pension plans for the elderly, the older population began to die at a very high rate – as part of *the Intensive Death Rate characteristic of intensive capitalism* (Gorbachev called it *"wildcat capitalism"*). The wide-spread unemployment also caused the women of child-bearing age to postpone having children, given the condition in which even their own survival was questionable. The dramatic increase in the death rate and the drop in the birth rate caused the population to decline, despite large-scale immigration of Russians from the other successor states of the Soviet Union. The total population of the Russian Federation had been going down by about a million persons every year since 1992. The population decline slowed down after the passage of about twenty years. The total population then stood at about 142 million.

Despite all the disasters faced by Russia, Putin kept his apparent promise not to try Yeltsin and his family for the super-corruption they had engaged in. He, also, tolerated some of the criminal and corrupt "self-made" billionaires, who had grabbed enormous amounts of Russian wealth by devious means in collaboration with Yeltsin, his family and the rest of his gang. After serving two terms as the President of the Russian Federation, Putin decided to assign the presidency to his Prime Minister, Medvedev, and himself became the Prime Minister. After one term, he stood for the position of President again and won. The conduct of those elections was criticized by many. He started his third term as President in 2012, since the Russian constitution had been amended for this purpose. The constitution of Russia was promulgated in December, 1993. According to Gorbachev, it gives autocratic power to the presidency. This was certainly Yeltsin's dream! Putin had only inherited it. Referring to the victory of Putin's political party, United Russia, Gorbachev called on the authorities to hold a new election, citing electoral irregularities, noting that in Russia *"politics is increasingly turning into imitation democracy with all power in the hands of the executive branch...to go further on the path of tightening the screws, having laws that limit the rights and freedoms of people, attacking the news media and organizations of civil society, is a destructive path with no future".*

In 2018, Putin was re-elected to his fourth term as President, Gorbachev characterized this "tandem" Putin-Medvedev rule as backsliding on democracy, encouraging more corruption and creating a dominance of security officers. According to him - *"The electoral system we had was nothing remarkable but they have literally castrated it."*

Under Putin's leadership, Russia has recovered economically and in terms of military strength. Also, by means of harsh measures, terrorism emanating from Chechnya was brought under control. However, the future of its political system remains a source of concern.

Consolidation of the New World Order

As the Soviet Union was dissolved and it fragmented into fifteen separate states, the United States saw a great opportunity in this development, which was totally unexpected for anyone, except Gorbachev and the few members of the Soviet Communist Party around him, who knew about Gorbachev's plans. When Gorbachev initiated his program called "Perestroika", he knew the risks involved, but, from his point of view, those were acceptable risks. His calls for "restructuring" were taken to mean restructuring of the Soviet Union and its economy, or its military posture. No one realized that Gorbachev intended to globally restructure all relations between the Soviet Union and all other states of the world, that he wanted to restructure the whole world's economic, military and political relationships, but this was what he intended and he had immediately embarked on this project after assuming the leadership of the Soviet Union.

When the Soviet Union fragmented into fifteen independent states and Russia withdrew all its nuclear and conventional forces from European states, from the Chinese borders and from all other foreign states, a dramatic shift occurred in the global economic, military and political balance. Sensing an opportunity to fulfill its original dream of global domination after the Second World War, the US initiated a series of reviews of its military and economic policies. The US dream of global domination had been frustrated by the emergence of the Soviet Union as a global power, and extension of its influence into Eastern Europe and, later on, into China. Now, it seemed possible that the original objective of the US may be achieved, but China was likely to become another obstacle in this regard.

Expansion of the Roles of NATO, UNO and IMF

One result of the review of US policies was that an expansion of NATO was planned and newly independent European states were encouraged to join the European Union and NATO, despite a promise made by the US to Gorbachev, not to move NATO forces "one inch" forward after reunification of Germany. This process resulted in a dramatic expansion of NATO and the European Union, with most East European and other European states becoming members of one or both these organizations. A new concept was developed for the expansion of Panamization of Europe. According to this concept, since the end of the Cold War, the world had come to face "complex new risks to Euro-Atlantic peace and security, including oppression, ethnic conflict, economic distress, the collapse of political order, and the proliferation of weapons of mass destruction." Thus, this "alliance", dominated and controlled by the US, committed itself to dealing with "all stages of a crisis – before, during and after" - enabling the US to Panamize, or dominate otherwise, any area of the world that posed a "problem" for it.

The US also became more active in the United Nations Organization. The UN supervised the 1993 elections in Cambodia and the 1999 referendum in East Timor and peacekeeping operations in Angola, Bosnia, Congo (Kinshasa), Eritrea and Ethiopia, Haiti, Kosovo, Mozambique, Sierra Leone, Somalia, Sudan and Chad. In addition to these peacekeeping operations, the US started to use this organization for furtherance of its own agenda. When it sought to Panamize Iraq, or Afghanistan, it used the UN to create a group of states, including those in NATO, to support its invasions, or economic blockades, thus claiming that the actions it was undertaking were supported by the "international community". Whenever, the members of the UN disagreed with its plans, it went on alone, or with the Panamized states covered by NATO, to invade and occupy areas of the world it coveted, for their mineral and oil and gas resources. This, especially, happened at the time of its invasion of Iraq and continues with regard to some of its actions against North Korea, Libya and Iran. The UN has also been active in setting standards of human dignity and freedom, such as in the Universal Declaration of Human Rights and the establishment of international labor standards, and has been a forum for discussion on some environmental issues. The US continues to use this forum for putting pressure on those states which were formally referred to as "Communist States" and which, in their efforts to provide security to and improve the lives of their citizens, end up putting limits on their freedoms of expression and association while guaranteeing their economic and social rights. These economic and social rights are naturally ignored by the US, although they are included in the *Universal Declaration of Human Rights*, since its system of economic exploitation, *benevolent capitalism*, cannot guarantee those rights.

The monetary system that emerged after the *"Nixon Shock"* is sometimes referred to as Bretton Woods II. The reason is that the economic relationships, that were established by the fixed exchange rates, imposed by the US through the IMF in the post-1945 period, were very difficult to change in favor of the "peripheral" countries. Thus, the international system that was composed of a "core" imperial power, i.e., the US, issuing the dominant international currency, and a "periphery" consisting of all other states which accepted, or were forced to except, fixed exchange rates, continued even after the *"Nixon Shock"*. The periphery was locked into export-led growth based on the maintenance of *undervalued exchange rates*. In the 1960s, the "core" was the United States and the periphery was mainly Europe and Japan. This old periphery then changed into Europe and the Asia-Pacific region. The Asia-Pacific region had become the center of economic activity at this time. The "core", of course, remained the same - the United States. Thus, the system of global trade remained "stable", under US dominance, although the exchange rates were no longer fixed. A "realignment" of currency exchange rates was also proposed, whereby deficit nations may devalue their currencies and surplus nations may revalue theirs upward. After the dismantlement of the Soviet Union and the revival of the economic market in China, the IMF progressively relaxed its stance on "free market" principles such as its guidance against using capital controls. Capital, thus, began to move freely to China, Russia, Brazil, South Africa and the Asia-Pacific region, causing increases in employment and wages there, while causing unemployment in North America, Europe and Japan. Many companies in Europe and North America began to engage in "outsourcing", trying to maximize their capital accumulation by using cheap labor in advanced feudal economies of Asia.

Global Trade and Finance

The World Trade Organization (WTO) was formally set up in 1995 after trade negotiations in Morocco, under its parent organization named General Agreement on Tariffs and Trade (GATT). Interstate tariffs on trade had already been lowered substantially during negotiations under GATT, between 1947 and 1994, when the WTO agreement was finalized. WTO deals with regulation of trade in goods, services and intellectual property between participating countries. It provides a forum for negotiating trade agreements and provides a dispute resolution mechanism for adherence to those agreements by participants. It oversees the implementation, administration and operation of agreements arrived at by its members. China joined the WTO in 2001, after a long period of negotiations. The fear of nationalization had disappeared with the Soviet Union. WTO allowed free capital flows between states, thus making it possible for the corporations in the US, Canada, Europe and Japan to freely invest their accumulated capital into Russia, Brazil, India, China and other states of Southeast Asia.

The US negotiated regional free trade agreements with its neighbors, Europe and some states of East Asia. In North America, the agreement between the US, Canada and Mexico is known as "North America Free Trade Agreement (NAFTA)". The Central America Free Trade Area (CAFTA) was an agreement between the US and six of its central American neighbors. It was later renamed CAFTA-DR, when the Dominican Republic joined the other six central American countries. The European Union (EU) and Association of South East Asian Nations (ASEAN) were such pre-existing agreements. These regional trade agreements further lowered import duties and other trade barriers, thus facilitating free movement of products in those regions and led to a great increase in inter-state trade. Trans-state corporations were given powers to lodge legal actions against states, thus greatly increasing their dominance of inter-state trade. However, the immediate effect of these regional trade agreements was to increase unemployment in the poorer states, because of the sudden increase in agricultural imports from North America and Europe, leading to lowering of local wage levels. Also, these agreements promoted the movement of industries from the US and Europe to the other regions. Increases in inter-state trade and foreign direct investment (FDI) have intertwined the economies of all states of the world.

The function of the IMF has not really changed in this regard, however, and it continues to manage interstate trade and negotiate currency exchange rates as before, under the new conditions - to impose devaluations, removal of subsidies, reductions in social support programs, etc. Capital of advanced countries, i.e., states at the stage of benevolent capitalism, has been invested in the emerging markets of Central Eurasia, South America, South Asia and South East Asia, instead of the US and other advanced economies. Trans-state corporations have transferred large parts of their production processes to affiliates in other states and intra-firm trade has become a large part of world trade. Exposure of trans-state corporations to native companies has resulted in inter-state investment and finance. In response to these needs commercial banks have also become globalized, issuing loans in different currencies. Investment companies and banks have also done the same[1] – courting foreign investments and investing in foreign companies, commercial banks and stock markets. A shadow banking system has also developed, avoiding most state regulations[2]. China has, however, received the largest amount of foreign direct investment. The result is that China and other recipients of FDI have been growing fast, while the economies of the US and Europe have been stagnating, or showing anemic growth.

Panamization of Iraq

After the Iranian revolution in 1979, the United States had encouraged the Iraqi dictator, Saddam Husain, in his invasion of Iran, which caused a long war over several years, causing the deaths of about one million Iraqis and Iranians. The US had supplied Saddam Husain with military equipment on a large scale, for this invasion. This had done immense damage to both Iraq and Iran – *which was the US objective, as it was in the fighting between Hitler's Germany and Stalin's Soviet Union, when either side was helped when it seemed that this action would maximize deaths and destruction on both sides.* In 1991, Saddam, angered by the rulers of Kuwait, by their behavior in OPEC, attacked and occupied Kuwait. The United States encouraged this invasion also, by diplomatic means. When Kuwait had been occupied, the US declared its opposition to this action, and attacked Iraq, with the intention of Panamizing it ultimately, since Iraq had much larger reserves of oil and gas. As a result of this attack, the Iraqi military forces were mostly destroyed, but, due to the danger of Soviet "interference" and the realization that Iraq was not yet "ripe" enough for Panamization, it was not fully occupied. It was subjected to sanctions and a "no fly zone" was imposed on it. The sanctions were designed to turn the Iraqi people against their rulers. It is estimated that besides hundreds of thousands of soldiers who had died in military action, more than a million civilians, mostly children, were killed due to the sanctions imposed by the US, resulting in denial of food and medicines to its people. The Soviet Union was in the last days of its existence at that time. After the Soviet Union disappeared, the project of Panamization of Iraq was taken up seriously by the US President George Bush and the US invaded Iraq again in 2003. This time, the excuse given for the invasion was that Iraq was developing "weapons of mass destruction (WMDs)", i.e., nuclear and chemical weapons – as if only the US and its allies had the right to do so! No such weapons were found, but Iraq was completely occupied, its army and its government and the ruling political party were disbanded. The US, then, started to set up new Iraqi army and police for its Panamization purposes. All opponents of the Saddam regime were collected together, as the new ruling class, and given government responsibilities, under US hegemony. All Iraqi oil companies and their assets were auctioned off to trans-state corporations. The ruling class of Iraq was manipulated into developing vested interests in these corporations and other US corporations to "harmonize" US and Iraqi interests, as usual. When the new military forces were trained and ready, the US forces withdrew, leaving a relatively small high-tech contingent in Kuwait, *just in case the new Iraq army got the wrong ideas.* Iraq was, then, ruled from the "US Embassy" (spread over a huge area, in the Green Zone of Baghdad.) The oil production of Iraq was, naturally, quadrupled. Iraq, thus, stood fully Panamized. So, it seemed. However, more conflict was to follow.

Consolidation of the Jewish Colony

Several Palestinian organizations had been resisting the colonization of Palestine and the systematic uprooting and expulsion of its inhabitants. The Soviet Union and the People's Republic of China had naturally sided with the Palestinians, although the Chinese support had not been very aggressive. As China changed its focus to peaceful economic development, keeping other disputes aside, and as the Soviet Union disappeared, the Palestinians found themselves without much support from other Arab states and the global community. The Palestine Liberation Organization (PLO) realized that its position had become much weaker in confrontation with the US-sponsored Jewish colony, which continued to be economically and militarily backed by the United States. It, thus, sought to obtain, from the Jewish rulers and their patrons in the US, whatever relief they were willing to give to the people of Palestine. On September 13, 1993, Israel and the Palestine Liberation Organization (PLO) signed the Oslo Accords (a Declaration of Principles). The principles established objectives relating to a transfer of authority from Israel to an interim Palestinian Authority, as a prelude to a final treaty establishing a Palestinian state, in exchange for mutual recognition. May 1999 was set up as the date by which a permanent status agreement for the West Bank and Gaza Strip would take effect. Next year, the two parties made some progress in terms of transfer of power to the PLO. The Jordanian Bantustan quickly took advantage of the opportunity to sign a Peace treaty with Israel, thus securing its continued existence and US patronage.

Prime Minister Yitzhak Rabin and PLO Chairman Yasser Arafat signed the Israeli–Palestinian Interim Agreement on the West Bank and the Gaza Strip on September 28, 1995, in Washington. The agreement was witnessed by President Clinton on behalf of the United States and by Russia, Egypt, Norway and the European Union. The agreement allowed the PLO leadership to relocate to the occupied territories and apparently granted autonomy to the Palestinians with talks to follow regarding final status. The agreement was opposed by Hamas and other Palestinian factions. Rabin had a barrier constructed around Gaza to prevent attacks. In the fall of 2000, talks were held at Camp David to reach a final agreement on the Israel/Palestine conflict. Ehud Barak offered to meet some of the Palestinian requirements in terms of territory and control of water resources, including some of the Arab parts of east Jerusalem; however, the effort to reach a settlement had to be abandoned, in the face of Israeli insistence on maintaining control of key economic sources and militarily significant aspects of the territory on the West Bank of the river Jordan, and the desire to retain territories in addition to what was allowed by the UN partition plan for Palestine. The Israelis were also not willing to allow the Palestinians to exercise their right of return to their home areas. If accepted, the Israeli proposals would have created two territories in which the Palestinian side was controlled by Israel and was not really sovereign - leading to "apartheid", South African style, not independence. The Palestinian territories would then have become providers of labor to the Jewish state, like the former Bantustans in South Africa, without human rights of a free population.

In September, 2000, the West Bank exploded in rage. This is referred to as the "Intifada". In 2005, all Jewish settlers were evacuated from Gaza. Disengagement from the Gaza Strip and the Northern part of the West Bank was completed in September, 2005. However, Hamas won the Palestinian legislative elections in 2006 and rejected all agreements signed with Israel. In June 2007, Hamas took control of the Gaza Strip, seizing government institutions and replacing Fatah and other government officials with its own. Thus, two authorities came into being on Palestinian territories. Throughout this process, the US maintained the position that the two parties should negotiate an agreement "without any pressure" from any other party, while the US was fully backing the demands of the Jewish colony. It seems the US wants to prolong the process of negotiations, while it completes the Panamization of the Middle East. Once it has achieved that objective, it would unleash the Jewish onslaught on the Palestinians and Arabs of other adjoining states, to achieve a "permanent" solution.

Panamization of Afghanistan

When the Soviet troops were withdrawn from Afghanistan in 1988 - 1989, the Afghan government was left alone to deal with the Islamic terrorists. It did continue, however, receiving some economic and military help from the Soviet Union. The Afghan government tried to start negotiations with the "holy" terrorists, but failed. The terrorist gangs continued their activities from across the border in Pakistan. The leadership of the "holy war" had been taken over by the extremist General Zia ul-Haq, who had already "achieved" the judicial murder of the Prime Minister of Pakistan, whose government he had overthrown, by setting up Kangaroo courts and manipulating them. The Afghan government collapsed in 1992, after Yeltsin discontinued all help to it. The US, then, quickly withdrew its financial and military support for the terrorist war, since its objective had been achieved. The Islamic terrorists (called "Mujahidin", or Jihadists) had been organized into gangs headed by warlords of various tribes of different areas of Afghanistan. Each warlord, now, wanted complete control over all of Afghanistan, just like bank robbers behave after a robbery. The terrorist gangs started fighting among themselves. At this time, the cities of Afghanistan were in quite good shape, as no lasting damage had been allowed by the Soviet forces. The savage gang wars between the various terrorist groups resulted in complete destruction and desolation of the Afghan cities. The population of those cities fled, mostly to Pakistan and Iran. The fighting between the holy warriors continued and Afghanistan quickly sank into a state of total anarchy, lawlessness and economic collapse. General Zia, the "chief holy warrior", was also killed when his plane crashed near Bahawalpur, Pakistan.

Zia had completely destroyed the educational system of Pakistan, among other institutions of the state. Government schools had been turned into marriage halls, grain storage centers and animal farms. Instead, Zia had set up a system of religious schools, known as Madrassas, as part of his plan of "Islamization" of Pakistan. The madrassas proved to be truly MADrassas, as they began to turn out millions of young men brain-washed with religious mythology and totally devoid of any scientific knowledge - incapable of any kind of constructive work or employment and good only for terrorism in the name of Islam. Many Afghan refugees were also "studying" in these madrassas. One group of such students set up an organization called the "Taliban", meaning "students", in Qandahar, Afghanistan. Mullah Omer, a holy warrior during the Jihad in Afghanistan, became their leader. The Pakistani ISI, sensing an opportunity to control Afghanistan, supported this organization. Even the US supported them and almost recognized them as the government of Afghanistan. The Taliban prevailed in their fighting against the other groups of Islamic terrorists in Afghanistan, with the help of the ISI, and captured most of the territory of Afghanistan, except the North-Western Tajik area, which continued to be ruled by the "Northern Alliance" of Tajiks and Uzbeks.

Under the conditions of complete lawlessness in Afghanistan, Osama Bin Laden, another "holy warrior", and his group set up an organization called *"Al-Qaida"* (meaning *"the Base"*) in Afghanistan and the tribal areas of Pakistan, bordering Afghanistan. While in Sudan, he had approached leaders of ignorant extremists all over the Arab world, had recruited them for a new holy project – creation of Al-Qaida. He moved to Afghanistan and set up Al-Qaida training camps. Recruits from all over the Islamic world began to arrive in these camps, to obtain training in conducting terrorist operations. Al-Qaida refers to these activities as "Jihad", or "holy war", against what it calls the "Crusaders". Several attacks were launched by Al-Qaida against US forces, in Yemen, East Africa, Indonesia, etc. The eleventh of September, 2001 terrorist attacks on the World Trade Center and the Pentagon demonstrated to the US and its allies that political disorder in distant parts of the globe could have terrible consequences at home. The Al-Qaida terrorist group had used Afghanistan as a base to strike at the civilian population in the US, adopting hijacked airliners as improvised weapons of mass destruction to kill thousands of civilians. This was a big event for the United States, since it had never faced such a military attack since the Japanese attack on Pearl Harbor. It was quite clear, at that time, that the US had created a Frankenstein monster by launching a war of terrorism against Afghanistan in 1978. The monster had, then, turned on its own creator! This war was the brain-child of Zbigniew Brzezinski, the National Security Advisor to President Carter at that time. Brzezinski continued to defend his recommendation till his death in 2017, with the absurd claim that it had led to the collapse of the Soviet Union. He refused to acknowledge that the "Mujahideen terrorists", who he had confused a good man like President Carter into approving, had *murdered about one and a half million Afghan men, women and children.* This was a great crime. However, to his royal Polish mind, it was a great achievement!

After the attacks on buildings in New York and Washington D.C., the United States decided to launch a full-fledged invasion as part of another project of Panamization, under the leadership of President Bush, who was driven by his desire of dismantling an "Axis of

Evil", i.e., move the projects of Panamization of Iraq, Iran and North Korea forward. The US forces landed in Afghanistan, flying over Pakistani territory and quickly took control of Afghan cities, while the Taliban retreated into mountainous areas. The US forces picked up, by helicopter, the future puppet "leader" of the Afghans, Abdul Haq, from the Pakistani territory near Peshawar and tried to use him to negotiate purchase of the support of some Afghan tribal leaders, against the Taliban. Instead of cooperating, the tribal leaders killed the new "president designate". The US, then, had to pick up a new puppet. This time, Hamid Karzai got lucky and became the 'front man" for the US forces and the "president designate". A puppet regime was set up in Kabul, with Karzai as the "President". The last King of Afghanistan was brought in from Italy, where he had been sent by his brother-in-law, Sardar Daoud. The process of Panamization was, then, taken up. (This process is generally referred to, by US administrations, as "nation building"). New systems of road transportation, tele-communications, broadcasting, education, economic management, etc., had to be set up. The water and electric power supply system had to be set up in all cities. A new currency was introduced along with a banking system. Thus, a state had to be re-created from scratch, while the Taliban started a guerrilla war against the NATO forces of occupation. The cost of this new project to the US was estimated to be about $70 billion per year, at its peak. This took more than twelve years, as a new Afghan army and police was trained to take over most of the military activities. Afghan immigrants, in the US and Europe, were brought back to Afghanistan to create a local and friendly ruling class. The process of Panamization is nowhere nearing completion, however. The withdrawal of forces of the US and its allies was almost completed, in accordance with the time-table of President Obama. As a part of the scheme of Panamization, a small but highly trained and well-equipped "residual" force of about 15,000 men was kept in Afghanistan. It was expected to stay in Afghanistan for a long period of time - to keep the Afghan military under control and to support its actions if necessary. The war against Al-Qaida, however, became a global war, since this organization had set up a global network of operatives and sub-organizations. Osama Bin Laden was found in hiding, in Pakistan, and killed by US forces and his position was taken over by his Egyptian deputy, Ayman Al-Zawahiri.

The Great Depression of Central Eurasia

The recession of early 1990 in the US followed a particularly long period of recovery and expansion from 1983 through 1988. It was caused, among other reasons, by several financial disasters in financial institutions, which had their origin in the Reagan administration's deregulation of the Savings and Loan Associations, which were specialized banks which issued mortgages and other consumer loans. The administration had also raised interest rates. The lack of regulation and oversight created an environment in which CEOs of S&Ls were motivated to create fraudulent schemes to enrich themselves at the cost of the customers of their organizations. They started inventing creative accounting strategies that turned their businesses into Ponzi schemes that looked highly profitable, thereby attracting more investors and growing rapidly, while actually losing money. Like all Ponzi schemes, these S&Ls also failed in large numbers, resulting in heavy losses to the government, while benefiting the fraudulent debtors and schemers in the ruling class. Also, social spending had been cut drastically during this period. When the recession arrived, it technically lasted about 8 months between July 1990 and March 1991. During this period GDP of the US declined by about one and a half percent and unemployment peaked at about seven percent.

The recovery of the US, after this recession, was fast due to the concurrent events in the Soviet Union. After Yeltsin had staged his coup in the Soviet Union and had taken over dictatorial powers with the help of the Russian military, he initiated his program of *"Shock Therapy"* for the Russian economy. Price controls were suddenly abolished, virtually all state institutions dealing with the economy were dismantled and Yeltsin's family and other gang members started grabbing former state property, creating a new class of *super-looteras*, or *oligarchs*. In this environment of total anarchy, gold ornaments, paintings and other decorations of Churches were stolen. Gold from mines and state banks was stolen by criminal gangs and transported to North America and Europe. About $160 billion worth of valuables are estimated to have been stolen this way. Human trafficking became a booming business. Soon members of the ruling classes of imperial powers arrived to "buy" state-owned factories and mines at cut-rate prices. They robbed wherever and whatever they could, with the help of the new "self-made" local capitalists, and dumped what they did not want. Yeltsin and company celebrated their new "freedom", while the people of Russia, and other former Soviet Republics, began to die of lack of food and medical care. Pensions of retirees evaporated with the Russian currency. Scientists began selling tomatoes to protest the closures of their research centers. Education and health care systems collapsed. *The GDP of the Russian Federation fell by more than 40%. The Great Depression of Central Eurasia was more severe than the well-known global Great Depression of the 1930s.* All this benefitted the US and other imperial powers as large capital flows occurred into these states. Thus, the recession in the US came to an end and, quickly, became a boom, in 1993, while the Great Depression of Russia was deepening. The worst stage of the *Russian Great Depression* was around 1997 when Putin took over from Yeltsin. Then, Russia and the successor states of the Soviet Union also began to recover. The Great Depression of Russia, and Central Eurasia, was, however, not a usual extreme recession of a market economy, but was due to the *sudden Looterization* (generally referred to as *"privatization"*) of Central Eurasian economy. The resulting concurrent boom of the US and European economies, that started around 1992, lasted till 1999.

Recession of Early 2000s

As the post-recession boom of the First World, i.e., North America, Europe, Japan and Australia, progressed during 1992-1999, a great upsurge of software development occurred. New technologies of computer networking, database storage and computer operating systems took hold. Internet users began to grow fast, especially in East Asia and Russia, and there was an explosion of cell phone and internet usage, triggered by laying of optical networks on a global scale. These technologies offered an opportunity to establish websites for businesses on a large scale. Venture capital was attracted by new companies developing these new technologies. These companies, themselves, had internet websites advertising their services. They began to be called dot-coms in popular discourse. At the same time, the disappearance of the Soviet Union had given a great confidence to the ruling classes of imperial powers, so that they started investing their hoarded capital in China, India and other emerging markets, instead of North America, Japan and Europe. This dramatically reduced investment in the local economies of the imperial powers, causing the local companies, especially the dot-coms, to move their operations to the emerging markets and closing their local operations, or establishing new factories, mines, call centers and maintenance centers in the emerging markets of East and South-East Asia, Australia and South America. Those companies which failed to adjust to these new conditions, lost their competitive edge and started failing in large numbers. The dot-com bubble also collapsed. Large numbers of workers, employed by local companies, became unemployed due to factory closings and outsourcing of labor-intensive projects. Under these conditions, many formerly high-paid manufacturing and professional employees were forced into much lower paid service positions. The extra spending on the global Y2K (i.e., "Year 2000") problem in computer software, also, stopped suddenly as the new millennium arrived. Unemployment increased to more than six percent. The September, 2001 terrorist attack on the World Trade Center in New York had some effect on consumer confidence and spending. The New York Stock Market crashed and did not recover till March, 2002. The US Federal Reserve lowered the discount rate in gradual steps and increased the money supply also to achieve a "soft landing". This recession, however, did not meet the strict requirement of two consecutive full quarters of negative growth, but there was a clearly noticeable downturn of economic activity lasting for more than one year. This recession and decline of 1999-2002 had occurred after a long time of continuous growth since the disappearance of the Soviet Union in 1991. Recovery from this recession was very slow, requiring low interest rates for quite a long time.

The Great Recession

An examination of the next major recession that started in the United States in late 2007, would clarify the details of the self-serving modifications to the Keynesian scheme of management, which have been developed, and practiced by the ruling classes of imperial powers[3]. This recession was a very severe one and had a great resemblance to *The Great Depression*, although it was better managed and thus did not lead to comparable results. However, because of its severity it is generally referred to as *The Great Recession*. This recession had one characteristic common with the Great Depression - It started with a collapse of the financial system. It was also a global recession because of the progress of globalization. Just before the recession started, the Chinese, Indian and Russian economies were growing, approximately, at 10 percent, 8 percent and 7 percent respectively. The US and the EU were also growing, approximately, at the rate of 3.5 percent and 2.5 percent, respectively. The *Great Recession* started in December 2007, in the United States, and took a particularly sharp downward turn in September 2008 and became global. The origins of this recession can be traced back to the following developments in the previous fifteen years:

- Government programs, in the United States, apparently promoted affordable housing for low-income families; and successively established more and more ambitious targets for institutions guaranteeing mortgages for this purpose. For the benefit of those in the ruling class of the US, whom he referred to as "his constituents", President George Bush's administration encouraged relaxation of regulations governing real estate mortgages with slogans like "We want all working people to become home owners". Consequently, those institutions were pressed to relax underwriting standards to meet their targets. People ended up purchasing mortgage loans for about 10 trillion dollars. Lessons of the 1989 Savings and Loan Crisis, caused by similar relaxation of regulations by the Reagan administration, were ignored. During the same period, recovery from the recession of the early 2000s had been very slow, requiring low interest rates for quite a long time. Home owners, especially low-income families, had to refinance their mortgages to survive, when unemployment remained high and family incomes remained low. Those home owners, thus, got trapped into adjustable rate "sub-prime" mortgages.

- A big trade deficit had developed in the US. As a result, an inflow of investment dollars occurred, mainly from the ruling families of US-controlled Bantustans and Panamized states, which had accumulated large amounts of cash, mainly from sale of oil and gas. This was needed to fund the US trade deficit, which had grown from less than 1% of GDP in the early 1990s, to about 6% in 2006. Much of that inflow of money was in the form of fixed income savings and went into dodgy, or "sub-prime", mortgages to buy overvalued houses, creating a housing bubble in the US. The value of these investments increased from around $35 trillion in 2000 to about $70 trillion (!) during this period. The "dodgy mortgagees" had developed due to the relaxation of loan standards by the banking system - to attract capital for investment. The main techniques were: overvaluation of houses and other real-estate by fraudulent appraisals, reduction in the percentage of down-payment required for mortgages (in some cases, this was reduced to zero), overstatement of buyers' income combined with understatement of their liabilities, in determination of affordability and introduction of heavily back-loaded adjustable-rate mortgages. These adjustable mortgages had very low initial rates of interest, which were to rise to very high rates step by step, in a few years. The knowledge of these "innovative" techniques of "creative financing" spread to Europe (especially Southern Europe after the creation of the Euro) and other unregulated parts of the world also, causing housing bubbles to develop. Concurrently due to refinancing, US household debt, as a percentage of annual disposable personal income, rose from about 75% in 1990 to more than 120% at the end of 2007. In other advanced economies, it soared to 130 percent and, in some states, to about 200 percent.

During the previous thirty years, or so, a *shadow banking system* had developed and expanded to rival, or even surpass, conventional banking in importance. All the protections developed during and after the *Great Depression* - the Federal Reserve as a lender of last resort, federal deposit insurance and a broad system of regulations, became ineffective against this shadow banking system. Thus, *blind and unscrupulous greed* had found a way through all those barriers against financial abuse: All the "achievements" of this shadow banking system were immune to the regulation that was applicable to regular banks. These achievements included the following:

- Repo Lending: Repurchase Agreements (REPOs) are a form of short-term borrowing for banks dealing in government securities. The dealer sells the government securities to investors, usually on an overnight basis, and buys them back the following day. For the party selling the security (and agreeing to repurchase it in the future) it is a repo; for the party on the other end of the transaction, it is a reverse repurchase agreement. The original seller is effectively acting as a borrower, using their security as collateral for a secured cash loan at a fixed rate of interest. Repos are usually used to raise short-term capital. But, using loopholes in accounting rules, they were used by shadow banks to hide huge debts.

- Mortgage Backed Securities: These 'securities" were shares, of pseudo "investment companies", that were being offered for purchase worldwide, with promises of very high rates of profit. They were also "opaque" since it was not clear to the investors, as to how well the relevant entities were actually performing, and there was no way to get at this information either. This was the latest creation of the shadow banking system to attract investment capital from ruling classes of Bantustans, Panamized states and other states with high rates of accumulation of capital, because these ruling classes, driven by their unquenchable greed, wanted still higher rates of profit, which were promised to them by unscrupulous "shadow bankers" who were driven by their own unlimited greed and lack of any ethics whatsoever.

- Credit Default Swaps (CDS): These "derivatives" are a kind of insurance on loans, which creditors can buy to protect themselves from defaults. They were invented in the mid-90s and have come into widespread use because of globalization of finance. They can also cover baskets of loans and mortgage-backed securities issued through pseudo-companies. Thus, even "sovereign debt" owed by states, may be covered. The CDSs may be bought by companies and individuals having no insurable interest in the debts. This can be done for speculation purposes only. Multiple entities can buy CDSs for the same debt, in which case, they are referred to as "naked CDSs". The buyers of CDSs have to pay a premium to the sellers. In case of default the sellers have to compensate the buyers generally by paying the "par value" of the debt. The par value of a debt may not be the nominal value, since a debt may have some market value, even after a default, which can be recovered.

- Hedge Funds: A hedge fund is an aggressively managed portfolio of investments that uses advanced investment strategies such as leveraged, long, short and derivative investments in global stock markets, with the goal of generating high rates of profit. These techniques are not allowed to be used in regular mutual funds. The hedge funds are, generally, unregulated and most often set up as private investment partnerships that are open to a limited number of investors and require a very large initial minimum investment. They are, thus, highly risky funds to satisfy the high level of greed of the ruling classes. The name "hedge fund" is a misnomer. Hedge fund managers make speculative investments and these funds carry much higher risks than the overall stock market.

With the marketing of these new kinds of derivatives, especially the mortgage-backed securities, a deluge of foreign investment followed. A multi-trillion-dollar repo lending market developed. It did not have any regulatory protections to prevent financial meltdowns. By the end of 2007, the outstanding CDS amount was more than 60 trillion! The investment banks, also transferred assets from their balance sheets to the stock markets and other unregulated markets, while rating agencies and regulators could not properly assess the risks involved in such transactions, making those banks more fragile and vulnerable. *The ruling classes of the imperial powers, Bantustans, Panamized states and other states that had accumulated large amounts of capital, had gone berserk with greed!*

When the interest rates on the adjustable rate mortgages began to increase, many home owners could not make the required payments and began to default in large numbers, especially those holding "underwater mortgages". Foreclosures and fire sales of houses resulted and the housing bubbles collapsed everywhere. When house prices declined, many households saw their wealth shrink relative to their debt, and, with less income and more unemployment, found it harder to meet mortgage payments. This accelerated the sub-prime crises. Ultimately, housing prices had fallen from their peak by 20-40%, in different areas of the world. This sub-prime mortgage crisis rendered the mortgage-backed securities worthless and, suddenly, a major financial crisis developed in the shadow banking system. The investment banks in this system reacted with a freeze on inter-bank lending and also stopped funding mortgage-related investments. The financial liquidity crisis started on 9 August 2007, at the interbank lending market when central banks had to step in with lending to the banking market. BNP Paribas temporarily had to block money withdrawals from three hedge funds - citing a "complete evaporation of liquidity". As share and housing prices declined, many large and well-established investment and commercial banks in the United States and Europe suffered huge losses and even faced bankruptcy. Many firms, by not renewing sale and repurchase agreements (repo) or increasing the repo margin ("haircut"), forced many banks into insolvency. Bank failures are highly contagious – domino effect of one bank failure causes many more due to loss of public confidence, resulting in "bank runs". This causes investment activity to stop suddenly, triggering a recession.

When massive defaults occurred on underlying mortgage securities, companies like AIG that were selling CDS were unable to pay their obligations and defaulted. The US government promptly paid about $100 billion on behalf of AIG. During 2008, three of the largest US investment banks either went bankrupt or were sold at fire sale prices to other banks. Some investment banks converted themselves into commercial banks, thereby subjecting themselves to more stringent regulation and becoming eligible for government help. The capitalist-controlled governments stepped in to protect the property of their class. Obama administration helped many banks with loans amounting to $20 billion - $100 billion each. The outstanding CDS amount came down by two-thirds in the next three years. As the stock market crashed, many manufacturing companies found themselves with big inventories and no buyers. This happened especially to car manufacturers who faced bankruptcy as a result. In the US the Obama administration helped them with large multi-billion Dollar bailout packages. In Europe, many banks were nationalized outright. Thus, governments immediately used their tax revenue to protect the ruling classes from large losses of capital. Naked CDSs were banned on the debt of European states. In the US, recession, technically, lasted for 18 months. It quickly spread to Europe and other parts of the world and became a global recession. Unemployment in the US and Europe peaked at about 10 and 12 percent, respectively.

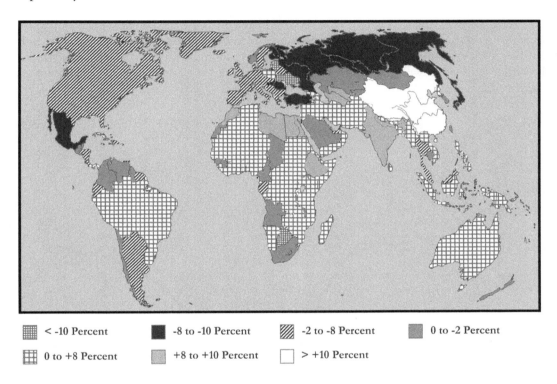

▦ < -10 Percent	■ -8 to -10 Percent	▨ -2 to -8 Percent	▪ 0 to -2 Percent
▦ 0 to +8 Percent	▨ +8 to +10 Percent	☐ > +10 Percent	

Figure 11-01: GDP growth/drop due to the Great Recession - 2009

Gdp_real_growth_rate_2007_CIA_Factbook.PNG: Sbw01f, Kami888, Fleaman5000, Kami888 derivative work: Mnmazur (talk) (https://commons.wikimedia.org/wiki/File:GDP_Real_Growth_in_2009.svg), „GDP Real Growth in 2009", color and grades of GDP changes by Tayyib A Tayyib, https://creativecommons.org/licenses/by-sa/3.0/legalcode. (Map based on world map at: https://www.d-maps.com/carte.php?num_car=13184&lang=en)

In addition to the bailout packages for financial and manufacturing companies, the US government announced a large stimulus package. Small amounts were paid to most tax payers and a large infrastructure building plan was taken up. Other governments of the world also took similar actions. Discount interest rates were reduced drastically and money supply was increased. In the US, the Federal Reserve Bank started injecting money into the banking system by buying back treasury bonds. Also, regulations of the stock market and the banking system were upgraded. However, due to the high household and government debt, the recovery was very slow. At the end of 2013, the unemployment rate in the US remained above 7 percent. That of Europe was about 12 percent. Southern Europe had much higher unemployment rates at that time. In Greece, Italy, Spain and Portugal, the rate of unemployment went up to about 20%. The IMF and the European Central Bank (ECB) had to give large loans to these states to keep their banking systems afloat. They, still, had great difficulties of adjustment to the terms of those loans. Specially, Greece went through a major crisis. In Europe, conservative political parties came into power and started austerity programs, further slowing the recovery. Even after the passage of about six years, the global economy had not really recovered. Also, a considerable risk of a relapse persisted. It took another five years for the US GDP growth rate to recover. In 2018 it reached about 4%, but for Europe it remained around one percent. China's economy did not go into a recession. It, however, did slow down but has recovered and is growing at the rate of about 6.2 percent. At present, the US and other European states are growing at a very slow pace - about one percent for Europe and about three percent for the US. The US unemployment rate has, however, come down to less than 4%, but unemployment in Europe remains high, especially in its Southern states.

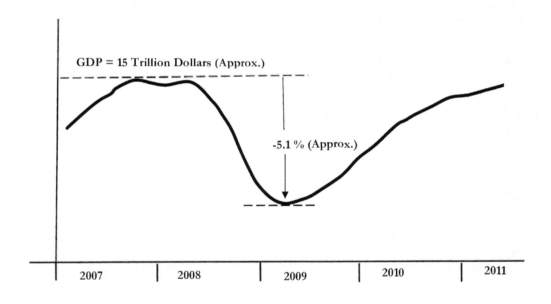

Figure 11-02: Changes in the GDP of the US during the Great Recession

Only a few states escaped ravages of the recession - China, India, South Korea, Poland, Vietnam, Laos, Ethiopia and some South American states were among them. The main reason behind this escape was state ownership of their banking systems and several sectors of their economy – removing run-away greed's control over those institutions, thus, avoiding risky investments and lending activities. Also, some had major investments and trade relationships with China and this acted as a cushion for their economies. Most states, however, had to face a considerable negative impact on their economies.

The Great Recession has demonstrated certain aspects of the Keynesian scheme that have become popular with the ruling classes of imperial powers. When a recession arrives, any financial or other corporations that come under stress, are bailed out of their predicament by government subsidies or outright state take-over (generally referred to as *"nationalization"*), if the owners are unable to manage these corporations. Taxes are reduced and some stimulus payments are made to tax-payers. In the spirit of capitalist-siders, some tax breaks are also given to the ruling class, even though they do not have any effect on the recovery. Major development programs may also be undertaken to increase employment and improve the infrastructure in general. Social spending, including that related to unemployment insurance, is also increased. Along with these fiscal measures, the central bank's discount rate is also reduced in stages and the money supply is further increased by government bond buyback activities, or reduction in the required reserve ratio of commercial banks, if required. When the economy begins to recover, these measures are gradually brought to an end. If the government debt has increased too far because of the increases in expenses and the bail-outs of corporations, then the subsidies and social supports of the people are reduced to control the debt that was incurred by government's help for the ruling class in the beginning of the recession. *Thus, the burden of the management of the business cycle is shifted from the ruling class to the people, as much as possible.*

During the growth cycle, the interest rates are increased gradually as and when the economy seems to be growing too fast for an ultimate soft landing. Also, other money creation activities are also curtailed. At a later stage of growth, the government increases its taxes, to "cool down" the economy, i.e., to maintain sufficient unemployment to keep wages down to a level that is tolerable to the ruling class. *In the US, one function of the Federal Reserve Bank is said to be to "keep inflation under control" – i.e., maintain sufficient unemployment and not let wages rise "too much". In fact, wage rises are considered part of "inflation". Thus, increases in the interest rates tend to constrain investment and employment and thus "achieve a sustained growth".* The tax increases to "control inflation" are also aimed at the people, not the ruling class, as much as possible. Since, in the globalized economy, recessions also tend to be global, actions are undertaken by large groups of states, in consultation with each other.

Consequences of Globalization

Initially, some work being done in advanced economies was "outsourced" to low-wage countries. Direct investment also grew dramatically. Due to this transfer of capital to emerging economies, under-developed states began to develop faster with lower unemployment and improvements in standards of living of their working populations. The advanced countries, on the other hand, began to experience lower wages and increased unemployment.

Big advances were made in reduction of extreme poverty, globally. However, most of this poverty reduction actually occurred in China. Large trans-state corporations began to make huge profits for their share-holders, resulting in increases in economic inequality – especially in the states at the stage of benevolent capitalism. The result has been a dramatic increase in social inequality in Europe and North America, especially in the US where ten billionaires own more wealth than the bottom half of the population!

In this environment of globalization, business cycles have also become global. Their management has also become much refined and predictable. The Keynesian scheme of fiscal and monetary measures has come to be universally adopted, with some *modifications to profit the ruling classes whenever recessions occur.*

As mentioned before, the core process of an economic recession, or depression, is the lack of aggregate demand for goods and services, when the business cycle is at its peak of economic activity. When production is at its highest in an economy and unemployment is at its lowest, but products cannot be sold because of high prices and the inability of the working people to buy those products because of their lack of sufficient income, inventories of businesses and corporations begin to grow. When such businesses notice the growth of their inventories, they tend to reduce their employment by laying workers off, reducing further investment and reducing prices of their products. *Further layoffs, and a halt to all investment, are resorted to when inventories continue to grow, because the layoffs have reduced aggregate demand further. These cyclical changes, ultimately, result in a contraction of the economy, leading to a full-fledged recession or depression.*

As mentioned before, a recession, however, can also be triggered by bank failures and a resulting financial crisis, which brings a sudden stop to all investment resulting in massive layoffs. This, then triggers the core processes of a recession. The great depression during 1929-1939, the recession during of 2000-2002 and the great recession during 2008-2009, were primarily triggered by financial crises.

The processes involved in global recessions have changed due to globalization. The main reason is that a large part of investment is being done by the trans-state corporations of the US, Canada, Europe, Japan and Australia. China and the countries of South East Asia are the main recipients of this foreign direct investment (FDI). Also, a large amount of this FDI is going into Mexico and countries of South Asia and South America. Wages, to varying degrees, are relatively low in all these recipients of FDI, but the output of those countries is exported to the rest of the world, in addition to being consumed locally. Thus, the increases in wages, e.g., in China, do not have much effect on the thinking of the management of the trans-state corporations. *Those corporations continue to make huge profits and, thus, do not reduce local employment, but may shift their factories to other countries of lower wages, or may direct their further FDI there. The aggregate demand and prices of their products remain high in the "home" countries in which those corporations are based. The employment in their home countries does not rise fast due to the lack of investment there, but their profits remain high because of high prices there. Thus, it takes much longer for their markets to be saturated and a decrease of global aggregate demand to arise.* This phenomenon has resulted in a long period of recovery/growth during 1992-1999. The long recovery/growth of 2009-2019 has, also, occurred due to the same reasons. However, the recovery of 2002-2007 was short lived, only due to the rash and abusive financial actions of the Administration of President Bush junior.

Another major factor behind the long periods of growth/recovery between recessions of the global economy is the role of China, *which tends to act as a stabilizer of global economic activity.* For example, China's stimulus package, during the recession of 2008-2009 and the following recovery, helped to reduce the severity and extent of the recession. However, China's focus on raising the standard of living of its population and increasing its domestic consumption, would tend to decrease this stabilizer effect in the future, by relatively isolating it from the global economy.

A new corona virus pandemic started in China in January, 2020. The virus was named Covid-19 by the World Health Organization. Drastic lockdown of Wuhan city was undertaken due to the fast spread of this infection. The pandemic started in China during the Chinese New Year, when the Chinese people travel in huge numbers to their ancestral villages and homes. The infection spread fast within China and, ultimately, to the rest of the world. Production and business activities were severely affected all over the world. The initial effects were on travel and entertainment industries. Airlines, hotels, restaurants were affected immediately. The tourist and film industries were also affected severely. All kinds of gatherings were stopped in country after country and full lock-downs are resorted to by governments in others, in an effort to slow down the infection and death rates. Hospitals and health care facilities have become over-burdened with patients needing help. In short, production and trade has suffered severely and a high level of unemployment is emerging at the time of this writing. Huge business losses and a drastic reduction of production and aggregate demand has developed. Global financial institutions are predicting a major global recession. It remains to be seen how severe this recession, or depression, would be.

Chapter-12

Inter-State Trade

Interstate trade, generally referred to as "international trade", was briefly referred to in previous chapters, but a detailed analysis of the basic processes of sale and purchase of products among states was postponed, due to the distortions of inter-state economic relations during nineteenth and twentieth centuries. Even under conditions of relative free trade among sovereign states, these processes can benefit one state far more than another, and may benefit others not directly involved in bi-lateral trade. To understand this phenomenon, we need to think of a world in which sovereign states are trading with each other. If imperialism is in existence, then the system becomes much more abusive and it becomes more difficult to see the real effects of trade. Since conditions during earlier times were such, we did not discuss inter-state trade in the chapter dealing with *intensive capitalism* in Europe. Inter-state trade at that stage was mostly inter-empire trade. Under *benevolent capitalism* in Europe and elsewhere, inter-state trade became somewhat free, although global *economic market subjugation* remained in control because of the imposition of the US dollar as the reserve currency and the emergence of the United States as a superpower with global reach. Trade relationships continued to remain distorted by the existence and further expansion in the number of *Bantustans* and *Panamized* states. Thus, our analysis, in this chapter, of inter-state trade in a globalized world of the present would remain applicable to the world of intensive capitalism and benevolent capitalism, provided we are conscious of the distortions caused by the imperial techniques and structures of *Colonization, Bantustanization, Panamization and Controlled Market Subjugation,* even in the present globalized world. Fair and free trade does not really exist even at this time! It is a figment of imagination of economists, who seek to justify the capital accumulation by the ruling classes of some powers and the resulting global economic and social inequality generated by the un-restricted operation of the law of supply and demand in some regions of the world, while such trade is highly regulated in other regions.

Within a state, products are expected to be traded or exchanged, mostly, on the basis of their values. Prices, as we have mentioned before, tend to fluctuate around the values of products. But this does not happen if some products, generally raw materials, have become scarce. Then, the perceived value of such materials, or their prices, tend to become very high and unconnected to their value. Recently, oil, gas, high-quality coking coal, uranium, gold, silver and quite a few rare metals have become scarce and their prices have become very much out of line with their values. Thus, in inter-state trade, the value of products is not the only determining factor of their prices. The labor market in each state is different. The social systems in different states are at different stages of evolution and wages of workers are generally very different, because of local labor market conditions. In a feudal society, wages are generally very low because of existing inequality and severe control by a class of landlords. People are denied education and knowledge and remain ignorant. When a society evolves into the early stages of intensive capitalism, real wages tend to fall further towards a subsistence level and continue to fall throughout this stage, for the working class as a whole. Only when labor shortages develop, when this society reaches the end of this evolutionary stage, wages stabilize and then tend to rise. Thus, the prices of most products are very different in different states and are determined by supply and demand on a global level. The ruling classes of both states, that engage in exchange of products, gain by this trade. However, the gain is not equal on both sides. Besides, people are not directly involved in such trade. Their wages are determined by local labor markets, which are not immediately affected by such trading, unless it is accompanied by changes in currency exchange rates.

All products produced in a state and all services provided by its population and corporations are not tradable. Tradable goods and services have been rising with time for most states, but are generally less than half of their output. In the US, the share is about 40 percent in terms of employment for production of such goods. At present, the trend of employment in services is such that it would eventually exceed that in manufactures.

Adam Smith[1] had justified inter-state trade on the grounds of absolute advantage – that each product should be produced in the state that can produce it at the lowest cost. David Ricardo[2] showed that unrestricted trade would benefit all states involved in it, even if one state had the absolute advantage in the production of all goods. All that is required is that trading take place according to *comparative advantage. However, "comparative advantage" had a special meaning according to Ricardo and this special meaning was based on the concept that every commodity has a "natural price", which is the value of that commodity. The most important commodity in production is labor. According to Ricardo the natural price of labor is the cost of commodities required by working people and their families for subsistence. Thus, Ricardo had assumed that working people would require and would be entitled to only subsistence wages, i.e., wages which would enable them to reproduce enough workers for the ruling class, in the future. In other words, he had assumed workers to be wage slaves.*

Absolute advantage is the capacity to produce more units of a product than a competitor can for any given level of resource use. Comparative advantage in cost of production is the relative advantage based on relative ratios such that either the absolute advantage is greatest or the absolute disadvantage is the lowest. As long as the relative productivities differ between states, they can engage in mutually beneficial trade.[3] In other words, a state has a comparative advantage in producing a good or commodity if the opportunity cost of producing it in terms of other products, or goods, is lower in that state than it is in other states. Trade between two states can benefit both if each state exports what it can produce at a comparative advantage.[4]

This does not mean that the whole population of a state benefits by trading in goods in which it has a comparative advantage. Inter-state trade is between the ruling classes of states that trade with each other. In fact, the whole ruling classes of the two states are not involved – only the exporting and importing corporations and their owners are involved on both sides. While such trade may benefit the consumers on both sides to some extent, it may benefit the exporters on one side, while causing losses to importers on the other side, because of increased competition with domestic producers.[5]

It is important to stress that trade between two states on the basis of comparative advantage does not mean that both sides gain equally, even under conditions of wage slavery as assumed by Ricardo and Adam Smith. At this time, wage levels are not uniform among states. Also, wages are not subsistence wages, especially, in states under benevolent capitalism and in those states, which are going through a managed transformation of their economies. Thus, at this time, equality of gains is not possible for both sides engaged in inter-state trade. The side with lower wage levels has a weak bargaining position. It loses due to its competition with other states with low wage levels. Also, the side with higher productivity or higher technology use has a stronger bargaining position in the global market and it gains more.[6]

In general, the state that is abundant in a certain resource, exports the products which incorporate more of that resource.[7] Inter-state trade results in gains to owners of the state's abundant resources and causes losses to those who own its scarce resources[8]

To maximize their gains from trade, states adopt certain policies. These policies include import tariffs, export subsidies, import quotas, voluntary export restraints and bans on import or export of certain products. These policies are generally meant to increase exports, increase government revenue, protect certain local industries from competition, or deny certain high-tech products, including military equipment, to certain states considered to be adversaries or competitors. Trade between otherwise sovereign states does not flow freely or smoothly due to such restrictions and may cause short-term or long-term losses to the states involved.[9]

In the globalized world of the present, we can divide the products involved in inter-state trade into four categories:

- Products that are produced widely by most states - with varying costs and values. The products in this category are freely traded, but their prices are negotiated, case by case. In the global market, these products have prices determined by supply and demand, although value does have a high influence on these prices. Generally, prices are determined by the lowest-cost producers, who tend to have some natural advantage like certain kind of weather for a specific type of agricultural crop, or the lowest wage levels in their local labor market and, thus, have a "comparative advantage".

- Products that are produced by certain states, or small groups of states, because of special weather conditions or availability of certain raw materials, e.g., coconut oil, certain types of flowers, oil, gas, uranium, etc. The prices of scarce products are determined solely by supply and demand and have virtually no relevance to their value. For example, oil in Saudi Arabia may be produced at a local cost of less than $5/Barrel, but may be sold at a price of $110/Barrel, since that may be the inter-state market price, at a certain time.

- Products which incorporate new and advanced technologies, e.g., nuclear power plants, wind turbines, liquefied petroleum gas (LPG), tunnel boring machines, high speed passenger aircraft, etc. This category included computers in the 1950s to 1980s, cell phones in the last two decades of the last century, solar panels and LED TVs in the 1990s and the first decade of this century and LED bulbs in the beginning of this century. The prices of such products have the tendency to fall with time.

- Products that are made scarce because of controls imposed by certain powerful states, which do not allow free trade in such products, e.g., uranium, nuclear power generating plants, nuclear reprocessing plants, nuclear enrichment plants, certain kinds of advanced offensive and defensive weapon systems, super-computers and certain kinds of software for computers, etc. Trade in such cases is only *controlled trade*. Prices involved in such trade are arbitrarily controlled by one or more powerful states, with "inter-state bully" status.

Role of Prices and Values of Products

To have a clear understanding of inter-state trade in the first category of tradable products, let us assume four states are trading with the US. Also, suppose these five states, i.e., US, China, Japan, Iraq and Congo, have a clear competitive advantage in production of corn, cotton, television sets, oil and copper, respectively, and are trading in these commodities only. As shown in Figure 12-01, the local costs of the commodities are $300 per ton, $45 per bale, $140 per unit, $30 per barrel and $800 per ton, respectively. The assumed global prices for these commodities are also shown. These are $500/ton, $100/bale, $250/unit, $100/barrel and $3,000/ton, respectively.

	USA	CHINA	JAPAN	IRAQ	CONGO
DESCRIPTION	CORN	COTTON	TV SETS	OIL	COPPER
Currency	$	YUAN	YEN	DINAR	FRANC
Currency Exchange Rate	1	9	110	1,200	1,600
GDP/Capita ($) PPP Basis	$60,000	$16,700	$37,000	$16,500	$800
Approx. Median Wages ($/Hour)	$25	$2.00	$18	$1.00	$0.50
Objectified Labor Used, Per Unit	$100	$15	$60	$20	$600
Total Living Labor Used, Per Unit	$400	$85	$190	$25	$2,400
Living labor - Paid, Per Unit	$200	$30	$80	$10	$200
Living Labor - Unpaid, Per Unit	$200	$55	$110	$15	$2,200
Total Cost ($)/Unit	$300	$45	$140	$30	$800
Sale Price/Unit	500	100	250	100	3000
Rate of Profit	67%	122%	79%	233%	275%
Rate of Exploitation (ROE)	50%	65%	58%	60%	92%

Figure 12-01: Issues of Production Related to Inter-State Trade

The five states are at different stages of their evolution. The US and Japan have reached the *benevolent capitalism* stage - with high wage rates, stagnating native populations with shortages of labor resulting in immigration from other parts of the world. They have highly-educated and trained manpower and have institutions developing new technologies. Their high per capita GDP could justify our assumptions of median wages of workers as $25 and $18 per hour, respectively. Based on these statistics, we assume that the objectified labor is $100 and $60 respectively, i.e., 33% and 43% of the cost of their products. This objectified labor is the labor contained in the fixed, facilitating and unfinished input capital of the respective production centers. Out of the living labor, 50% ($200) is assumed to be paid for in the US and 42% ($80) in Japan, because of its relatively lower real wages for its workers, after adjustment for local prices of consumer goods. The unpaid labor is, thus, $200 and $110, respectively. It becomes part of a product, i.e., becomes objectified during the production process. These assumptions may be considered reasonable for illustration purposes, based on current approximate GDP levels, per capita income and wages in these states on the Purchasing Power Parity (PPP) basis. The numbers for the other states involved have also been assigned with similar assumptions.

	----- EXPORTS -----			---------------------------------- IMPORTS BY EACH STATE ----------------------------------									
	Price				USA		CHINA		JAPAN		IRAQ		CONGO
Country	Per Unit	Units	Total Price	Units	Total Price	Units	Total Price	Units	Total Price	Units	Total Price	Units	Total Price
USA	$500	340	$170,000		$0	80	$40,000	40	$20,000	100	$50,000	120	$60,000
CHINA	$100	400	$40,000	400	$40,000								
JAPAN	$250	80	$20,000	80	$20,000								
IRAQ	$100	500	$50,000	500	$50,000								
CONGO	$3,000	20	$60,000	20	$60,000								
Total Exports of each State					$170,000		$40,000		$20,000		$50,000		$60,000
TRADE BALANCE (Unpaid Labor)					-$14,300		$6,000		$800		-$12,500		$20,000
TRADE BALANCE (Price)					$0		$0		$0		$0		$0

Figure 12-02: Transfer of Unpaid Labor in Inter-state Trade.

We are assuming that products are exchanged at or near their value for all states except Iraq. The price of oil of Iraq is assumed much higher than its cost of production, because oil prices are generally high due to scarcity of this commodity. Trade on the basis of values, of course, does not normally happen. Also, wages of workers are very different for different states. For example, workers are paid lower wages in China as compared to Japan or the US. China is now considered a middle-income state and the lower wages are due to the relative unfavorable local labor market for workers as compared to the US or Japan. Both Japan and the US are at the stage of *benevolent capitalism*. Iraq is a *Panamized state*. It has vast oil and gas resources, but it has faced devastating invasions by the US, which have crippled its economy. It is going through a rebuilding process, but the wages of its working class are quite low even compared to the wages of its workers three or four decades ago. The prices of oil and gas have gone up in recent years, although the cost of production of oil has not increased as much. The high price of oil is due to its global scarcity. Iraq may be considered to be at the stage of *advanced feudalism*. The Congo, i.e., Congo-Kinshasa is a very impoverished state that has faced intense colonial abuse by Belgium and devastating invasions and political instability after independence. Its per capita GDP is only about $800, although it is rich in copper and many other minerals. It has a tribal society, so we consider it to be at the stage of *primitive feudalism*. We have, thus, a variety of states to illustrate the shortcomings and inequities of the inter-state trading system, from the point of view of the working people of the world.

The prices we are using are not the current global prices for these products, but have been assumed on the basis of rough correspondence. The assumed amounts exported and imported by each state are shown in Figure 12-02. In this example, we see that, *although, the trade is in balance for all states, but still, three states end up exporting high levels of unpaid labor, i.e., its equivalent value to the US and Iraq.* Thus, US and Iraq gain enormously even in the small-scale trading shown in this example. The gain by Iraq, however, does not become the gain of its ruling class, because Iraq is a Panamized state. Part of this capital gain ends up in the hands of the ruling classes of the US and its allies. Because of the low wages in China, Iraq and the Congo, the local rates of profit are very high and the rates of exploitation are also very high in these states, reaching 92 percent in the case of the Congo. The trade is, of course, between the members of the ruling classes or their corporations for all states engaging in it, except China and other states ruled by their communist parties.

IMF and its Role in the New World Order

Currently, major currencies are free floating. These include the British Pound, the Canadian dollar, the Japanese Yen, the Brazilian Real and the Euro for countries in the Eurozone. Most countries have exchange rates fixed to the US dollar, but managed by their central banks. China and India are among such countries. The IMF continues to monitor the economies and exchange rates of all states. The work of the IMF can be better understood by examining its relationship to one underdeveloped country with a floating currency exchange rate. The underdeveloped states are, now, referred to as "developing countries", which were former imperial possessions or colonies. Argentina, Egypt, Pakistan, Indonesia, or any other former imperial possession may be considered for this purpose. Virtually all states, now, have their central banks with authority to create money. Figure 12-03 shows the financial organization of a state and its relationship with the IMF.

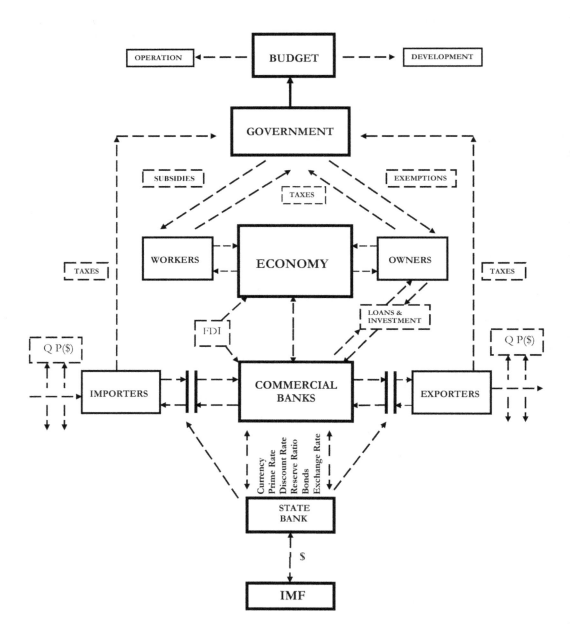

Figure 12-03: IMF in Action

For any state, the central government formulates an annual budget on the basis of expected tax revenues in the coming year and its expected or planned expenditures. The planned expenses may be for its social programs and the normal functioning of its various departments and ministries, or for its economic development activities. The central government revenues come from income taxes on companies and on individuals. Import "tariffs" and export "duties" are other such taxes. There may be sales tax or value-added tax revenues, or other taxes on inheritance and capital gains. There are two kinds of tax-payers - businesses and their owners form one category. The other category is of individuals who are employees only, and are not owners of any kind of business. Both these categories of tax-payers may have subsidies and exemptions allowed to them by their government.

The owners of industrial or agricultural businesses and those who own export and import companies, have to deal with commercial banks to obtain credit or to make payments in local as well as foreign currency. The state's government-controlled central bank controls the commercial banks. It can increase or reduce the money supply by selling or buying government bonds or foreign currency. It can change the reserve ratio for commercial banks to adjust the amount of loans they can issue. It can also change the *discount rate* it uses to issue loans to banks, which has an effect on the *federal funds rate*, as it is called in the US, that banks charge each other for inter-bank loans. This, also, has an effect on the *prime lending rate* of commercial banks, to which most consumer loans are linked. Also, the ruling class makes its investments through the commercial banks. Foreign direct investment (FDI) is also routed through these banks.

In a state that is tightly controlled by its ruling class, as in a developing country, the imports tend to be more for the consumption of luxury goods and to meet the needs of the local industry for plant and machinery. If the imports keep on growing as the demands of the ruling class keep going up, then a situation can arise such that the balance of payments goes negative and the foreign currency reserves of the state may not be enough to support the achieved level of imports. Also, if the tax revenue of the government is reduced because of tax reductions for its corporations or higher income individuals, or if the government has initiated new expensive social programs, its central bank may not be able to maintain its managed exchange rate with the US dollar, because it may be running out of foreign exchange reserves. Another reason for this may be the capital flight due to financial activities of its ruling class or emigration of its members, causing reduction in the foreign currency reserves of the state. In such a case, the IMF is willing to issue a foreign currency (i.e., US Dollar) loan to the state bank, so that the state may not default on its loans and would have enough reserves to meet its import requirements. In return for this favor to the ruling class and its government, the IMF normally demands that the government devalue its currency. The devaluation makes exports cheaper for foreign companies and imports more expensive for the local population. Thus, exports tend to rise and the imports tend to decrease. This tends to improve the balance of payments situation in the state's favor, increasing its foreign currency reserves, but since more products are exported than usual, shortages of products may develop in the local economy and their prices would, then, tend to rise, i.e., inflation results. To deal with the expected inflationary pressures, the IMF generally demands that the state cut its expenses by decreasing or eliminating the subsidies and exemptions it has given to the people or its companies - more so for the people. In negotiations, the representatives of the government of the ruling class in the state also tend to favor the cuts in subsidies for the people, rather than their companies. IMF may further demand an increase in government taxes on individuals and companies so that the budget deficit is kept within a specified limit. In general, an agreement is reached to eliminate or reduce subsidies and social programs, i.e., to reduce government expenses. This is referred to as a *"reduction of budget deficit"*, or *"reduction of deficit financing"*. Also, indirect taxes, that mainly affect the people, may be increased. *Thus, the burden of fixing the economic situation is shifted to the people while the trading partners within the developing state are protected to some extent.* They may gain because of the jump in the level of cheaper exports due to the devaluation. The importers may however, lose because of the reduction of imports due to their higher prices in terms of dollars. Further, for a state that has state-owned enterprises, sale of these enterprises to private owners is another favorite recommendation of the IMF. This tends to give greater control to the local ruling class over those state-owned enterprises. It leads to its rich members becoming their owners, providing them the opportunities to move the resulting accumulated capital to the developed countries. Also, this de-nationalization, or "privatization" makes those enterprises the targets of future foreign direct investment and acquisition. Normally, the reason given for this recommendation is that privately owned businesses are more efficient.

The general effect of the IMF's activities is that the standard of living of the people of the developing country falls. As inflation develops, strikes and protests of the people, especially government workers and employees of the former state-owned enterprises, result. People are laid off and unemployment and inflation increase. This constitutes the political cost to the government of the ruling class in the target state. The ruling class and its government have to face and "manage" its people. Since, a devaluation reduces the competitive value of the local currency, the local account holders tend to move their money into foreign currency and foreign accounts. Thus, local interest rates are generally increased to control this tendency. However, this also tends to fuel inflation further, because individuals and institutions of foreign states, also, find the increased interest rates attractive for short-term investment of their financial reserves and tend to move their bank account balances to the banks in the developing country – thus increasing the money supply in the local currency. Also, to deal with the immediate problem of foreign exchange shortage, the government of the developing state may engage in further direct borrowing from other states. This, also, may put inflationary pressure on the local currency. The IMF has acted this way since it was established after the Second World War, during the pegged rate period and, now, during the time of free-floating and "managed" fixed exchange-rate currencies. This is exactly what it was designed to accomplish, as an *Imperial Monetary Fund (IMF)*.

Effects of Devaluation of Currency

To understand the real effects of the devaluation of a state's currency in more detail, we can look at the last example again. However, we assume that, among the four trading partners of the US, China has devalued its currency by thirty-three percent, while Japan, Iraq and Congo have devalued theirs by fourteen, seventeen and thirteen percent respectively. The devaluation does not immediately change the cost of the fixed capital, already in existence and the facilitating and unfinished input capital already in inventory of local businesses of the developing country. There is no immediate effect on local wage rates. Workers are paid the same wages as before the devaluation. Thus, business owners feel that they have to pay less in local currency in terms of dollars, for products to be exported. Thus, the local rate of exploitation increases due to the devaluation, although the workers may be unaware of this situation in the short term. Also, due to the decrease in the dollar price of paid labor, the unpaid labor increases in dollar terms. This results in a decrease in the dollar cost of production of goods in the developing country, in the short term, as indicated in Figure 12-04. Because of the devaluation, the export prices go down in the developing country, in accordance with the rate of devaluation. The result is that demand of those products increases in importing countries. Thus, more products are exported by the developing country as is the objective of the devaluation.

	USA	CHINA	JAPAN	IRAQ	CONGO
DESCRIPTION	**CORN**	**COTTON**	**TV SETS**	**OIL**	**COPPER**
Currency	$	YUAN	YEN	DINAR	FRANC
Currency Exchange Rate	1	9	110	1,200	1,600
Rate of Devaluatioin	0	33%	14%	17%	13%
Exchange Rate After Devaluation	1	12	125	1,400	1,800
GDP/Capita ($) PPP Basis	$60,000	$16,700	$37,000	$16,500	$800
Approx. Median Wages ($)/Hour		$1.34	$13	$0.83	$0.44
Objectified Labor Used, Per Unit	$100	$15	$60	$20	$600
Total Living Labor Used, Per Unit	$400	$85	$190	$25	$2,400
Living labor - Paid, Per Unit	$200	$20	$69	$8	$175
Living Labor - Unpaid, Per Unit	$200	$65	$121	$17	$2,225
Total Cost ($)/Unit for Exports	$300	$35	$129	$28	$775
Sale Price/Unit for Exports	500	$67	$216	$100	$2,625
Rate of Profit, for Exports, in Local Curren¢	67%	186%	94%	253%	287%
Rate of Exploitation (ROE)	50%	76%	64%	67%	93%

Figure 12-04: Effect of Devaluation on Production of Goods

However, supposing that a rough financial balance is still maintained for the all the trading partners, the results would be as shown in Figure 12-05. We can see that for China, Japan and the Congo, exports had to increase from 400, 80 and 20 units to 597, 93 and 23 units, to maintain the virtual balance of payments for all the five states. However, the US receives more imports than before and the *unpaid labor transferred to it increases* from $14,300 to $41,733. The unpaid labor transferred to Iraq decreases somewhat, as the price of oil is not changed. However, the effect of Panamization, on Iraq, remain unchanged. If other states, besides the

US, are also importing products from China, Japan and the Congo, their balances of trade would also be affected. If those are developed countries, whose currency exchange rates are linked to the US dollar, or are determined by "managed" money market conditions, then, their imports would also increase sharply, because the currencies of China, Japan and Congo would have depreciated in terms of their currencies also. *Thus, a large transfer of unpaid labor would occur to those states also. The benefits to those states, in addition to the US, justify their membership and contributions to the funds of the IMF.*

Since many products, not just one, are actually exported, the real effect of the devaluation cannot be calculated in this manner. The long-term effects, on the economy of the developing state, are widespread. In general, the trade balance, in dollar terms, does not stay the same. Exports of products, from the states devaluing their currencies, increase considerably and the balance of trade tends to shift in their favor in terms of dollars and their foreign exchange reserves tend to increase. However, export of unpaid objectified labor from those states is greatly increased, in accordance with the objectives of the "Imperial Monetary Fund" (IMF). The imports of the developing states tend to decrease after devaluation, due to their resulting higher dollar prices. This process helps in creating a trade balance and tends to increase the foreign exchange (i.e., dollar) reserves of a developing country/state.

	EXPORTS			IMPORTS BY EACH STATE										
	Price			USA		CHINA		JAPAN		IRAQ		CONGO	
Country	Per Unit	Units	Total Price	Units	Total Price	Units	Total Price	Units	Total Price	Units	Total Price	Units	Total Price
USA	$500	340	$170,000		$0	80	$40,000	40	$20,000	100	$50,000	120	$60,000
CHINA	$67	597	$39,999	597	$39,999								
JAPAN	$216	93	$20,088	93	$20,088								
IRAQ	$100	500	$50,000	500	$50,000								
CONGO	$2,625	23	$60,375	23	$60,375								
Total Exports of each State					170,462		39,999		20,088		50,000		60,375
TRADE BALANCE (Unpaid Labor)					-41,733		22,805		3,253		-11,500		27,175
TRADE BALANCE (Price)					462		-1		88		0		375

Figure 12-05: Transfer of Unpaid Labor in Inter-State Trade, due to Devaluation

As the exports grow from a developing country/state after a devaluation, shortages of exported products develop in the local economy, giving rise to inflation. The increases in prices of essential consumer goods, if exported, directly affect the working people. Also, the budgetary cuts in spending on social programs and subsidies on food and services, like electricity supply, gas supply, housing rents and education fees, etc., affect them. The transfer of financial assets, by the local ruling class of the developing state, directly profits the ruling classes of developed states and tends to reduce the foreign currency reserves of the developing state. The privatization of state-owned enterprises results in layoffs and the increases in indirect taxes, like sales tax, mainly affect the working population. As the standard

278

of living of the working-class falls, it creates demands for wage increases by the workers. These demands are normally resisted by the ruling class of the developing state. Protests and demonstrations by workers result and, generally, lead to suppression of such demands by police violence. Thus, the working people become the main victims of the devaluation.

Effects of Unfair Trade on Developing Countries

Developing countries have learned their lessons after the recession of 1980-1983 accompanied by the debt crisis of 1980s to 1990s[10] and the Asian financial crisis of 1997[11] and most have accumulated high levels of foreign exchange reserves to avoid dictation by the IMF. Further, these states continue to maintain positive current account balances in their inter-state trade.[12]

Despite these precautionary measures, developing countries, continue to face high level of exploitation under the New World Order. Financial management of Inter-state trade under the control of the IMF (which is really the "Imperial Monetary Fund"!) is not the only cause of this exploitation. Large amounts of capital are also being transferred to imperial powers, as "tribute" by the ruling families and tribes of Bantustans and by the ruling classes of Panamized states created and maintained by those imperial powers. Yet, even "liberal" economists seem to be puzzled by this phenomenon. They refer to this huge uphill flow of capital from poor countries to rich imperial powers as a "paradox" and invent elaborate theories to explain its causes – ignoring the obvious imperial techniques.[13] Further, what is not realized is that only the financial uphill flow of capital is being accounted for. *The far greater flow, in terms of unpaid labor and value, is ignored!*

When former imperial possessions and colonies were declared independent by their imperial powers, the ruling classes of those imperial possessions inherited all the institutions of control and abuse from their former rulers. One technique that continued to exist was the exercise of power by big land-holders, who were formerly patronized by the imperial powers like Britain. These big land-holders, or pseudo-rulers were rewarded with large "free" agricultural lands as rewards for services to the ruling power. The inherited legal systems continued to be heavily tilted in favor of the moneyed classes, with virtually no property title system to protect small land holders and dependent peasants. The "elections" held by Britain in India had heavy property requirements even for voting, thus enabling landlords in selecting other landlords as "representatives" of the people! The same applied to the thirteen British colonies in North America that fought for and achieved independence from Britain. As power was concentrated in a few hands in the former colonies, it generated extreme cases of corruption in the local population in accordance with the principle that – *"power corrupts and absolute power corrupts absolutely"*. Under those conditions in the newly independent countries, state ownership of resources, protection of new "import substitution" industries, establishment of financial systems for business and trade, etc., were also forced into corruption and resulting inefficiencies. Many of the legitimate criticisms of developing states tend to ignore this fundamental phenomenon.[14]

Some developing states have succeeded in developing their economies and trade relationships. South Korea and Taiwan may be included in this category, but their apparent success lies in Panamization and the creation of the US controlled *"Greater East Asia Co-*

Prosperity Sphere" - in which, initially, the US had to invest heavily to rehabilitate the war-ravaged economies. China, India, Indonesia, Malaysia, South Africa, Brazil also fall in the category of successful states. Their success is based on the fact that they focused on development of their human resources, or "human capital", by investing in education and health care and promoting population control. They, also, developed their infrastructure and high-tech industries in parallel with development of their human capital and protected those import substitution industries till they took off. Further, they carried out land reforms to varying extents and developed state-owned banking and insurance systems. Most of such successful states have also reformed their judicial systems and have tried to stamp out corruption – in this aspect they are still struggling as corruption is part of capitalism. Those states are developing while capital accumulation continues with exploitation based on the "market mechanism". Some of that accumulated capital is used by companies for bribes to government functionaries, to obtain tax exemptions and other favors

Chapter-13

The Semi-Global Empire

Origin and Evolution of Empires

When the ruling class of a state is not satisfied with the accumulation of capital by exploitation of the population within its boundaries, then its intense greed drives it to seek accumulation of wealth by exploitation of other states and their peoples. Since a society is divided into two classes at all stages of evolution of a state or its society, this motivation is developed by the ruling class of a state at any stage. However, this desire for establishment of an empire is most acutely felt at the two evolutionary stages of intensive and benevolent capitalism. Here is a summary of this evil phenomenon at all stages of evolution of mankind:

At the stage of primitive feudalism, societies consisted of tribes, in which the ruling families accumulated wealth that their tribes were able to gather, or loot, from other tribes, including land, animals, food-grains, slaves, etc. Complete freedom of the jungle existed at that time. There were no restrictions on the "economic market", i.e., the jungle. The "law of the jungle", or the "law of supply and demand", was supreme at that time. Tribal wars were frequent and whole tribes could be wiped out by bigger tribes. However, since the loot was not substantial, it was mostly consumed. Only land, the ruling family's habitation, its slaves, and some valuables could be passed on to the next generation of rulers.

When the social system evolved and nations emerged, by conquest of many tribes by one or more tribes, or by voluntary fusion of tribes, resulting in creation of one central authority, the accumulation of capital increased substantially. Nations existed as advanced feudal nations. Wars against tribes and other nations continued resulting in mass murders on a large scale. Whole tribes and nations were enslaved, or effectively enslaved by being forced to pay "tribute" to the dominating nations, e.g., the nations conquered by the Roman Empire had to pay "tribute" to the Roman Emperor, the Caesar. Complete freedom of the jungle continued at this time. No restrictions were imposed on the "economic market", and the "law of the jungle", or the "law of supply and demand", continued to prevail. Thus, any individual, tribe, or nation could be robbed whenever possible. The rulers of these nations inherited much more than the tribal "elders" used to inherit in tribal times. This was the time when civilizations emerged in Egypt, Mesopotamia, the Indus Valley, around the Yangtze river and Greece. Still, the accumulated wealth, or capital, was not huge, because most of it was used on construction of what look like wasteful royal and religious structures, today.

With the beginnings of development of intensive capitalism in advanced feudal societies, states and empires, factories and mines were developed in Europe and North America, mercilessly abusing the working people by creating and maintaining a "freely" abusive labor market, leading to fast accumulation of capital in the hands of the ruling classes of these regions, and the resulting Intensive Death Rate of their people. As huge amounts of capital began to be accumulated by the owners of factories and mines, the insatiable greed of the ruling classes overflowed their factories and mines and even their states. The ruling classes embarked on subjugation of peoples living on large sections of the world and began to create empires for loot and plunder of those peoples, in addition to those at "home". The Great Genocide was a concurrent phenomenon in the territories newly occupied by Europeans. The United States, meanwhile, broke away from the British Empire and started the creation of its own empire, occupying more and more land of North America and killing its native peoples, during the *Great Genocide*. It then began to occupy lands outside North America and declared South America its "backyard", which only the US had the "right" to exploit.

As the whole world was covered by new empires, and the insatiable greed of their ruling classes collided, large scale wars began among their imperial masters. These imperial wars increased in intensity, till the empires destroyed each other in the two World Wars. The United States used these wars as an opportunity to establish its hegemony even over the existing imperial powers and to start towards its ultimate objective - establishment of an empire over all of humanity. Only the Soviet Union resisted the power of the US, leading to the destruction of virtually all previously existing empires and also bringing the process of colonization to a virtual halt. Thus, the US could not succeed in establishing its virtually global hegemony, although it did succeed over a large part of the world.

With the disappearance of the Soviet Union, the US has, become, apparently, the unchallenged imperial power of the world. It has established its hegemony over all other imperial powers and many other states. Only China and, basically, the successor states of the Soviet Union remain independent. Also, some other states, like Brazil, India and Vietnam, are still maintaining their independence to some extent. The US is, however, continuing to expand the semi-global empire it controls. To maintain its virtually global imperial hegemony, it has deployed its military on all oceans, and is maintaining more than 600 military bases, spread all over the world.

Tran-State Corporations

States and Corporations are the two most important types of social groups that dominate the world's political development today. A large number of trans-state global corporations have developed, mostly centered in the US and Europe. These trans-state corporations (generally referred to as multi-national or trans-national corporations) have grown enormously in size and power. They operate in all regions of the world with more or less freedom and exercise influence on the states of those regions. Many smaller states are helpless in facing the enormous power of these trans-state corporations, which exploit them with the "golden" principle of intensive capitalism – *"Maximum rate of intensive profit (Maximum RIP)"*.

The most important global financial institutions, at this time, are the International Monetary Fund (IMF), the World Bank and the Asian Development Bank (ADB). These financial institutions were set up by the United States and are also controlled by it to a great extent, directly and indirectly. The US uses these institutions to further its global imperial hold. There are numerous trans-state corporations based in the US - General Electric, Chevron, Exxon Mobile, ConocoPhillips, Wal-Mart, Boeing, McDonnell Douglas, McDonalds, Burger King and PepsiCo are some of the American trans-state corporations.

Similarly, Mitsubishi, British Petroleum, Siemens and AREVA are Japanese, British, German and French trans-state corporations, respectively. These corporations are owned and controlled by their "home" states, and their "home" ruling classes, although they have global operations. The capital that they accumulate by their wide-spread operations ends up in the hands of their share-holders, who are, mainly, citizens of their home states. Since capital flows in the post-Soviet era have been freed and the shares of these corporations are available to all investors, a multi-state and multi-national ownership of these corporations has emerged. However, since the US ruling class is the biggest class of owners of business and industry, or oligarchs, in the world, the ownership of these corporations has come to be dominated by the US ruling class and is becoming more so, as the capital accumulation continues globally.

A corporation is purely an economic social group. It is a powerful private tyranny and its purpose is maximum economic exploitation of its workers and customers, or maximum profit, no matter how and where it is engaged in production or distribution of goods and services. It always tries to maximize its capital accumulation and makes the capital gains available to its share-holders. Because it is the most organized and most flexible institution that capitalism provides, it has a great influence over the relevant state itself. Corporations gobble up other corporations and keep growing in size and power. At this stage, this power has grown to such a level that many corporations exercise a high level of control over their "home" states and the other states where they operate. The reason is very simple - the corporations are owned and controlled by the ruling classes and the governments of most such states are also controlled and collectively owned by the same ruling classes. Deceptive claims of "democracy" and "human rights" are routinely made by these governments, but these corporations actually promote the aims and interests of the ruling classes only. Unintended consequences of their activities may be that the people of their home states, or other states, may also benefit. Those states which control the bigger corporations, or are controlled by these big corporations, are the ones which end up controlling other states, indirectly, through the use of government and corporate power. As the members of the ruling class in a state want more, *in fact maximum*, profit, the corporations owned by this class also work for this objective, guided by the unquenchable greed of their owners. This economic power, with unlimited greed, is now in the process of transforming the world into, basically, one empire. The semi-global empire has a large number of corporations. These corporations are all interwoven in terms of ownership, but are controlled, effectively, from one center of management, the "center of gravity" of this empire - in the United States, at this time.

Structure of the "New World Order"

What Genghis Khan and Hitler only dreamed about, has been achieved by the United States by the end of the twentieth century. Its virtually global hegemony is consolidating itself and is deepening the exploitation of all of humanity. The United States is also trying its best to expand its hegemony to cover areas of the world still outside its military and economic control. It is easy to see the structure of the emerging, virtually global, empire. The United States has used all the three techniques used by other imperial powers over the previous centuries and the technique of "Panamization" that it has developed and used, first, in the creation of Panama. The approach, now, is much more sophisticated, since the population of the whole world, no longer just of the United States, has to be deceived into believing that the

United States is the "leader of the free world" and stands for human rights and justice, as a "shining city over a hill" for all of humanity, while its ruling class engages in the exploitation of the people of the whole world, including the people of the US. This grand deception has succeeded, so far, in deceiving the people of the United State, who are one of the most compassionate and large-hearted people in the world, believing in justice and fair-play. But it has also succeeded in confusing most of the other people of the world, because of the control over the mass media that the ruling classes have, globally. We can see all the four techniques at work in the "New World Order", or its Semi-global Empire, that the United States, in reality, has created, and is still trying to expand.

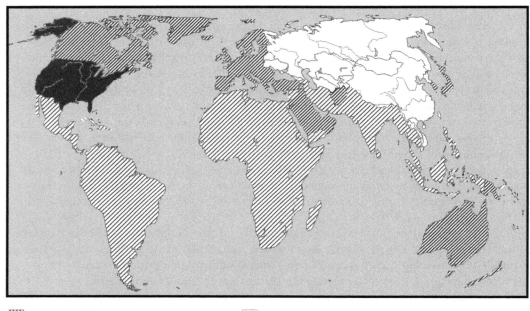

| | Panamized/Bantustanized areas | | Areas resisting imperial control |
| | Areas subjected to controlled-market subjugation | | |

Figure 13-01: Structure of the Semi-global Empire

(Map based on world map at: https://www.d-maps.com/carte.php?num_car=13184&lang=en)

Panamization

The United States has Panamized whole of Europe, North America, Australia, Japan, South Korea and Taiwan. Panamization of Iraq, Libya and Afghanistan is in progress. Corporations of the US have set up extensive operations in these Panamized regions, in addition to the Jewish colony in Palestine. Thus, the United States controls the Panamized regions of the world, including the other "allied" states of Europe and Japan. It is, now, imposing Panamization onto new states, especially in the Middle East.

At this time, the focus of the US is on completion of the Panamization of the independent oil-rich states of the Middle East. Most of this region has already been Bantustanized. Iraq has been the one of the latest states to be subjected to this process of Panamization starting with the 1991 invasion. Panamization of Iraq is now almost complete and most cities of Iraq are in ruins. More than a million people of Iraq have been killed in this process. Afghanistan is, at present, going through this process. The objective of the US in Afghanistan is not the control of natural resources of this state. Afghanistan is not a resource-rich state, though it does have some copper, iron and precious metal reserves. In Afghanistan, the US goal is to use this state as a spring board to access the oil and gas resources of Central Asia and Iran and to deny these resources to China. Elimination of the nuclear power status of Pakistan is another important objective, since Pakistan, as a nuclear power, is considered a threat to the US domination of the Middle East. Thus, Panamization of Afghanistan is to help in the "containment" of China, exploitation of oil and gas resources of Central Asia and destruction of Pakistan as a nuclear state, in addition to suppression of the current terrorist activities of Al-Qaida and its allies. In March, 2007, General (Ret) Wesley Clark of the US Army, gave an interview on "Democracy Now" - a television station. A video of this interview was posted on YouTube also. In this interview, he explained the plans of the US military, in late 2001, to invade seven countries in the Middle East in the coming five years – Iraq, Syria, Lebanon, Libya, Somalia, Sudan and, then, Iran. He thought that the main reason for these attacks, was the fact that these countries have large oil reserves. According to him, these countries would have been treated like countries in Africa, if they did not have oil.

Bantustanization

The United States has not created many Bantustans outside its territory, except those in Nicaragua and Guatemala. These two Bantustans, headed by two families, have been dismantled by their people. Those within its territory are referred to as "Reservations". It has, however, taken over the British-created Bantustans of the Middle East. Its oil companies also dominate the British-created Bantustan of Brunei, in South East Asia.

Colonization

The United States is giving lower priority to the expansion of the Jewish colony over Arab lands, because of the new project of Panamization, from which it expects to increase its "tribute" from the region as a whole. Once, the Panamization of the Middle East region is complete, then, the United States may increase its collaboration with the Zionists for further expansion and consolidation of the Jewish colony. However, the revolts of the people of Arabia as part the "Arab Spring", against their ruling dictatorial regimes and the ruling families, tribes and classes of the Bantustanized and Panamized states, created by Britain and the United States, may jeopardize those plans for further Panamization. Israel is the only state that is trying to swim against the flow of historical forces and is trying to create a racist state based on apartheid, under the protection of the US. Ultimately, it would have to face the power of the Arab people and the rest of humanity. But, in the short-term, Israel suites the imperial plans of the ruling class of the US, as a "Manhattan" of the subjugated Middle East.

Controlled Market Subjugation

The US and other "allied" powers collectively exploit South America, non-Arab Africa, South Asia and South-East Asia by using the technique of controlled market subjugation, with the help of the IMF, WTO, the World Bank and the Asian Development Bank. Part of the capital, accumulated by the European sub-imperial powers, Australia and Japan, gets transferred to the US, through its trans-state corporations - thus, the US obtains "its share" of the "tribute", from these "allies" also. In 1987, Gorbachev discussed the realities of the abusive relationships between the imperial powers and developing countries as follows:

> **While we do not approve the character of current relations between the West and the developing countries, we do not urge that they should be disrupted. We believe these relations should be transformed by ridding them of neocolonialism, which differs from the old colonialism only in that its mechanism of exploitation is more sophisticated. Conditions are required in which the developing countries can be masters of their own natural and human resources and can use them for their own good, rather than for somebody else's.[1]**

> Quoted from *Perestroika* by Mikhail S. Gorbachev. pp 139. Copyright© 1987 by Mikhail Gorbachev. Used by permission of HarperCollins Publishers.

After 2008, Gorbachev called for a "Perestroika" of the societies around the world, starting in particular with that of the United States, because the 2008 recession precipitated by a financial crisis showed that the development model under the new world order was a failure that will have to be replaced, sooner or later. According to Gorbachev, countries that had resisted the dictates of the new world order and its institutions, like the IMF, the World Bank and the Asia Development Bank, had done far better economically on the whole and achieved far more fair results for their citizens than those countries which had given in to the demands of these institutions. He included China and Brazil among those states which had resisted those institutions.

Developments within the United States

The emergence of the United States has been discussed in Chapter-6. After the Second World War, as the US initiated the creation of its global empire, an unexpected development created a serious difficulty in achievement of this objective. This was the emergence of the Soviet Union. The US embarked on the Panamization of Europe, Japan and South Korea, but to face the Soviet threat, concessions had to be made to the working people of North America, Australia, Europe and other Panamized states – to prevent the people of these regions from staging rebellions because of the "communist" ideas being propagated by the Soviet Union and its allied states. A progressive tax structure was adopted to finance the required social security programs and the standard of living of the people of the US improved greatly. In the 1980s it was noticed that the economic and military policies of the Soviet Union had landed it in great difficulties and its defense burden had become unbearable. With this realization, the post-war policies of the US and Europe were reversed by their ruling classes. Taxes on businesses and the richest members of the ruling classes were reduced and cuts in social programs were initiated. The result was that the standard of living of the people began to fall. This process was accelerated after the disappearance of the Soviet Union. The standard of living of the working people has continued to fall. Strikes and demonstrations by the affected people have increased with time - both in the US and in the European Union. As a result of the growing inequality and lack of social protection, recessions have also grown in duration and intensity, as described in Chapters – 7 & 9. Howard Zinn, a historian, has described the conflict between the "establishment" and the people of the US in detail, even in recent times. [2]

Trying to understand political events from the point of view of the citizens of the US, it is strange to see the level of trust placed on television news, TV programming and media in general. Most people seem to think that whatever is said, or displayed on television, is the absolute truth, especially, if an event in a foreign country is being discussed. Of course, they never get to know what has not been displayed or discussed on TV. Generally, if two sides of an argument are presented, they are not radically different. It takes time to understand the reasons for this much power of television. What this hold of the "media", or the "soft power" of the US ruling class through its ownership of all media, on the minds of the American people means, needs some explanation:

- Children grow up watching TV programs, like "Sesame Street" and learn how to spell, what to say and what not to say. As they grow up, they see advertisements and TV programs, in which new models of cars, new mansions, boats, motor-bikes, etc., are depicted. Pretty girls and handsome young men are inevitably associated with such ads. Young people learn to covet such things and their standards of beauty and goodness are formed by such ads. In other words, TV has become a teacher and a parent for most American children. They are trained by the Almighty TV and its programming. No wonder, they learn to trust whatever it says!

- What applies to pre-school children at home, also applies to teenagers in school. The Almighty TV continues to work on their minds. Teachers, and the textbooks and videos they use at school, build on what has already been started by the television programs. Virtually no high school teaches world history to its students, as a required subject. In many schools, teaching of the law of evolution and evolutionary history is forbidden. Haphazard courses are offered by most schools about history of some regions of the world, which the students can choose to take, or not, at their own discretion. History of the US is, of course, a compulsory subject and so are videos of Hitler's persecution of Jews, without any mention of persecution of Palestinians by the Jews. The result is that it seems to most students that history started in 1792 and Jews are the only people who suffered at the hands of Hitler! Economics that is taught to students is heavily weighted in favor of "business" as an honorable profession, which can do no wrong! Virtually no geography is taught and most students are, thus, ignorant about geography and geology of the rest of the world.

- Through books, newspapers, movies, television and radio programs and the internet, covering all kinds of cultural activities, even adults are constantly bombarded with ideas equating "capitalism" with democracy, glorifying "free trade" and "free enterprise", without any analysis of these economic issues. The "American way of life" is also glorified without specifying what it means. All this ends up reinforcing what has been learned in school and watching TV as children and teenagers - that the US and all this is great! Everybody begins to dream of becoming a "great business person" and making it to the "American Dream", i.e., becoming rich and famous members of the ruling class! Wars and bombings of other peoples and countries look like video games to most Americans. When the US kills hundreds of thousands of people in other countries - the victims seem to them like pseudo-robots in video games, deserving no sympathy. Justifications for such wars by "experts" on the TV screen automatically look valid to them. In fact, most Americans seem to think that they are waging those wars - and not those in power.

Given this power of the media in American society, it is not difficult to imagine that what is being presented here, in this book, may be treated with skepticism or even with hostility. *Such is the "cultural hegemony" of the ruling class of the US – as Antonio Gramsci would have described it.* Most Americans think of the US as a "great" country, or an ideal state. It has certainly become great in some ways, but not in every way to be desired. Specially, its treatment of its own minorities and other countries has been terrible.

Developments Outside the Empire

China, Russia, India and other Central Asian successor states of the Soviet Union, are not within the stranglehold of the United States and its Panamized sub-imperial "partners", or their corporations. Also, included in this category are other smaller states like North Korea, Vietnam, Mongolia, Kazakhstan, Cuba, Laos, Angola and Mozambique, etc., which have managed to stay independent of foreign dominance.

China

As described in the last chapter, China is fast developing its infrastructure and key industries, under an ambitious plan of social and economic development. It, now, has the largest economy in the world in GDP terms, on Purchasing Power Parity (PPP) basis. It is likely to become the largest economy of the world on the exchange rate basis also, in about a decade. Because of the large state-owned sector of its economy and its social controls to achieve wide-spread and stable well-being of its population, *China has achieved in terms of economic and social development, what no other state has achieved in history.* Since after the 2008 financial crisis, China has also embarked on restructuring its economy from export-based growth, to growth based on domestic consumption. It has dramatically changed its focus on pollution of its ground water and air. It has started moving towards renewable energy resources instead of use of oil and gas for electric power generation and electrical transport vehicles, etc., so as to decrease its CO_2 emissions that contribute to global warming. Its other environmental protection policies are also being enforced strictly. It is working on increasing the energy efficiency of its industries and is moving its economy toward higher technology, including artificial intelligence and robotics. These transformations would result in a healthy and efficient high-tech economy which would result in higher standards of living for its population. These are dramatic changes but the transition may take decades.

Among the larger economies, China fared the best during the Great Recession. Its GDP growth rate, however, decreased to about 6.2% in 2019. It should be noticed that despite introduction of the economic market into its economy, China's major industries remain heavily state-owned. Its health care, education and financial systems have remained almost fully state-owned till very recently. All land is owned by the state. Thus, China has remained immune to the global financial crises and its banks have had no involvement in the foreign mortgage financing or stock market activities. It has, thus, established an economic system devoid of recessions and depressions. The global economy continues to face recessions every few years, but *China's economy acts as a stabilizer, or a "flywheel", of the global economy. Its growth rate does go down if the rest of the world is going through a recession, but it does not go into a recession.* For the last thirty or so years it has behaved this way, even though it has been focused on export-led growth. Now, after the planned shift towards internal consumption, the Chinese *economy would be further isolated from the ups and downs of the rest of the global economy.* Thus, it would continue to have a higher level of long-term efficiency compared to the economies of states under *intensive capitalism,* or even *benevolent capitalism.*

In the future, Taiwan's reunification with the mainland is a foregone conclusion, as China becomes the center of a powerful super-state. Japan, Korea and other South-East Asian states are also likely to become closely intertwined with this super-state. During the 1980s, Taiwan and the People's Republic of China had public contacts with each other and cross-straits trade and investment has been growing ever since. Although the Taiwan straits remain a potential flash point, regular direct air links were established in 2009 and a process of de-Panamization of Taiwan is currently underway and a similar process is likely to transform Japan and Korea also, in the future. Meanwhile, their governments are officially adhering to a "One-China policy."

China has also become a major investor in African and Asian states, especially in the oil, gas and mining industries. Further, it is importing key minerals on a large scale from Australia and South America and is developing economic relationships with major mining companies. Its own mining and oil and gas producing companies are also growing and expanding their operations to South-East Asia and Latin America, including Cuba.

In 2013, China proposed a huge infrastructure plan to boost trade with Europe and all over the Eurasian land mass and Africa (referred to as the "world Island"). The plan is referred to as the Belt and Road Initiative (BRI). It is presented as a new "silk road" between China and the rest of Eurasia and Africa and implies creation of an economic belt of relationships between China and other states of this huge region. It is meant to directly connect China with its Asian neighbors, Europe and Africa by means of roads, railways and seaports. Dramatic progress is being made in this long-term project. An additional motivation behind this project is the security of China's trade routes, since China does not have a big navy at this time. The project would, thus, dramatically transform trade relationships of most of the Asian and European states. Even African states and states on other continents would be affected, since the impact of this project, which may take about 50 years for its completion, would be *global*. China has also launched the Asian Infrastructure and Investment Bank (AIIB), which would also facilitate this project. Since China has become a big importer of raw materials, especially minerals, it has developed close economic links with African and South American states also. It has established the National Development Bank (NDB) with the BRICS (i.e., Brazil, Russia, India, China and South Africa) group of countries and has formed the Shanghai Cooperation Council with its Asian neighbors, to discuss economic and security matters. One objective of these activities is to introduce the Yuan as a currency for trade transactions – ultimately leading to adoption of its currency as a reserve currency, at least for export/import transactions with its neighbors.

Around 2015, China announced a plan to dramatically improve its manufacturing sectors. It is already the biggest producer of steel in the world, producing about half the steel produced in the whole world. It is also a large producer of aluminum and copper. The new plan is called "China 2025" and describes China's intensions to focus on high technology development of its industry. Under this plan China aims to be in the forefront of global industrial development in ten sectors, which it considers crucial for its future. The ten sectors are:

- Ship building
- Aviation
- Agricultural Machinery
- Robotics and Artificial Intelligence (AI)
- Internet Technologies
- New Materials
- Medical Devices
- Power Generating Equipment
- Railways
- New Energy Vehicles

China is already ahead of all countries in manufacturing railway equipment and construction of railway lines. It already has the longest high-speed railway-line system in the world, utilizing the magnetic levitation technology for most of its railway lines. It, also, has become the largest manufacturer of ships, especially container ships, for shipping lines. It is ahead of all other countries in new energy vehicles, especially electric and hybrid road transport vehicles. It has recently produced a medium-range civilian transport aircraft, named the C919 and is working on a larger aircraft in cooperation with Russia - the CR929.

China has gone far ahead of other states in 5G technology for its computer and telecom networks. It has developed high powered super-computers and is, already, ahead of the US in terms of power and number of its super-computers. Its internet-based companies have also become quite large in terms of global market share. In this area, China intends to become, globally, the dominant state. China is already quite advanced in the field of robotics and expects to be one of the front-line states in this area. What this means is that China intends to introduce high-level automation and robotics in its manufacturing industries to meet the shortage of manpower that is expected to develop in the near future.

In the area of renewable energy and production of electric vehicles, China is already far ahead of any state of the world. It intends to further develop its renewable energy resources, including hydroelectric, solar, wind and nuclear power industries. The development of solar cell and wind turbine technology, has reached a point where these two sources of electric power generation are becoming less expensive than generation of electric power using fossil fuels, i.e., using coal, oil or natural gas. This, in addition to its stricter air and water pollution standards, would increase its energy-use efficiency and would help in its contribution towards controlling the dangers of global warming, while improving the environment in its cities. China, also, has a large program of nuclear power generation and is helping other countries in this area. Under this plan, China also expects to become a leader in production of agricultural machinery, in development of biotechnology and medicines, medical equipment and in developing new materials, based on nano-technology, for industrial use.

It is expected that China's economy would grow to three to four times its present size in terms of its GDP, by the middle of the 21st century. By that time, its economy may be equal to the economies of the *US and Europe combined*! The standard of living of the average Chinese citizens would improve accordingly. At the same time, most of its population would have moved to urban areas, resulting in a dramatic increase in urbanization that is already clearly visible in Chinese cities. The current rate of urbanization of its population is about 3 percent. However, due to aging of its population and transformation of its economy towards domestic consumption, higher technology, environmental improvement and higher standard of living of its population, the rate of growth of its economy is expected to decrease - this process is expected to be spread over the next three or four decades, as its population levels off and then decreases. *Its effective rate of growth is, however, likely to remain much higher than the economies at the stages of intensive or benevolent capitalism, since it is not likely to go through recessions in the future – like in the previous four decades.*

China's government is not the government of its exploiter class. So, naturally, it has shown no imperial appetite to conquer foreign territories and peoples. In fact, its foreign policy is focused on helping other developing states in achieving self-reliance and independence, especially those in Africa and South-east Asia. China is also helping states of these regions in terms of acquisition of military equipment for improving their defense capabilities. This is bound to raise its global political stature.

Russia

Russia's economy has begun to develop after going through a destructive phase during the Yeltsin years, when the command system of economy was abruptly changed into intensive capitalism, naturally causing an *unusually intensive mortality rate* and *intensive drop in birth rate* of its population because of the sudden change. The processes involved at the stage of intensive capitalism have continued to work in Russia and its population has been decreasing despite large-scale immigration from other successor states of the Soviet Union, as the old and sick keep on dying due to lack of proper nutrition and medical care. Recent policies of the Putin government have, however, resulted in an increase in birth rates. Despite the drastic reductions in its nuclear arsenal, Russia remains a nuclear superpower and has started taking part in resolution of global political problems like global warming, nuclear fuel disposal and global terrorism.

It is also participating in efforts to resolve conflicts like the Israel-Palestine problem and Iran's conflict with the US regarding its uranium enrichment program despite the multilateral agreement between Iran, the US, Russia, China and several other states of Europe. The US withdrew from that agreement in May, 2018. North Korea's regional conflict because of its nuclear program, is another issue in which Russia has become involved. Russia's involvement has the effect of restraining the US in its aggressive military adventures.

Russia has also developed a strategic military and economic relationship with China because of the complementary nature of the economies of the two states. The European Union is the main trading partner of Russia, but it has become a major supplier of military hardware and technology to China and the trade between the two neighbors has been growing geometrically. Russia is the richest country of the world in terms of natural resources. Since it is also the largest country of the world in terms of land area, and is sparsely populated, the twenty-five years, or so, of continued reduction of its population has made matters worse. On the other hand, China is an over-populated state, which has shortages of oil, gas and other strategic minerals. Thus, the economies of these two states are complementary to each other in terms of population, natural resources and the sectors in which they have, already, made technological progress. Both the states are facing economic and military pressures from the US.

India

The main mistake that India made at the time of its independence is that it forcibly occupied most of the territory of the then "Princely State" of Jammu and Kashmir, against the wishes of its Muslim majority population. Later on, it refused to negotiate and demarcate its border with China, considering Ladakh (The Tibetan part of Jammu and Kashmir) and South Tibet (which it now calls "Arunachal Pradesh") as its "imperial inheritance" from the British Raj. The British had conquered these Tibetan territories when they invaded Tibet in 1906. The McMahan line was drawn to indicate the new border of "British" India with Tibet. Both these mistakes have cost India dearly, but its ruling class continues to stick to its ambitions of re-creating the "Greater India" of Hindu mythology. Hindu Imperialism has led to a war with China in 1962 and three major wars with Pakistan, in addition to other minor conflicts. The subjugated nations of Kashmiris and Baltis, in the Indian occupied part of the "Princely State" have continued to struggle against occupation of their territory and have consistently demanded their right of self-determination. In 1989, the struggle changed into guerilla fighting. The terrorists of the Indian Army have responded with mass murders of civilians and gang rapes of their women in the Serbian style. More than 100,000 men have been killed and thousands of women have been gang-raped. Several mass graves have been discovered as reminders of the war in the Balkans and atrocities similar to the *Srebrenica massacre*. As the fighting continues, the Indian government has started labeling the guerilla fighters as "terrorists"!

India developed a mixed economy during the cold war period. The state sector of its economy remains quite large even though its economy has been "liberalized" to some extent. Its banks, airports, seaports, motorways, railways and airlines remain mostly in state hands. Similarly, a large segment of its mining, oil and gas production, steel production, etc., are still in the state (or "public") sector. It has a system of public education for children. Many state-owned universities have been set up, where higher education is subsidized. There is, also, a food security program for the poor. However, the religious discrimination against Muslims and Christians; and against its "lower class untouchables" continues, affecting its peaceful development. Its GDP growth rate has picked up steam, after its economy was opened up for foreign direct investment and trade. Like China, it did not go through a recession during the global Great Recession, although its growth rate went down during that period – more so than China.

India has remained independent and continues to be so, even after the disappearance of the Soviet Union, its main source of military equipment and economic help during the cold war. However, now, the US is preparing it for its policy of military "semi-containment" against China. For this purpose, the US has extended military and nuclear support to it. It remains to be seen how far India would become an ally of the US in this regard. India has an economy that is equal to that of Japan and Russia combined. It is likely to become a military and economic super power in the more distant future, as compared to China and Russia. The US is not likely to achieve Panamization of India, however. Instead, India is more likely to remain the power center of the South Asian super-state. Its stature and major power status would rise if it is able to settle its territorial and national disputes with China, Pakistan and the Kashmiri nation.

Middle East

During the Trump administration, the US withdrew from the Iran nuclear deal and re-imposed sanctions on Iran with additional political and military demands. The Iranian government has, so far, refused to enter into new negotiations. Movements of US and British naval vassals into the Persian Gulf and the activities of unmanned "drone" aircraft near the Iranian border has greatly increased tensions in the area. *Once, Iran is "ripe" for a conquest, the US and its "allies" are likely to launch the final invasion to initiate the process of its Panamization.* But the project is being delayed because Panamization of Afghanistan is taking much longer than expected. Only when the Afghanistan project has been completed and the full Panamization of Iraq and Libya has been completed, the invasion of Iran may be launched to *"prevent it from developing weapons of mass destruction"*, as usual! At present, the theocratic regime in Iran is doing everything possible to make it possible for the US to conquer Iran - by denying basic human rights to its citizens, converting the state into a theocratic state, in the first place, instead of separating state and religion, like the rest of the civilized world. In short, the Iranian regime is doing virtually everything that an ideal state would not do and is, mostly, not doing what an ideal state would do, thus creating the conditions which may, finally, lead to the subjugation of its people by the United States. However, Russia and China are trying to strengthen Iran's economic and military posture. Things have changed since the negotiations were started regarding Iran's nuclear program. Iran has received military equipment from both Russia and China. China has also started several economic projects in Iran and is a big customer of

Iranian oil. Both China and Russia are trying to mediate between states bordering the Persian Gulf. It seems that a successful conquest of Iran has become a much more difficult project than before.

As can be clearly seen, like Iran, the regimes of the Arab states have adopted the worst forms of political systems to resist their subjugation. Thus, the subjugation is succeeding in state after state. The Bantustans of Arabia, of course, are already subjugated by the United States and Britain. Those military regimes, that seized power in the 1950s and 1960s, consolidated their hold on their states during the 1980s and have been in power until recently. The military regime in Egypt collapsed after the people rose against it in rebellion and the dictator, Mubarak, and the military had to give in. A representative government was set up, but because of its "Islamic" ambitions, it could not succeed in controlling the military and was overthrown. Similarly, the people of Yemen rose in a rebellion against their dictatorial "president", Salah, who had to flee to Saudi Arabia. He died later on, in Yemen. A civil war in Yemen is in progress, with Saudi Arabia trying to install its favorite regime there. The people of Jordan, Morocco, Algeria and Kuwait had also risen against their regimes. The Tunisian regime was overthrown, but the other regimes, hurriedly, made political and economic concessions to their people and managed to stay in power, for now. The people of Bahrain also staged a rebellion, but the rulers were able to violently suppress this rebellion with the help of Saudi Arabian military forces.

In Libya, Colonel Gaddafi's regime also faced a rebellion, which started in the Eastern Libyan city of Benghazi. Gaddafi had been a dedicated leader of his people. His regime, when it came into power in 1969, had nationalized the British and American oil companies and removed the British puppet "King" of Libya. His regime had, then, used all the oil income of Libya to create the infra-structure for a modern state. Roads and air-ports had been built and a state airline was created and expanded. Hospitals, houses and schools were built, sufficient for the needs of the whole population. Free education, up to the highest level a student could reach, and medical care, were guaranteed. Modern Army, Air Force and Navy were created. Education was made free and compulsory for all children. Military training was also made compulsory for boys and girls alike. When under-ground reservoirs of water were discovered in the process of oil exploration, a huge project was launched to pump the water to the Northern cities of Benghazi and Tripoli and the Libyan coastline in general, creating a green belt where only desert existed before. The work on this "great man-made river" was completed at the cost of about $ 40 Billion. Gaddafi was not driven by greed for wealth and lived a simple life, donating even his own house to the army. He, however, had one weakness in common with other Arab rulers - He wanted to be King and wanted his sons to be in power after him, just like the other rulers of Arabia. After 42 years of his rule, the people rebelled and, sadly, he was violently killed by his opponents, who were organized and aided by the "intelligence" agencies of the US and its "allied" powers, using a large-scale bombing campaign focused on Gaddafi's assassination.

At present, Syria and its dictatorial regime are under attack, because the regime has lost the support of its population, while the US and its "allies" have tried to take over control of this state by secretly supporting and "managing" the opposition to the regime. Russia has also become involved. The fighting has resulted in destruction of several cities, large numbers of deaths of civilians and movement of millions of people out of Syria. Most of such refugees have either found shelter in Turkey, or in the European Union, especially Germany. The dictator, Bashar Al-Assad is the son of the previous military ruler, Hafiz Al-Assad, who set up his "kingdom" in Syria, during the 1960s.

Ever since its territory was taken over by Britain and France, from the Turkish empire after the First World War, the Arab nation has been badly divided and the divisions along tribal lines have grown. Also, the national and tribal rights of the Kurdish nation and the Berber tribes have not been paid attention to. Arabs face a bleak future unless a broad and comprehensive strategy for national survival is formulated.

Imperial Policies of the US

The United States has continued to pursue its imperial policies that it adopted after the Second World War. These policies constitute an in-depth and very well thought-out, plan for creation of a global empire. The US mass media continue a campaign depicting the US as the "symbol of freedom and leader of the free world". The whole world is continuously lectured on human rights, with reference to only the rights of freedom of speech, freedom of conscience and freedom of assembly. Mass murders of the native people of North and South America, Australia, New Zealand, South Africa and the denial of basic human rights of the people colonized by the US and its sub-imperial powers are never mentioned. The denial of the basic human rights of the native people of Palestine is also not considered important.

"1. Everyone has the right to a standard of living adequate for the health and well-being of himself and of his family, including food, clothing, housing and medical care and necessary social services, and the right to security in the event of unemployment, sickness, disability, widowhood, old age or other lack of livelihood in circumstances beyond his control.

2. Motherhood and childhood are entitled to special care and assistance. All children, whether born in or out of wedlock, shall enjoy the same social protection."

Universal Declaration of Human Rights, Article 25

What is important to the ruling class of the US, is that it continue mass murders of the people of the states it desires to Panamize, while claiming to promote "democracy". It is, also, important to the US that the governments of China and Russia, and other states of this kind, allow the unrestricted freedom of maligning their governments and state institutions, to their opponents. Since, restrictions are placed by such governments on the exercise of such "rights", in the interest of their populations, by protecting them from the "cultural hegemony" of the US, these have to be condemned loudly, while the US itself refuses to accept the basic human rights defined in article 25 of the Universal Declaration of Human Rights, while the citizens of this rich state languish on the streets of San Francisco and Los Angeles, under bridges of New York city, or in card-board boxes on the streets of Washington D.C. The people of the US continue to be denied universal health care, a minimum living wage and equal pay for its female citizens - unlike other states in the world of benevolent capitalism. The economic disparity between the oligarchs and semi-oligarchs of its ruling class and its people has been growing ever since 1980. No end to this process is in sight

The ruling class of the US continues to lecture other states about the rights of nations to determine their own future, while invading other states at will and establishing its subservient regimes there and while keeping quiet about the denial of the same rights to the people of Palestine and Kashmir, etc., because talking about such matters does not suite its strategy of extracting the maximum "tribute" from mankind - including from the people of the United States, who are becoming poorer and poorer while the ruling class is growing richer and richer. This propaganda by its mass media (TV and radio stations, movies, books, the internet), schools and universities and government agencies, serves the ruling oligarchic class of the US very well, while it pursues its fundamental objective - subjugation of mankind for extraction of *maximum "tribute"*. We can summarize the current US strategy as follows:

- Claim the "moral high-ground" by raising slogans of "human rights" against opposing states, while "forgetting" the history of slaughter and immense abuse of human beings in the past, especially during the Great Genocide.

- If people, of states with oppressive dictatorships, rebel against those regimes, set up secret civilian and military organizations to achieve "regime changes" in those states and establish Bantustans in those states or Panamize them, depending on their level of social development.

- Continue the imperial relationships with all Bantustans. If and when the populations of these Bantustans rebel against their conditions of subjugation, "help" the ruling families, or tribes, of these Bantustans, in adjusting to changed conditions. If the population exerts too much pressure on the ruling tribes, and a change becomes unavoidable, transform these Bantustans into Panamized states.

- Panamize all the oil and gas-rich advanced Arab states (excluding those already Bantustanized) and other non-Arab states, including Iran. If the people of a Panamized state rebel against their conditions of subjugation, adjust the system of Panamization to meet the demands of the people to the minimum extent possible while maintaining the system, as long as possible.

- Change the military aspect of the containment of the Soviet Union into a system of containment of China and Russia and the successor states of the Soviet Union aligned with Russia, as far as possible

- Continue economic cooperation with China and Russia, to take full advantage of their relatively cheap labor and resources. Continue to work closely with the business classes of these states, with the hope that, one day, they may be able to take over their states and become dependent on the ruling class of the US, thus enabling the US to extend Panamization to these states also.

- Continue the use of the IMF, World Bank and Asian Development Bank to maintain the system of controlled-market subjugation of those peoples and states which are not already subjugated by colonization, Panamization or Bantustanization.

- In all the regions of the world, continue to promote its cultural hegemony – as related to global culture and depiction of historical events, concepts of economics, trade, military power use and inter-state relations in general.

- In the worlds of advanced and primitive feudalism, continue to Panamize those resource-rich states, occupation of which can be expected to increase "tribute" substantially

Nothing less than a global empire is the objective of these policies. If Russia, China and the super-states developing around them, or the other states of Asia, Africa and Latin America are not able to maintain their independence, the US would succeed in its ambition to create a global empire under its hegemony. It is difficult to decide whether the US would succeed in subjugating China, Russia and states closely related to them, though both these states continue to strongly resist US ambitions – and seem to be succeeding at this time.

Only time would tell whether the US would actually succeed in establishing a global empire and how long that empire would last. As described before, during the evolution of mankind, the subjugation of many tribes, with similar racial, linguistic and religious characteristics, by one tribe has generally led to evolution of one nation consisting of all such tribes. Similarly, the domination of many nations, by one nation, has led to creation of empires, and ultimately, multinational states. The domination of Europe and North America by the United States has created the European Union and the effective Economic union of the three states of North America. Thus, we can expect that the creation of a global empire by the United States may have the un-intended consequence of hastening the process of emergence of a global system of closely-knit states - or a global state!

The Moral Record of the United States

Driven by the insatiable desire of its ruling class for more and more "tribute" from the people of the world, the United States has subjugated state after state and has accumulated a terrible record of abusive behavior towards other states, while always clothing its real aims in slogans of "human rights" and concerns about "democracy". Professor Carl Boggs, of National University of Los Angeles, has described the inter-state behavior of the US, in great detail in his book entitled *"The Crimes of Empire"*. A summary, of the treaties and laws that the US has violated, is also given in his book.[3]

The Resistance

The most important development during the second half of the twenty-first century, would be the growth in the economic and military strength of China and Russia and the consequent emergence of closely integrated super-states of East Asia and Central Eurasia. China expects to move into the category of high-income states, during that period, and is likely to achieve the status of an economic and military super-power with global trading and investment relationships and a world-class military.

Russia is already a military super-power and is likely to continue to develop its economy, especially in cooperation with China. Both China and Russia are already helping the states of the Middle East, economically and with military equipment for defense. Russia has increased its political support towards Iran, Turkey, Syria and Egypt. Economic and military cooperation of both these states would thus tend to help the Middle East in resisting further Panamization by the US.

China's navy is likely to expand into the Western Pacific and the Indian Ocean. China has a long-standing policy of non-interference in the internal affairs of other states and it has consistently promoted win-win cooperation with all states, including the US. The increase in its military power and its global economic reach would tend to help the states that have been Bantustanized, or Panamized and would tend to strengthen the independence of states which, presently, are not dominated by the Semi-Global Empire. China is setting up independent financial institutions and is trying to raise the status of its currency in the global trading system. These developments would tend to support those states which are subjected to a controlled market. China's model of economic development is already being adopted by most states which are ruled by their communist parties. It would become more attractive to other developing countries with the passage of time.

While China has adopted economic policies which have the effect of isolating its economy from the rest of the global economy, in terms of boom and bust cycles, the US has moved in the other direction – adopting policies which would increase economic inequality in its population – making the boom and bust business cycles much worse in the future. India is likely to cooperate with the US and Japan in resisting the expansion of the Chinese navy into the Indian Ocean. Despite its rapid growth, however, its economy is likely to continue to bear a heavy burden of military confrontations with its neighbors – especially China and Pakistan.

Thus, the Semi-Global Empire is likely to face increasingly stiff resistance and an uncertain and bleak future, because of policies of the US government. This, of course, is good news for the future of mankind.

Chapter – 14

Threats to Existence of Mankind

Two existential threats to mankind have emerged during the twentieth century – nuclear weapons and global warming. It is quite clear that they would continue to grow and threaten our continued existence on our planet. Let us look at the basics of these two threats, with a quick review of how they have developed.

Nuclear Weapons

As described in Chapter-6, the United States became an *imperial power* right after it won independence from the British Empire. It continued the slaughter of the native population of North America as part of the *Great Genocide* and continued to conquer territories from the other imperial powers by using its military forces and financial incentives, till it reached its present extent. Its ruling class, then, developed the desire to create a hegemonic empire over all of North and South America. This desire to dominate the Americas, then, grew into a dream of a global empire. The project to create a global empire was launched during the First World War, but had to be postponed due to the resistance of the European imperial powers and the difficulties created by the unexpected Russian Revolution and the resulting state of the Soviet Union. An invasion of the Soviet Union, in 1918-1920, could not eliminate the threat to the plans of the ruling class of the US, despite the active support it managed to obtain from all European imperial powers and Japan in the effort to crush the Soviet Union.

The Second World War created a much more favorable atmosphere for US dominance. The war brought the Great Depression to an end and led to a dramatic increase in GDP of the US. Production of armaments was greatly expanded and these arms helped the European imperial powers in thoroughly destroying each other, making it possible for the US to achieve its war aims. The IMF, World Bank and the United Nations were created and the Panamization of Western Europe was initiated with the creation of NATO. In the Asian theater of the war, the US developed and dropped two Atomic bombs on Japan in 1945 to ensure its exclusive occupation by the US forces. In this process, the US also achieved the dubious distinction of becoming the only state in human history to engage in nuclear terrorism. The Soviet Union was behind the US in developing an atomic bomb. Then, the US developed the hydrogen bomb and the Soviet Union followed suit shortly afterward. The US had threatened the Soviet Union, North Korea and China with nuclear weapons during the Korean War. As a result, all three states have also developed nuclear weapons. The Soviet Union, especially, developed a huge stockpile of conventional and nuclear weapons, far in excess of its defense needs, and became a big obstacle to the dream of the creation of a global empire by the US. After the dissolution of the Soviet Union, its place has been taken over by China to some extent – especially in the economic field, as the economy of the US has begun to lose steam relative to the other emerging states besides China. Both the US and Russia have cut their nuclear forces down to smaller number of weapons, but the danger of accidental or intentional nuclear war still remains very high, since the number of states with nuclear war capability has increased substantially. Britain, France, Israel, India and Pakistan have also become nuclear powers. The limits to use of nuclear weapons in a nuclear war, between major powers, were dramatically described by Mikhail Gorbachev as follows:

> **"Even if one country engages in a steady arms buildup while the other does nothing, the side that arms itself will all the same gain nothing. The weak side may simply explode all its nuclear charges, even on its own territory, and that would mean suicide for it and a slow death for the enemy."** [1]

Quoted from *Perestroika* by Mikhail S. Gorbachev. pp 219. Copyright© 1987 by Mikhail Gorbachev. Used by permission of HarperCollins Publishers.

A global nuclear war can certainly bring an end to the human species. The US is, however, continuing its efforts to Panamize all the oil-rich states of the Middle East. A semi-global empire has come into existence. It is currently expanding in the Middle East, but is facing resistance by China, Russia and other smaller states. It remains to be seen whether the US ruling class would succeed in its vision of establishing a truly global empire ruled by it, while the ruling classes of its Panamized states play the role of junior partners in it.

Although the dangers posed by nuclear weapons are widely understood, there is another aspect to possession of these weapons that is not fully understood by many. In a confrontation between two nuclear-armed, but unequal, military powers, the possession of such weapons by the otherwise-weaker state makes its position more secure because if it is attacked and responds with a nuclear counter-attack, the militarily stronger state finds that it is not willing to accept the resulting extensive damage – and, thus, avoids attacking the weaker state. The behavior of India and the US, against Pakistan and North Korea, respectively, has amply demonstrated this fact. This situation makes it essential to resolve the underlying political problems between militarily unequal states, before meaningful universal nuclear disarmament can be realistically expected. Of course, as long as the desire of establishing a global empire, motivated by blind greed coupled with duplicity, remains alive, the oft-repeated goal of global nuclear disarmament would remain a hollow and deceptive slogan.

Global Warming

Scientists have noticed, for quite some time, that the climate is changing. The existence and duration of the last ice age has been the basis of research related to the peopling of the planet (Ref. Chapter-1). It is now widely understood that ice ages and warming periods have been occurring on this planet for a long time. *Ice ages have occurred roughly about every hundred thousand, or so, years, always interrupted by much shorter warming periods.* Our current warming period began approximately twelve thousand years ago. As the previous warming periods generally lasted about fifteen thousand years each, we can say that our planet is moving towards another ice age.

The exact causes for the alternating extreme warming and cooling of the planet are difficult to determine, since several factors like solar activity, shifting of continental plates and ocean currents come to mind in this respect. The changes in the ocean currents seem to be the most important – since the other factors are unpredictable, although they can cause abrupt changes when they coincide with this basic phenomenon.

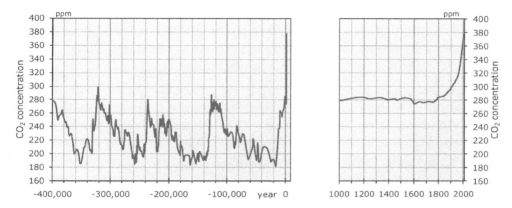

Figure 14-01: Relationship Between CO2 Content of Atmosphere and Recurrence of Ice Ages over the Last Four Hundred Thousand Years.

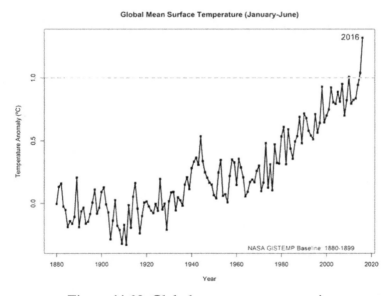

Figure 14-02: Global mean temperature rise

What is the evidence for these conclusions? In a way, the ice sheets on the planet's poles and high mountains have stored the history of its climate. Where there is permanent ice, fresh snow falls on the ice surfaces every year. Considerable quantities of air are trapped between the ice crystals. Thus, new masses of snow form each year, over the previous layers of snow. Under the weight of more and more snow, that at the lower levels, changes into ice with small bubbles of air trapped in it. Drilling into the ice reveals the history of the planet over hundreds of thousands of years and the carbon dioxide content of its atmosphere at those times. Such drill core studies are unanimous in their conclusions that the concentration of carbon dioxide before the period of current industrialization was never more than 300 ppm as indicated in Figure 14-01. These studies have also shown that the proportion of carbon dioxide in the atmosphere is higher today than at any other point in the previous period of more than six hundred thousand years. This leads to the conclusion that global warming is not an unusual phenomenon. However, what is clearly unusual is that the current relatively abrupt changes in climate are being caused by human beings.

During the last ice age, currently densely populated areas of Europe, Northern parts of Asia and North America and the extreme Southern parts of South America and Australia, were covered with thick layers of ice - like Antarctica is at this time. During the previous hot periods those areas, where our current coastal cities are located, were many meters deep under the oceans. If, now, similar major changes to climatic conditions reoccur, they would certainly have a devastating impact on the face of the earth, especially on our seaports.

The increase in temperature during the last 50 years was twice as high as during the last 100 years. Figure 14-02 indicates the Global mean temperature rise since 1880. Glaciers are shrinking worldwide, as are the ice sheets on Greenland and Antarctica. During the 20th century, sea levels have risen by about seven inches. Frequency of rains, droughts and extreme temperatures has increased and tropical cyclones and hurricanes have become much more intense. The temperatures in the polar region rise even more quickly than in the rest of the world. The ice coverage in the Arctic has decreased dramatically. Hurricane Katrina alone, which devastated the city of New Orleans in 2005, caused about 125-billion-dollar worth of damage and cost the lives of about 1300 people. Recent hurricanes in the Caribbean and in Florida caused about 200-billion-dollar worth of damage. US and British Virgin Islands and Porto Rico were devastated.

Without the protective influence of its atmosphere, earth's temperature would be about -18° C and this planet would be ice bound. Some natural trace of gases in the atmosphere - such as steam, carbon dioxide and ozone - prevent the earth from emitting all incoming solar energy back into outer space. As in a greenhouse, these gases radiate part of this energy back to earth. This natural phenomenon is referred to as the *"greenhouse effect"*. It causes the average temperature to settle around +15° C. It is the basis for life on earth. A balance has been created in the level of trace gases in the atmosphere, during the last few thousands of years and it has enabled life in the form that we know today. For a long time, skeptics even questioned whether climate change was really taking place. Now, no one can seriously claim that it is not happening. Also, there is universal scientific agreement that the increase in carbon dioxide and methane emissions are the main causes of global warming.

Figure 14-03 illustrates the greenhouse effect. It is based on average data that is a few years old. The numbers shown indicate rates of energy exchanged per square meter of earth's surface, in terms of power, i.e., Watts. They correspond to energy that is received and radiated back by earth and its atmosphere. Thinking in terms of average changes related to the surface of our planet, at any moment, the atmosphere may contain energy corresponding to 452W per square meter. Energy received, per square meter, from the sun is 235 W. Of this amount 67W is absorbed by the atmosphere, raising its accumulated energy to 519 W. 168 W goes directly to the surface of the earth (both land and oceans). Of the 519 W accumulated by the atmosphere, 195 W is radiated back by the atmosphere and the rest, i.e., 324 W is absorbed by the surface of the earth. The total energy absorbed by the surface of the earth, i.e., 492 W, is returned to the atmosphere. Of this amount, 39.2 W is directly radiated back to space and 102 W is retained by the atmosphere. In addition, 350 W of energy is reflected by the outer layer of the atmosphere and is retained by it in earth's atmosphere. Thus, the energy retained by the atmosphere continues to increase by an amount of 0.8 W, or so, per square meter. Of course, these numbers are based on recent scientific studies. The energy retained by the atmosphere of the planet is a quite large amount, considering the total surface area of the planet. This is the phenomenon that is leading to global warming. [2]

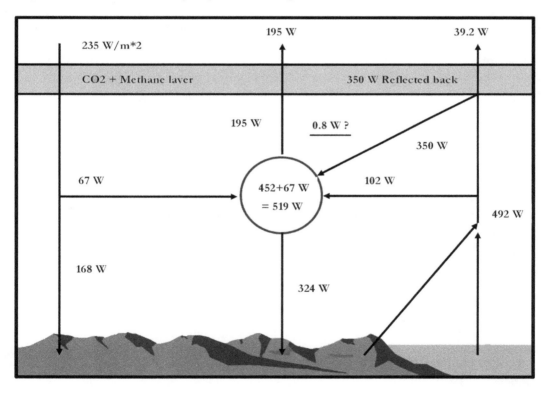

Figure 14-03: The Greenhouse effect

At the same time as climate change has become a serious threat to the survival of mankind, the rising oil and natural gas prices, and estimates of global reserves, show that the supplies still available will not be enough to cover our requirements for much longer and that other alternative sources of energy must be exploited as soon as possible. Scientists have been aware of the relationship between energy use and global warming, for quite a long time. In Europe, experts have been calling for speedy restructuring of energy supplies and some European governments have started taking actions and setting goals for use of renewable energy resources, instead of coal, oil and gas. There is a growing awareness that climate change has already begun. If we do not take this emergency seriously, the catastrophic consequences of climate change will far exceed even our powers of imagination. The awarding of the Nobel Peace Prize to Vice President Al Gore, and the Intergovernmental Panel on Climate Change, was an indication of how seriously the problem is being considered by the global community. Renewable energy could completely cover all our energy supply needs within a few decades, in an environment-friendly and sustainable way. This is the only way to end our dependence on energy sources like oil and uranium. These "traditional" resources are not only costly in financial terms, but they greatly damage our environment.

In December, 2015, 196 countries signed an agreement in Paris, France, dealing with greenhouse gas emissions mitigation, adaptation and finance starting in the year 2020, after lengthy negotiations. The objective of this agreement, known as the Paris Agreement or the Paris Climate Accord, is to respond to the global climate change threat by keeping a global temperature rise this century well below 2 degrees Celsius above pre-industrial levels. Also, the intention is to make efforts to limit the temperature increase even further to 1.5 degrees Celsius. In the Paris Agreement, each country determines plans and regularly reports what contribution it intends to make in order to mitigate global warming. There is no mechanism to force a country to set a specific target by a specific date, but each target is expected to go beyond previously set targets.

In June 2017, US President Donald Trump announced his intention to withdraw the US from the Paris Agreement, causing widespread condemnation in the European Union and many organizations, states and cities of the United States. Under the agreement, the earliest effective date of withdrawal for the US is November 2020. All other countries of the world, however, announced their intentions to continue their efforts in this regard.

In mid-2017, France's environment minister announced France's five-year plan to ban all petrol and diesel vehicles by 2040 as part of the Paris Agreement. He also stated that France would no longer use coal to produce electricity after 2022 and would invest considerable amount of revenue in boosting energy efficiency. Other European and Asian states, especially China, continue to make similar efforts. China has a huge renewable energy generation program and has become the global leader in this regard. Similarly, India is also making considerable efforts in shifting from generation based on coal to renewable energy generation. However, at this stage, the global efforts seem to be insufficient to keep the average global temperature rise to less than 2° C. Hopefully, sanity would prevail in the end and the threat of irreversible global warming would be avoided by mankind, otherwise a catastrophic disaster awaits our future generations.

Chapter-15

Stages of Human Evolution - Summary

Evolutionary Stages of The Distant Past and the Present

We are social beings and the level of socialization of our social groups has grown with time. Looking back at the evolutionary history of mankind, we can, now, identify the following stages of human development:

1. Family - during the age of savagery, engaged in hunting and gathering. Always in conflict with other families in competition over habitat and resources.

2. Extended Family - during the age of savagery, engaged in hunting and gathering and domesticating and using animals. Always in conflict with other families over habitat and resources.

3. Nomadic tribe - in the age of savagery and the age of civilization, engaged in hunting, gathering and animal-farming for food and using animals for transport.

4. Tribes settled on land - initially engaged in agriculture. Sporadically coming into conflict with adjoining tribes and, in later stages, trying to establish hegemony over them.

5. Nations. Settled tribes merging into an advanced feudal nation. Nations trying to establish hegemony over other nations and tribes - thus establishing multi-national empires, or states.

6. Nations, multi-national states, or empires, developing a modern *intensive capitalist society*. Given an opportunity, trying to establish hegemony over other nations and tribes, or extending their hegemony over more nations or states and, thus, establishing, or extending multi-national empires, or states

7. Intensive capitalist society of a state becoming benevolent capitalist society. Nations, national states or multi-national states trying to establish empires over other nations or states.

As long as the above social groups existed as independent groups, we can refer to them as states. For the first three stages, independent groups do not exist anymore. Extended families, living close together, do exist in societies of South Asia, Arabia and Africa. Very few nomadic tribes exist now and their numbers are negligible. Families, extended families and nomadic tribes do not exist as independent social groups. Some nomadic tribes inhabit areas around the Sahara Desert in Africa. They also inhabit parts of Eastern Afghanistan and Western Pakistan and move between these two states as weather conditions change.

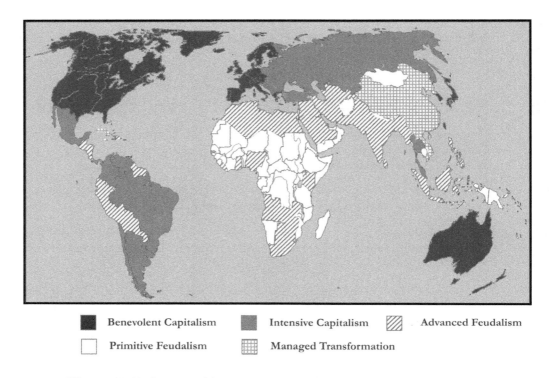

■ Benevolent Capitalism	■ Intensive Capitalism	▨ Advanced Feudalism
□ Primitive Feudalism	▦ Managed Transformation	

Figure 15-01: Stages of Socio-economic Development (Four Worlds)

(Map based on world map at: https://www.d-maps.com/carte.php?num_car=13184&lang=en)

The remaining four stages of evolution of societies do exist at this time, even as independent states, although the very primitive customs of openly practiced slavery and concubine-ship have disappeared. Also, at present, genocidal tendencies of tribes and nations as modern states, with clearly demarcated boundaries, have virtually come to an end. In all these stages, societies are divided into two main classes - a class engaged in economic abuse and a class being abused. *Thus, exploitation of human beings by human beings is virtually universal at this time. It has always been this way ever since mankind developed the evolutionary economic skills of production and exchange of products. We refer to this characteristic as "lootera-ism" and the two classes of abusers and abused as the "ruling class" and the "people".*

We refer to the present-day social system of a tribal society, with each tribe claiming a discrete tribal territory, as *primitive feudalism*. A national state may consist of one or more tribes, or some or all tribes may have merged into an advanced feudal society with no discrete tribal territories. Thus, a national state may have a tribal society at this stage of social development or it may have developed further. We refer to the social system of a national society that is no longer tribal, as *advanced feudalism*. Nations in a multinational state may also be at either of these two stages of evolution.

Intensive capitalism denotes a social system based on the factory system and a relatively free market economy. This system came into being in Europe after the French Revolution and was given the title of *"capitalism"* by Karl Marx. He coined this term to denote blind accumulation of capital by the ruling classes of Europe at that time, resulting in exploitation and intense abuse of the people and the working classes. The system has, since, spread into other regions of the world, while Europe's social system has evolved into a higher level, i.e., *benevolent capitalism*. Societies of South America and the successor states of the Soviet Union are, basically, at the stage of intensive capitalism at this time. The rest of mankind, excluding China and other states ruled by their communist parties, is at the stage of *advanced feudalism* or *primitive feudalism*.

We can, thus, divide the world of today into five Sectors, or five "worlds", each of these worlds consists of one or more regions of the world. There are four "worlds" of normal evolution, as follows:

- *The Fourth World, or the World of Primitive Feudalism.*

- *The Third World, or the World of Advanced Feudalism.*

- *The Second World, or the World of Intensive Capitalism.*

- *The First World, or the World of Benevolent Capitalism (or "Capitalism with a Human Face").*

In addition to these four worlds, as described above, a *fifth world of managed transformation* exists, consisting of states ruled and managed by their communist parties. These states are at one or another of the four stages of evolution as indicated above, but do not really fit into any one of those stages, because of their unique management. In the following sections, we would summarize the characteristics of these five worlds, which correspond to the last four stages still in existence, of social and economic development of mankind, along with a world of managed transformation, with a focus on the mode of production and population dynamics.

The World of Primitive Feudalism

This stage of human evolution was described in some detail in Chapter-2. If a society consists of tribes claiming separate tribal territories, we refer to it as a primitive feudal society. Primitive feudalism exists mainly in Arabia and non-Arab Africa. In non-Arab Africa, South Africa has moved towards advanced feudalism to some extent. It also has a substantial population of European colonizers, which is almost as advanced a population as that of Europe. *Intensive capitalism* is developing fast in this state. In Arabia, only Egypt, Syria, Iraq and Lebanon have progressed somewhat beyond primitive feudalism although tribalism still remains part of their societies. Besides these two major regions, tribal societies also exist in Afghanistan, Western Pakistan and Laos. Some tribes of the native "Indians" exist in North and South America also, but these are very small minorities and the societies on the two continents have evolved far beyond primitive feudalism. Bantustans, i.e., states dominated by single tribes, and controlled by an imperial power, also exist, mainly on the Arabian Peninsula. The political structure of Bantustans was described in chapter-6. Kuwait, Oman and Saudi Arabia, are some such subservient states. These have been created by imperial powers for their own purposes and are not a normal evolutionary phenomenon. In general, the world of primitive feudalism has been affected by its colonial experience. Slavery is not an open practice in this world and some industries have emerged here which would, normally, not be expected to develop in a society of this kind. Mining and production of energy resources of oil, gas and coal has been developed in this world. Also, copper, gold, diamonds and other precious metals are being mined here. Small scale consumer industries have developed and, in some states like South Africa and Nigeria, large industrial establishments have also developed. But, the society, basically, remains primitive and tribal. The conditions in Arabia are somewhat better, as education, and knowledge of modern sciences has spread. Most states of Arabia have developed modern militaries, but non-Arab Africa remains backward in this respect also. Trans-state corporations of imperial powers operate everywhere in this world, and exploitation of the local populations continues, by their subjugation to controlled markets, although the degree of subjugation is less severe now as compared to the colonial times of the recent past. In the Arabian Peninsula, the discovery of large reserves of oil and gas has proved to be both a good fortune and a curse, as the imperial powers have created Bantustans to rob the region of these resources. Iraq, at this time, is being fast developed into a Panamized state, by the United States. The reason for the establishment of a Panamized state, and not a Bantustan, is that most of the population of Iraq has advanced beyond primitive tribalism and has made significant progress in terms of education and political consciousness. Thus, the establishment of a Bantustan in Iraq is not a feasible project.

In a society at the stage of primitive feudalism, the population cannot grow fast, because of inter-tribal warfare and a primitive level of health care. However, as tribes settle down into a state of peaceful coexistence, and even tend to fuse into each other, the population begins to grow fast. This is now happening in this world, despite the HIV pandemic, which has caused huge numbers of deaths, especially in non-Arab Africa.

World of Advanced Feudalism

In many regions of the world, primitive feudalism, consisting of many tribes, has advanced to a stage where tribes have merged into communities which retain the basic characteristics of feudal societies - extreme class-consciousness, violent intolerance and superstitious belief systems. Separate tribal identities have tended to disappear in such states and their societies. We refer to the social system at this stage of evolution as advanced feudalism.

This process of tribes settling down on land and fusing together into a much bigger community, took a long time and happened when the mode of production had already changed to agricultural production and its peculiar relationships. With time, tribes and communities grew in size, accompanied by changes in social relationships which made large scale agricultural production possible. *Slave ownership* grew as a practice. Local "kings" or land-lords emerged equipped with small armies to impose their will on the people and to grab land from other individuals and tribes. Kings emerged to exercise control over the local "mini-kings", or landlords and to seek tribute from them. Emperors emerged to control the kings and to demand *tribute* from kings of far-away lands. This stage of feudalism was described in Chapters-2. The ancient Mesopotamian civilization and the civilizations on the river Nile and the Indus River, with time, evolved into advanced feudal societies and they, basically, remain so even at this time.

Advanced feudalism of today retains the characteristics of violent intolerance and superstitious belief systems, to some extent. Iran, Turkey and South-Central Asia, South Asia, South-East Asia and Central America have societies of this kind. But, due to their colonial experience these societies also have substantial communities who have had access to modern scientific knowledge and technology. Thus, in these regions, agriculture is being mechanized and modern manufacturing and service industries are developing. In fact, some states, like India and Thailand are, also, developing modern computer software and telecommunication industries. Thus, in the "womb" of this advanced feudalism, a new system of intensive capitalism is also developing.

As advanced feudalism develops, and tribalism dies out along with its inter-tribal wars, the level of violence begins to decrease and, because, of wide-spread agricultural production, the availability of food increases. Because of these developments, the population, even if it is very poor, has easy access to food because of direct contact with food-crops. The population of such societies tends to grow faster and faster as the system develops. This phenomenon can be easily observed in the history of South Asia, Japan, China and other South-East Asian states. In 1850, the population of South Asia was about 100 million. Now, it has increased to more than one and half billion - more than fifteen times in about one hundred and fifty years! In the beginning of the nineteenth century, the population of Japan was about 20 million. By 1840, it had increased to about 35 million. In the beginning of the twentieth century, the population had grown to about 80 million and by the end of the Second World War it had reached 100 million, when the stage of advanced feudalism had passed. In 1600, the population of China was about 100 million. In 1950, it was 600 million. In 1970, it reached 800 million. Now, it is about 1.4 billion after the passage of about forty more years, despite the extra-ordinary population control program adopted by the Chinese government.

World of Intensive Capitalism

As intensive capitalism begins to develop in an advanced Feudal society, the rate of growth of the population begins to decrease. The death rate begins to climb as the people are forced to move to the cities after losing their jobs due to mechanization of agriculture and being dispossessed of their small landholdings by powerful landlords. In the cities they are exposed to unhygienic conditions and they begin to die of hunger and lack of medical care. The cities, in effect, become concentration camps where the people are pushed together to die. Under these conditions, first the death rate of the population increases, then the birth rate begins to decrease, because more and more people realize, despite their ignorance of socio-economic processes, that it is not wise to have children when they cannot even feed themselves. At present, these processes are underway in South and South-East Asia. In the 1960s, scientists were talking about a *"population explosion"* in these regions. Now, the talk is about a *"population implosion"*, as the rate of growth of the population is steadily decreasing.

Population growth rates of India and Pakistan were about 2.4 and 3.5 percent respectively in the 1960s. Now they are about 1.2 and 1.4 percent.

In Britain, population statistics are available for the nineteenth century, but not the eighteenth century. Rest of Europe does not seem to have any such statistics for the two centuries. In Britain, Thomas Malthus, wrote his book "An Assay on the Population Principle" in 1798. His impression was that Britain's population was growing at "a geometric rate", such that it was doubling every 25 years. He thought food production could not keep pace with this population growth, because it could only grow at an "arithmetic rate". He, obviously, did not know about how intensive capitalism was affecting the population growth, since the new system was in its very initial stages at that time. His observation about the growth of population, however, does provide confirmation of the high rate of growth of 3-4 percent per year in Britain, which we can compare with a rate of about 0.5 percent in the late nineteenth century, from official statistics.

As was described in Chapter-6, Europe went through a stage of intensive capitalism during the seventeenth to mid-twentieth century. The main development took place in the nineteenth century and reached its peak in the middle of the twentieth century. During this stage, European societies gradually developed from advanced feudalism to intensive capitalism, as modern industry developed. The old social system gradually faded away and the new system took hold. This process is, now, underway in Eastern Asia (excluding China and other states ruled by their communist parties), Central Eurasia and South America. China's population growth has gone down due to its population control program.

During this stage of intensive capitalism, society goes through a massive change. Economic forces cause the people of rural areas to move to the cities in search of jobs. As the people are crowded into cramped and unhygienic living quarters, contagious diseases spread fast among them and they die of lack of food and medical care. Political consciousness of the workers develops fast under these conditions and they try to organize themselves into labor unions and political parties. The hold, of religious and other superstitious ideas in their minds, breaks down and they are able to see the greed and violence of their abusers, in the mines, factories and service centers of various kinds that they have to work in. The result is that the population growth slows down and, ultimately, stops. It may actually reverse itself. This is what happened in Europe during the nineteenth century and the first half of the twentieth century. This was the time of colonization. Some of the population did move to the colonies in North and South America, South Africa and Australia. These were the people who had some means of staying alive and they could afford to pay for their transportation to the new lands, but the rest of the people were forced to live under terrible conditions in the cities of Europe. They died of hunger or lack of medical care, as the new system became the dominant system in Europe. More than 70 million were killed in the two World Wars. In the mid-nineteenth century, Marx had noted that intensive capitalism, or "capitalism" as he called it, was, above all, producing its own grave-diggers - the desperate and abused industrial workers in the cities. He predicted that the system would be eventually buried by these "grave-diggers", but the system proved to be too strong for the working class, at that time. There was violence and protest throughout the nineteenth century in Europe. Two periods, around 1830 and around 1848 were especially full of social explosions in many countries. There were protest marches, rebellions and bloodshed all over Europe. Working people could not succeed in burying, or even drastically changing, the economic system and the exploitation and abuse of working people. Thus, the system ended up burying most of the rebellious workers and their families, instead.

At this time, advanced feudalism is being transformed into intensive capitalism in Southeast Asia, Central Eurasia and South America. The pace of this transformation is very fast. Central Eurasia consists of the successor states of the Soviet Union, Iran and Turkey. In this region, population has been decreasing since the end of the Soviet Union in 1991, as intensive capitalism was reintroduced. The population growth in Iran and Turkey is also slowing down. In South America, also, population growth is slowing down fast. These three regions - Southeast Asia, Central Eurasia and South America, are not exactly at the same stage of development, but the social system of intensive capitalism is nearing the point where it would dominate society and the state, in all three regions. Like in Europe, this may take about one hundred, or so, years for this process to reach its peak.

World of Benevolent Capitalism

After the Second World War, Europe, North America, Japan and Australia developed further into societies, and states, where a "kinder, gentler" form of capitalism developed. The late President Bush Senior would have, probably, called it a *"kinder gentler system with a thousand points of light"!* We refer to this stage as *benevolent capitalism* - or *"capitalism with a human face".* There were two reasons for this change. First the working population of Europe died off due to hunger and lack of medical care. Much of the rest were killed in the two world wars, or immigrated to the new European colonies in North and South America, Australia, New Zealand and South Africa. Thus, a labor shortage developed in Europe, forcing the employers and governments into making concessions to the people. The other reason for this change was the emergence of the Soviet Union, after the Russian Revolution. As much of Eastern Europe fell into the control of the Soviet Union after the Second World War, and as China also went through a revolution, the fear of another revolution in Europe, or elsewhere, began to haunt European and American ruling classes. Thus, concessions were made to the people of North America, Europe, Australia and New Zealand. The same happened in the newly Panamized states of Japan and South Korea. As a result, the working conditions and well-being of the people improved in these regions. They were given access to educational institutions for their children. The health care system was improved. The war-ravaged economies of Europe were given injections of substantial amounts of capital, by the United States. A period of rehabilitation and reconstruction was initiated and by the 1960s, Europe had basically recovered from the economic disaster caused by the Second World War. The people of those enumerated states saw a clear improvement in their lives as their housing, food supplies, transportation and other services improved, resulting in a higher standard of living.

This system has continued to evolve slowly, but, with the disappearance of the Soviet Union, this process has slowed down. In fact, there has been a reversal of this process for some time, as the ruling classes, in the states of most regions of this world, sensed that the threat of the Soviet Union had disappeared and have tried to reverse the concessions that had been made to the people. The people of the First World have, by experience, learned to be careful in planning their lives. They tend to acquire knowledge and skills by focusing on education – thus, improving their "human capital". They tend to acquire capital for their housing and save for hard times which inevitably come due to the boom and bust cycles of the economy of the system. Also, as family relations become, more and more, money relations, they also tend to have smaller families. Having children is no longer a priority. They tend to engage in more activities of personal enjoyment and pleasure than try to raise large families. Thus, the population growth of the various regions of the First World has drastically slowed down. The improved health care has resulted in increased life-expectancy and the population has begun to age. Since industry continues to grow, the demand for more labor has outstripped supply. Thus, the states of this World have been forced to liberalize their immigration policies. Since the 1960s, the United States and Canada have begun to welcome immigrants from Asia, South America and even Africa. The same has happened in Europe. Australia resisted this trend for some time, but it has also liberalized its immigration policies. The same is now happening in Russia also. The non-European segments of the populations of the United States and Canada are likely to grow and may overtake the European population. By the end of the twenty-first century Europe may be about twenty percent non-European. Perhaps, one-quarter of the population of Russia would be non-Russian. Even the population of Japan is likely to be about ten percent non-Japanese by that time, although the Japanese are very nationalistic and do not welcome immigration. Thus, the population of the First World is likely to become more and more diverse with time.

The World of Managed Transformation

This is a unique world ruled by its communist parties. China, Vietnam, Laos, North Korea and Cuba are included in this world. Revolutions occurred in these states when they were, basically, at the stage of advanced feudalism, while intensive capitalism was developing in their economies. Now, economies of Vietnam and Laos have become mixed economies and are developing fast. North Korea and Cuba are not far behind, although they have special problems which hinder their growth. Their human capital has developed fast because of their stress on education, health care, land reforms and development of basic infrastructure. Industries in these states are also developing fast along with development of technology and the income level of their people is growing along with drastic reductions in inequality and poverty. Strict controls on operation of the economic market have been introduced in these states, putting an end to intensive capitalism and promoting welfare of their populations instead.

As mentioned before, people of China have, mostly, escaped the cruelties of the system of intensive capitalism, because its government had developed a system of population control. Since 1980, it has prevented the birth of about 500 million children. Industry in China is growing very fast. The population is moving fast to the cities. There are large numbers of migrant workers. But, despite these changes, people are not dying of hunger. In fact, due to the improvements in health care, the people are living much longer. The population of the aged is growing. The birth rate has dropped and the population growth rate has decreased dramatically.

Population Dynamics

When we look at the four stages of economic, social and political development as described in the previous section, we can see that economic activity is the basis of social development. Economic development leads to changes in society and, as a social system develops into a higher level of organization, this has its concurrent effect on the development of social institutions also. Thus, economic, social and political developments are closely linked and the *basic form of economic activity, or the mode of production,* forms the core of all changes in the society of a state.

Hunting, gathering and grazing of livestock remain a substantial part of the activities of tribes of the society of primitive feudalism - they are the legacy of the stages of the age of savagery and nomadic existence. But agriculture becomes the main mode of production of such a society. We can see all these activities in existing societies of non-Arab Africa and, to some extent, Arabia and parts of Asia. Thus, control of land and protection of agricultural crops are matters of the highest priority at this stage. Intertribal warfare occurs, primarily, because of the need for control of agricultural land. Due to this inter-tribal warfare, and due to the scarcity of food and lack of health-care, the population of such a society cannot grow fast. However, as tribes begin to fuse into each other and the level of violence decreases, the rate of growth of population begins to increase and becomes very fast as the transition to advanced feudalism proceeds. Parts of Laos, Afghanistan, the Arabian Peninsula and most of non-Arab Africa have reached this stage. In Figure 15-02, we indicate non-Arab Africa nearing the end of this stage of development. We also indicate that Arabia, as a whole, has moved into the beginning of the stage of advanced feudalism. By Arabia we mean all the territories and states inhabited by the Arab nation.

The stage of advanced feudalism is reached when tribes merge into other tribes and the tribal control of land begins to disappear. Tribal identities may remain, but they no longer lead to violence over control of land. Thus, large tribes develop due to the fusion of smaller tribes and these large tribes, with common control of land, a less homogeneous but still with a common genetic heritage, a common language and religious beliefs, develop and become so large that they include all the neighboring populations with these characteristics. At this stage, such huge tribes, with little in common with neighboring population, can be referred to as nations. As nations develop their land resources, violence in society tends to decrease and practices like slavery begin to disappear. Population continues to grow fast, but a new mode of production begins to take hold, as new skilled artisans develop, new methods of production of goods are discovered or acquired from inter-action with neighboring nations, as trade of agricultural commodities with neighboring nations, develops. Thus, a new mode of production based on a large number of skilled workers in a workshop, begins to be established in society. This, ultimately, leads to establishment of large-scale workshops and factories. This system of concentrating a large number of workers in a small area to produce, through mining or manufacturing based on agricultural or mined products, leads to dramatic changes in society. As a result of this new mode of production, agriculture is mechanized, more and more of the population of the countryside is rendered surplus and is forced to move to large towns or cities, where most workers are unable to find means of subsistence, and are exposed to hunger and disease. Thus, the death rate of the population begins to climb

313

up, ultimately leading to a decrease of the birth rate also. Thus, the rate of growth of the population steadily goes down as industrialization increases. Figure 15-02, shows Arabia, South Asia and South-East Asia going through this stage of development. These regions are not developed to exactly the same degree, but the basic processes of transformation of society are the same in all these regions. Arabia is at the initial stage of advanced feudalism. South Asia and South-East Asia are more advanced. In fact, South-East Asia is nearing the end of this stage of development.

Figure 15-02 shows the initial graph splitting up into three branches. One branch, the one at the top, applies to "normal" societies, states or regions. This is the path that Japan has taken and Arabia, South Asia and South-East Asia are likely to take. "Normal" development, here, means the absence of large-scale inter-state warfare, absence of emigration of population due to the opportunity of colonizing new lands, or the absence of a drastic birth-control program like the one China has adopted.

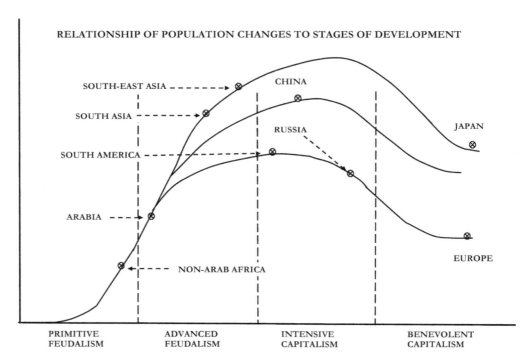

Figure 15-02: Socio-economic Development and Population Change

The second branch of the graph applies to China only, at this time. Chinese society developed like a "normal" society till about 1980. The population of China increased dramatically as 1980 approached, especially after the Second World War and the revolution in China. Because of these events, the population growth was not exactly like a "normal" society. However, the population continued to grow fast till the one-child policy was adopted. Initially, even this drastic population control program did not have much effect on population growth, because the health care and food supplies continued to improve for the population, resulting in increases in life expectancy. But, now, after thirty years, this birth-control program has succeeded in preventing more than 500 million births and is bringing the rate of population growth down. At this time, the population of China is growing at the rate of about 0.4 percent. This rate is likely to become negative sometime within the next three decades. Thus, China's population growth curve has become unique. So far, no other state has adopted such a population control program. China has reintroduced a form of tightly controlled economic market in its economy. Malnutrition and rampant disease and lack of medical care, which would normally occur under intensive capitalism, are not occurring in China, because the population is protected from such consequences by the state. However, large-scale migration to the cities is still occurring, while being properly managed by the state, and fast urbanization of society is underway, as agriculture is mechanized and fast industrialization of the economy is in progress. The government of China encourages the rural population to move to the smaller "secondary" and "tertiary" cities, instead of Beijing, Shanghai, Guangzhou, etc., which are referred to as "primary cities". This way, congestion in the primary cities is avoided. Thus, China has managed to avoid the large-scale *"Intensive Death Rate"* that takes place in the stage of intensive capitalism. Its population is expected to reach a peak of about 1.5 billion around the middle of this century. In the second half of this century, the population would decrease and is likely to stabilize at a level about 30 percent below its peak level. The one-child policy has been changed into a two-child policy for couples, at least one of whom is an only child. At what level stabilization is achieved would depend on how long the current system of birth-control is continued without change and, if it is changed, then how fast it is changed - i.e., at what stage controls are removed completely. Most forecasts expect a population of about one billion by the end of the current century.

The experience of Europe has, also, been unique in terms of population growth. When Europe had reached the stage of advanced feudalism, it became so powerful as compared to other states and societies on other continents that it began to colonize distant lands and grab land from natives of those areas. This explosion of a genocidal frenzy resulted in the virtual extermination of the native people of North America and large-scale mass-murders of the populations of South America, Australia, New Zealand and southern Africa. As these territories were colonized, a large-scale emigration of population took place from Europe to the colonies. Thus, the population of Europe could not grow as fast as would have been expected. Also, Europe went through a long series of wars related to the imperial ambitions of the European states. Tens of millions of people died in these wars which continued till the middle of the twentieth century. The mass-killings during these wars also had their effect on the growth of European population and slowed it down to a considerable extent. At the same time, during the stage of intensive capitalism, the "social murder" of a large section of the European population took place, despite the emigration of people of some means. Due to these reasons, we show a separate branch of the population growth trend graph for Europe, i.e., its lowest branch. Russia, and the former Soviet Union, also went through a similar sequence of events. Thus, Russia and, by implication, all the successor states of the Soviet Union, are also shown on the same graph.

At this time, the native population of Europe is not growing. In fact, it is decreasing. As a result, the governments of most European states have liberalized their immigration policies. The flow of immigrants has, thus, stabilized the European population. These trends are likely to continue for the foreseeable future. The situation of North America, Australia and New Zealand is quite similar. The population of Japan is also decreasing, but Japan has, so far, resisted liberalizing its immigration policies, because of highly nationalistic nature of its society. This, resistance is, however, likely to be overcome by the rulers of Japan, since, otherwise, Japan would have to undergo an increasing level of de-industrialization, as local industries are moved out to other states.

Former colonies of North and South America, Australia and New Zealand are not indicated on the diagram. The native populations of these colonies were mostly exterminated during the *Great Genocide*. The subsequent colonization of these territories, and the growth of populations there, does not follow the "normal" path, as conditions in these colonies were very different from the states with thousands of years of history. Also, the colonizers were from societies which were already well-developed.

Looking back at the territories of the five worlds of today, we can see that the population growth of the native populations of the first world has stopped as this world has reached the stage of *"benevolent capitalism", or "capitalism with a human face"*. Because of the aging of the native, or original population before 1950, or so, the death rate has increased. The population of this world is being stabilized by liberalization of immigration policies. Immigrants are moving into this world, mainly, from the third and the fourth worlds. The native population of the non-Muslim majority, successor states of the former Soviet Union, as a whole, has been decreasing. As most of the elderly and sick people have already been killed because of lack of nutrition and medical care, the decrease has slowed down. This is happening even in Russia, despite the inflow of large numbers of Russian-speaking immigrants. Russia has initiated a program to encourage couples to have more children, but it and the other states in this group would have to develop an extensive social protection network, like in the other European states, and would have to adopt liberal immigration policies to deal with this trend. Similarly, China is nearing the stage where its population would, first, level off and then drop dramatically by one-third or so. China would not have to adopt new immigration policies, since it would still have a very large population, but it would have to further liberalize its system of social security and population control. It is already doing this to some extent. Similarly, South America is also coming close to the point where its population would stop growing and would start going down. It would also have to adopt policies similar to other regions in the second world, i.e., Central Eurasia and the world of managed transformation, especially China.

The populations of the third and the fourth worlds are continuing to grow at this time, although the rate of growth is slowing down in the third world. These worlds would also go through the same changes as the first and the second world have faced. The population changes in the world of "managed transformation" are likely to be faster, but similar to those in the four worlds of evolution, depending on what level of development and transformation a certain state has reached. In the meantime, movements of population from the third and the fourth worlds into the first world would continue. This may be seen as a kind of reverse-colonization, and would result in increasing diversity there. The result would be that, in the next two hundred, or so, years the population of the planet would first stabilize, and then would decrease. This process would continue till non-Arab Africa joins the first world. At that time, it would no longer be possible to stabilize the population of states, or super-states, without drastic improvement of basic social protections of the population, i.e., without the creation of a highly just social system, to replace the inhuman and abusive system of loot and plunder that has dominated our history so far.

Condition of the Working People

As was pointed out before, the conditions, under which the people worked, did change when a society evolved from one stage to another. The changes that occurred in the first three of four stages of evolution, currently in progress, have been described briefly but with sufficient detail. The changes leading to benevolent capitalism from intensive capitalism have not been described with accuracy by some authors. In some cases, this seems to be deliberate misrepresentation. Let us deal with one important aspect of this evolution.

Under intensive capitalism, workers are paid subsistence wages because there are huge numbers of unemployed workers in the labor market and they cannot bargain for higher wages. This has been described in detail in Chapter-7. In each sector of the economy, new machinery and more and more automation is introduced to increase the productivity of labor, i.e., the output per worker employed, in an effort to increase market share and profits. As automation and mechanization grow in one sector, the rate of profit of that sector tends to decrease at a faster rate with time. The reason is that with lower number of workers, the total unpaid labor decreases, relative to "objectified labor" in the form of buildings, machinery, etc. Also, competition increases among businesses. The rate of profit falls forcing the owners to transfer new investment to new sectors of the economy, as described in Chapter-7. As higher levels of automation are introduced, larger numbers of workers are laid off accordingly. The number of workers laid off is much greater than those who are retained. Thus, the workers laid off and their families end up starving to death, due to lack of food and medical care, and the condition of those retained also does not improve much since they can be retained with minimal increases in wages, due to the glut of unemployed workers. This is what happens during intensive capitalism. Marx has referred to this phenomenon as the *"law of the falling tendency of the rate of profit"*. When a labor shortage begins to develop because of growth of industry and the high death rate of workers, due to what Engels has called *"social murder"*, or what we call *Intensive Death Rate*, conditions do change. Workers are then able to bargain for somewhat higher wages. When the shortage of labor becomes acute, the bargaining position of the workers also becomes better. The system, ultimately becomes benevolent capitalism and the living conditions of the working class begin to improve. Starvation is then, no longer, the only option for them. Still, under benevolent capitalism, some workers continue to suffer as before – as we can currently see in the US and in Europe. If we only look at the improved living conditions of workers and imagine that this has happened despite introduction of machines in production and increase in automation in general, forgetting the workers rendered jobless before and during the transition to benevolent capitalism, we can come to the erroneous conclusion that automation improves the condition of the working population. This is what some theoreticians have done. In general, it is correct to state that under intensive capitalism, due to introduction of machinery and automation, there is a tendency of profits to fall and for the working population to initially die at a higher rate, till a labor shortage develops.

When a social system has developed into benevolent capitalism, the tendency, of the rate of profit of corporations to fall, continues. However, it does not lead to reduction of wages to starvation levels. It only tends to increased unemployment, since more workers are laid off and less are hired when new machinery and automatic processes are introduced in a factory or corporation. Under normal conditions, the laid off workers are able to find jobs in other sectors of an economy – after a period of unemployment. Also, when the profits have fallen for most companies in a certain sector, companies tend to move to other sectors in pursuit of higher profits.

A good example is that of personal computer manufacturing sector. IBM had produced the first computer in the 1950s. Its computers were very expensive and only large companies could afford them. When other companies started producing computers, prices started to go down. A company named "Apple Computers" started to produce personal computer, due to development of microprocessors and other advances in technology. Soon, IBM and other companies, also, started producing personal computers. Prices of personal computer went down dramatically and, ultimately, IBM decided to sell off its personal computer production division to a Chinese company. Apple, also, stopped producing personal computers and went into cell-phone production and, because of its high rate of profit in this field, it became the largest corporation in the world. Now, many other companies are producing cell phones, which are getting cheaper and cheaper due to global competition. The profits per cell-phone have been falling, but the total quarterly profits of Apple Computers had not fallen till recently. Now, they have.

Stages of Political Development
Tribal States

In the early stages of tribes settled on land, the tribes could be referred to as "states", although the physical boundaries of their tribal territories were not clearly defined and were always subject to change due to tribal warfare. Some single-tribe states continue to exist till now, e.g., Lesotho, Swaziland (Now called the "Kingdom of Eswatini") and Kuwait. Multi-tribal states exist in most of non-Arab Africa.

National States

When tribes of contiguous tribal territories and with similar genetic heritage, language and religion united, a nation was born. Originally, nations had no clearly demarcated territorial boundaries. Later on, generally when intensive capitalism had started to develop in a state, the boundaries got well-defined. Also, as described in Chapters 6 and 7, intensive capitalism and the market mechanism created the irresistible desire in its ruling class to subjugate other states. Some national states, thus, became imperial powers and established their empires. In Europe, this phenomenon led to the establishment of empires by most European states during the seventeenth to early twentieth century. When most of the rest of mankind had been subjugated or slaughtered, the imperial powers started planning to grab parts of each other's conquered territories – leading to increasingly intense wars among them. The two world wars were the ultimate results of this conflict among imperial powers of Europe – motivated only by the insatiable greed of their ruling classes.

Multi-National States

During intensive empire-building, propelled by the new mode of production of intensive capitalism, several new multi-national states came into being, in North America, South America and Australia. These were mainly multinational colonies which, later, changed into independent states of today. United States, Canada, Mexico, Brazil, Argentina and Chile are some of these states. Some other multinational states emerged in the areas subjugated by the imperial powers, where the native people had managed to survive. South Africa, China, Bolivia and Guatemala are some such multinational states. The break-up of the British Empire resulted in the creation of several multinational states. India, Pakistan, Malaysia, Iraq and Britain itself are among such states, which are fragments of the former British Empire. The Soviet Union and Yugoslavia emerged as multinational states, through forcible conversion of empires. These have not survived as such and have, basically, fragmented into national states. But due to the basic good intentions of the original multinational state systems and the lack of extraction of tribute by one or more states from the rest, the seeds of eventual convergence into viable multinational states have been planted in the former territory of those multinational states; tolerance and recognition of common culture and cultural values are thus growing with development of civilization Yugoslavia was a multinational state, and United States is a multinational state, but there's a difference - United States was also a colony. In a colony, the colonizing people do not quickly develop the attachment to the colonized land and, so, there is very little resistance, if any, to immigration from other nations and states, and to the colony becoming a multinational colony and state.

It is very clear that a very high level of civilization is necessary for a multi-national state to develop and the new state structure has to be very tolerant of the nations involved and autonomy of each nation has to be guaranteed in matters of national importance. If there are tribes still existing on national territories within the state, it has to be tolerant of the tribes involved and has to guarantee a level of autonomy to such tribes in matters directly related to tribalism. This tolerance can only develop if a clear understanding of tribal and national rights develops in the political culture of the state. The emerging European super-state has reached that level of civilization.

Many states of today are national states, i.e., they basically consist of one nation only. As mentioned before, there are also many multi-national states in existence. Multi-national states are of two kinds - non-colonized multinational states and former colonies which have become independent states. The nations, in most non-colonized states, have clearly identifiable traditional territories. For example, Britain, at present, has four nations included in it - the English, Scottish, Welsh and Irish nations. Pakistan has ten nations – five large ones and five smaller ones. Similarly, India has about thirty nations. Since Africa was divided by imperial powers among themselves, ultimate colonization being the main objective, the boundaries of non-Arab African states, at present, do not reflect national or tribal boundaries. Non-Arab,

African societies are, in general, tribal societies and the states in this region are basically multi-tribal states. Some African tribes have, however, grown to a level such that they could be referred to as nations.

Some states, like the United States, Canada, Mexico, Brazil, Australia and New Zealand, are former colonies, where the native tribes, if any, and the colonizing nations are mixed up into one more or less diverse population, where none of the colonizing nations can claim a territory traditionally inhabited by them. For example, the United States, has people of all nations of the world in its population, but since these individuals, come from other states, they have their traditional territory in some other state than the United States. Some big communities in the United States are Germans, Italians, Chinese, Japanese, the Irish and the Africans. Their traditionally inhabited territories are in Germany, Italy, China, Japan, Ireland and Africa. Since the Africans were brought into the United States as slaves, their tribal origins are not identifiable. They have even lost their languages and religious belief systems. There is no traditionally German, or African, territory in the United States. Thus, stable multi-national states do exist today, but these are mostly former colonies, where different nations do not stake out territorial claims.

It has been very difficult for traditional multi-national states to survive, since it is very difficult to create a system of political management for them. Traditional multi-national states, have, thus, tended to disintegrate into national states. Japan has, over time, abandoned all ambitions of creating a multi-national state, as an empire, or a *"co-prosperity sphere"* in Asia. It has failed in all such adventures and has returned to the status of a national-state. Similarly, virtually all multi-national states in Europe, and all European empires in Europe or on other continents, have disintegrated. Only a few multi-national states exist today in Europe - Britain, Switzerland and Bosnia, etc. Similarly, the Soviet Union broke down into its national republics, which have become fifteen national states, except those whose national boundaries were not clearly demarcated. Conflicts have developed in such cases, e.g., in Ukraine and Chechnya. Also, some inter-national conflicts do exist in Western Europe and Eurasia, and can be traced to mismanagement of national issues in multi-national states. For example, violence has continued in the Basque and Catalan regions of Spain, in "autonomous region" of Chechnya, in Northern Ireland "province" of Britain, and threatens to start again in Bosnia. But most such national conflicts have been settled in Europe and Eurasia by creation of national-states. Thus, this process of national conflicts within multi-national states, or among such states, is continuing and, in most cases, is likely to create more national states in Eurasia and Africa.

Super States

During the First World War, the United States embarked on the most ambitious project of its existence - creation of a global empire. After the initial non-successful attempt during the First World War, this project was again taken up during the Second World War, with careful preparation. As described in Chapter-6 and Chapter-8, this ultimately led to the creation of the European Union, i.e., creation of a common market with a common currency for *free movement of people, goods, services, and capital within its boundaries*. Also, NATO was created. It was also dominated by the US and still is. These changes put an end to the constant military and economic conflicts among the European states. However, due to the emergence of the Soviet Union, states of Eastern Europe could not be included in this arrangement, but after the end of the Soviet state, this became possible and, now, Europe as a whole has come under US control. By this time, most multi-national states of Europe have broken down into national states and they have become part of a continent-wide super-state. The process of economic, military and political cooperation is continuing at this time, despite the planned British exit.

The US has also created a "common economic space" by finalizing *the "North American Free Trade Agreement (NAFTA)"* with Canada and Mexico. This has greatly increased trade and investment in the region. This can be seen as the basis of a virtual super-state of the future. The Arab states have created the *Arab League* and all the African states have created an organization for economic, military and political cooperation known as the *African Union*. Similarly, South East Asian states have created *"Association of Southeast Asian Nations (ASEAN)"*. China has developed significant economic relations with this organization. South Asian states have created a similar organization named *"South Asian Association for Regional Cooperation (SAARC)"*, although it has not made much progress in promoting economic integration. The successor states of the Soviet Union have also created a common economic space and have created *"Eurasian Economic Union (EAEU)"* and are increasing their economic and political cooperation in this respect. Economic and political cooperation is also growing among the South American states, based on a customs union. All these economic and political organizations are groups of states whose common interests have grown with time.

About two hundred states exist today, some large and some very small. Basically, about half of them are big states. Others are small and insignificant in terms of economic or political power and, naturally, their influence on global events is insignificant. Social groups, like tribes or nations, grow, or become smaller, as the struggle between them continues. States, despite being highly organized social groups, also grow or break down into smaller ones. States may grow by gobbling up other states. They fight each other, or compete with each other in other ways, as they always have. States are now highly organized, so the violence between them is also very organized and very well developed scientifically. The fight is fundamentally, the same old fight - over habitat, resources and reproduction; over who would survive and who would not.

Along with the process of creation of national states, the reverse process of coming together of national states into new groups of states is also in progress. This happens when nations feel secure within their national boundaries, which have become their state boundaries also. The feelings of security are further enhanced by the development of a system of social security. As a result of this feeling of security, nations, and their national states, become willing to engage in developing cultural and economic relationships with other national states. Similarly, multi-national states emerging from former colonies can have the same tendencies, if national conflicts do not exist on their territories. With the creation of the European Union and signing of NAFTA free movement of populations within the European Union and, possibly, within the NAFTA area in the future, the development of continental-size super-states has become a possibility. The United States, being a global superpower in terms of economic and military capabilities, has become the center of power around which these two super states have begun to emerge - North America and Europe. Europe has its big imperial powers, which continue to exercise imperial hegemony over some parts of the third and the fourth worlds – specially their former imperial possessions. But Europe itself has gone through a mild Panamization by the United States. The United States has established a permanent military presence in Europe to dominate the local military establishments of the European states. It has closely intermingled the ownership of US and European corporations, thus unifying the ruling classes of the two regions – while the US ruling class remains dominant in all economic matters. It has, also, established global financial institutions to control economic activity throughout the world.

The US had included European imperial powers in the new global financial institutions and obtained their agreement to end direct imperial control over their territories in Asia, Africa and Latin America. This was done to deal with the threat posed by the Soviet Union, which threatened complete destruction of their empires. In this process, the United States, succeeded in creating, effectively, a virtual near-global empire dominated by it. By means of economic measures it usually uses for creating a "community of interests" between its ruling class and that of a state it Panamizes, it created a Panamized Europe also. Thus, the two super-states of North America and Europe are emerging with time, with the United States being the center of power.

Similarly, the Soviet Union has disappeared but powerful economic, cultural and military forces still exist for development of a super-state in Central Eurasia. Some of the successor states of the former Soviet Union had already come together in what was called the Commonwealth of Independent States (CIS). This has changed into a Eurasian Union which is expanding with time. Special relationships exist in this Eurasian Union regarding movement of products and people. Also, military relationships between the successor states of the Soviet Union have continued even after its demise. Russia remains a big nuclear power with global reach. Also, powerful linguistic and cultural relationships continue to grow between Russia and the other members of the Eurasian Union. Russia is the center of power of this region and a new super-state is clearly emerging around it. Similarly, China is the center of power around which economic, and in some cases military, relationships are developing in East and South-East Asia. Thus, the contours of a future super-state are also visible in this region.

In South Asia, India is the center of power. Britain had divided South Asia, or "British India", into seven states - three Buddhist majority states of Burma (now "Myanmar"), Bhutan and Sri Lanka, two Hindu-majority states of India and Nepal and two Muslim-majority states of Pakistan and Maldives. Bangladesh has since separated from Pakistan. India's relations with Bangladesh, Sri Lanka and Pakistan are strained because of India's short-sighted policies of threatening its neighbors and trampling on their rights. It occupied the land of the Kashmiri and Balti nations of Pakistan and continues to subjugate these nations. Because of this behavior, which is based on Hindu imperial ambitions, South Asia has not been able to develop strong regional economic, military or political ties, although a forum for the promotion of economic cooperation among South-Asian states has been created. With time, religious hatreds would die down as this region goes further into an advanced stage of intensive capitalism and moves towards completion of the resulting economic and social transformations. When the rights of the Kashmiri and Balti nations have been restored to the satisfaction of these nations, a South Asian super-state would begin to emerge because of geographical compulsions, genetic homogeneity of the population and economic advantages of cooperation of all the states of the region. This virtual super-state would, naturally, be centered on India, although strong links between Pakistan and China are highly likely to survive this development.

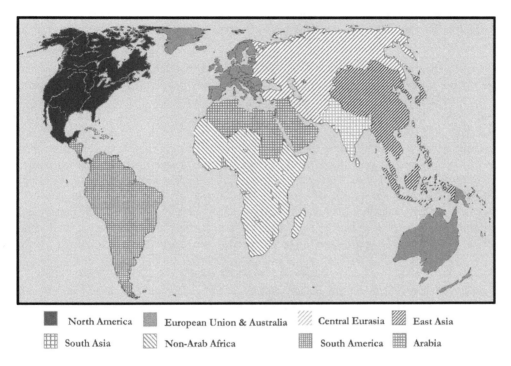

▮ North America	▨ European Union & Australia	▨ Central Eurasia	▨ East Asia	
▦ South Asia	▨ Non-Arab Africa	▦ South America	▦ Arabia	

Figure 15-03: Virtual Super-states of the Future

Map based on world map at: https://www.d-maps.com/carte.php?num_car=13184&lang=en)

Two more super-states are also emerging in Arabia and in non-Arab Africa. The Arabs are a primitive nation and, due to their tribal society coupled with imperial manipulations, they have not been able to form one national state. A national organization, i.e., the Arab League, already exists. Because of its national genetic homogeneity, language, religion and culture in general, the growth of national cohesion is inevitable. As this region goes through two more stages of economic, social and political development, basically, a national super-state would ultimately emerge. Most likely, this super-state would be centered on Egypt, but this is not necessary. The biggest danger to the development of an Arab super-state is from the United States, which is trying to completely Panamize all, oil and gas rich, Arab and non-Arab states of the Middle East. If the United States succeeds in completely Panamizing the Middle East region, or only the Arabian Peninsula, this may greatly delay the emergence of an Arab super-state.

The situation of non-Arab Africa is quite similar to Arabia. There is no single all-powerful state dominating the whole region at this time. Thus, a powerful center of power may not develop in this region for some time. Power would, thus be diffused throughout the region and this may delay the emergence of a super-state. However, with the existence of the African Union, the desire and need for such a union is highly likely to grow. The only region that may not develop into a super-state is the continent of Australia. Australia has enormous natural resources, but its population is too small for a powerful state. Australia is, at present, a Panamized state of the US.

Thus, already, we can see that eight virtual super-states are emerging in the world and would become more and more powerful as the five worlds develop further. Not only are the super-states emerging, but a global economy is also emerging. This process has dramatically speeded up after the disappearance of the Soviet Union and the creation of the World Trade Organization (WTO). Fears of state take-over of industries has decreased, also new "free-trade", unrestricted free-exploitation agreements have been negotiated between the ruling classes of the imperial powers and the developing states. Because of these developments, the imperial powers and their trans-state corporations have become very confident of their future profits. Thus, a semi-global empire has emerged and it is backed up by global imperial institutions - the International Monetary Fund, the World Bank and the military muscle of the United States. The process is being, euphemistically, referred to as "globalization" and is characterized by free inter-state direct investment and free inter-state extraction of profits, in the name of "free trade". This empire is clearly an empire of the US ruling class and its oligarchs, and the ruling classes of European, South American, Asian, Arab and African states are its junior partners. Only China, Russia, the Central Asian states and, to some extent India, are not yet part of this empire. In other words, the center of gravity of this semi-global empire is in the United States. However, movement of tourists and immigrants is also growing among the eight super-states, promoting tolerance and diversity everywhere. Military strength of the states forming the centers of power of these super-states is also growing. Thus, the "global village" is also coming together with time.

There is more to the emergence of eight super-states in the world than globalization of economic activity. Looking at what is referred to as the "global village", now-a-days; we can see that United States dominates North America. Economic relations between it and the other North American states are growing fast. Rate of migration of population among the states of North America has also grown. Thus, a diverse European, Native American, African and Asian population is developing in North America. What this means is that prides and fears of nationalism are dying as populations of different states of the world come together in North America. Also, the rate of growth of the population of the North American continent, north of Panama as a whole, is tending towards stability despite immigration. This means that the stresses that population growth produces, in the absence of full, or near-full, employment, are also becoming less severe. Social security of the people is also growing as better health care, unemployment insurance policies are adopted by the states of this continent, and as the social security net is strengthened in general, especially in Canada. In the US, the reverse is happening but it is not expected to continue for hundreds of years. Thus, what is happening in North America is what has already happened in Europe - the standard of living has tended to converge to a certain level for the whole continent, diversity has increased throughout and conditions of social security have also tended to converge to one level. This has given the ruling class of Europe a powerful incentive to demolish boundaries hindering easier and deeper economic relations between the states of the continent, including the movement of labor within Europe. Of course, the purpose is to increase capital accumulation throughout the continent, but it is giving the people of the region an opportunity to organize themselves to demand a uniformly higher standard of living. Thus, the social life of the population has improved since 1945, but is deteriorating since 1980. However, the standard of living of the population is expected to improve over the coming centuries, as political consciousness spreads among the people. National fears have died down and, thus, a new European mentality is taking hold. Thus, there are four factors that combine to create conditions of creation, and further development of relations, in a super-state - i.e., economic interdependence, converging standards of living, native population stability and social security safety-net.

The initial step, for consolidation of territories or unions of states, is taken by the ruling classes of such states to maximize their capital accumulation. The initial cause of creation of national states was the same. A dominant tribe tried and succeeded in subjugating many other tribes, which, if the control lasted long enough, created a nation. Similarly, multi-national states like Britain were created for the same reason by the ruling class of one nation subjugating other nations for maximization of capital accumulation. Such territory became a multi-national state if the control lasted long enough, as in the case of Britain. The initial cause of creation of large states like the United States, Brazil, Russia, or China, has been the same. This continues to be the primary reason even today as globalization is pushed by the global ruling classes and "free trade areas" are created over large territories involving many states, resulting in consolidation of groups of states into embryonic super-states.

Chapter-16

Human Conflicts

To understand human conflicts over habitat and resources, of the past and the present, we have looked at the evolution of mankind. The development of the human brain has been the basis of this phenomenon. We need to look at how evolution has resulted in the development of the human mind, or personality, in relatively recent times. To find out how to manage human conflicts, the limits of management have to be understood. In this chapter, we would look at the structure of human personality as it is shaped by individual, tribal, national and class influences. We would, also, look at resources over which conflicts arise. Then, we should be able to see what could be done to manage those conflicts, so that violence between individuals and social groups can be avoided, or reduced, as far as possible.

Human Growth and Personality

We are born with a brain and a set of instincts, which are determined by our genetic heritage. The instincts are unconscious behavioral patterns that we inherit. Our minds and personalities develop later. We can think of our brains as the "hardware" and our minds as the "software" that develops with our experiences as living beings, as we grow up. Another way of saying this is that our "personalities" develop as we grow up. Thus, by a person's mind and personality, we mean the same "software" that grows and consists of our beliefs and habits of behavior, during our lifetimes. One aspect of human life is that our biological development and development of our personalities is a very long process. We don't really see this in other animals. Also, our conflicts are unique in the sense that they are, to a great extent, conflicts of the mind, rather than simple physical conflicts, like other animals have. Many of our conflicts arise out of subjective reality in our minds. Ways have to be found to eliminate those conflicts or, at least, to control them. Many other conflicts arise due to objective causes that have a scientific basis. Even these conflicts can be, and have to be, resolved or minimized, if we hope to create a set of ideal states ensuring security and well-being of all their citizens, without getting into conflicts and wars with each other. Thus, it is important to investigate various aspects of the development and structure of human personality, at least those aspects which have close connection to the social and political existence of human beings. We shall be doing this, briefly, in this chapter.

As human beings develop from conception to adulthood, their genetic heritage, or their DNA, determines many aspects of their behavior. When a child is conceived, it goes through a series of changes, in the embryonic stage, seemingly repeating the stages of evolution through which mankind has evolved. Conception is the combination of two sets of genes, constituting their DNA, from two different human beings. The result is the creation of a new DNA, or the "blueprint" of a "program", by which the new life has to be born and has to develop physically, in the future. The two sets of genes have been developed by evolution, which means that the new individual has a new path given to him, for this physical development, from which he has virtually no freedom to diverge, except that environment may alter it somewhat. The survival instinct is the fundamental instinct passed to a child and it becomes active at birth. Sexual instinct is another, but it is inactive at birth and becomes active during puberty.

As a species, we reproduce the genetic heritage, as passed on to us and modified by the environment, by a long series of past generations and also the knowledge passed on to us. When we come to our full potential as adults, we tend to add to that knowledge and the skills passes on to us. We are the most social species on earth and, thus, as individuals we are very insignificant compared to the whole species, which is constantly expanding in numbers and skills. The child of a human being is a very helpless individual. It is said that the higher the level of evolution of an animal, the more dependent is its offspring to that animal. A chicken may take 10 weeks to raise a little baby to become an adult, and kick the adult out to lead its own life, but a human being takes twenty or thirty years to develop its offspring. It takes a lot of time because the knowledge and the skills to be conveyed are enormous. And that is why the offspring is so much dependent. Seals and otters train their babies to swim and to eat and to hunt, and the birds teach their babies to fly, but it doesn't take 30 years, it doesn't take that much effort. But human beings take a lot more time to acquire knowledge and understand relationships and the tricks of survival that we learn from our parents and others around us. Being very social beings in all our activities, when we have developed as mature adults, we are deeply connected to others around us. A developed human being has a big social side to him along with a personal one. A newborn child has an instinctive inner self which gives rise to an additional personal self of an adult. Our relationships with other living beings, especially other human beings are more mental than physical. Also, we are becoming more and more social beings and less and less individual beings. Our needs are becoming social too. Food, clothing and shelter are our basic needs, which are now produced by social institutions, like factories. But we have other social needs, both physical and mental, which have to be met to give us a sense of security. As we grow up, we develop family, tribal, national, state, religious and class loyalties and these become part of our personalities. Thus, to summarize, we can say that a mature human being has an inner self, a personal self and a social self. The more civilized a society an individual is born into, the more extensive and important his social self becomes. The inner self is inherited by a human being at birth and gets modified to some extent in later life. It is basically part of the unconscious mind. The personal self is inherited, to some extent, by an individual but basically develops as the personality develops. Part of the personal self is also unconscious, but another part is not. The social self is also partly conscious and partly unconscious. Thus, a human adult is a very complex organism. Some behavior of a human being can be called rational behavior, i.e., it is based on reason and logic. We also exhibit behavior that cannot, really, be reasoned with. This kind of behavior is mostly based on what is instinctive to us - our sense of security and emotional well-being, and is influenced by feelings in our minds, of which we are not conscious. Within an individual, there is always tension between the rational and the irrational mind, and these tensions get amplified under stress.

The science of the mind, or Psychology, became established as a science in the late nineteenth century. Sigmund Freud is considered the founder of this science. Since the human mind is very complex, this science also consists of very complex concepts. Sigmund Freud's concept of the human personality and its sub-processes remains widely accepted by Psychologists to this day. Some of the details have come under criticism, regarding their relative importance, but these do not affect us since we are concerned with, mainly, the political and social behavior of human beings and not so much with the details of the development of human beings and their normal and abnormal bilateral relationships with other individuals. The following sections describe the development of the human mind and personality, based on Freud's theories. [1]

Structure of Personality

As described in the previous section, the human personality, or the human mind, may be considered to consist of three major systems. These are called the "Id", the "Ego" and the "Superego", referred to as the "inner self", the "personal self" and the "social self" in the previous section. In a normal and mentally healthy person these three systems work together harmoniously. By working together co-operatively they enable the individual to carry on smooth and satisfying interactions with his environment. The purpose of these interactions is the fulfillment of a human being's basic needs and desires. Conversely, when the three systems of personality are at odds with one another, the person is said to be maladjusted. He is dissatisfied with himself and with the society around him.

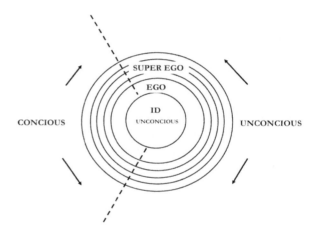

Figure 16-01: Structure of the Human Mind

Id

The sole function of the Id is to provide for the immediate relief from disturbances caused by internal or external stimulation. This function of the Id fulfills the primary principle of life which Freud called the *"pleasure principle"*. The aim of the pleasure principle is to rid the person of tension, or, if this is not possible, to reduce tension to a low level and to keep it as constant as possible. Tension is experienced as discomfort, while relief from tension is experienced as pleasure or satisfaction.

There is a tendency found in all living beings to maintain a constant, peaceful state in the face of internal and external disturbances. In its earliest form the Id is a *relaxation system* that, by physical movements, immediately gets rid of any sensory excitations reaching it. Thus, when a very bright light falls upon the retina of the eye the eyelid closes and light is prevented from reaching the retina. Consequently, the disturbance that was produced in the nervous system by the light, stops and the organism returns to its peaceful state. A human being is equipped with many such "reflexes", which serve the purpose of automatically getting rid of any bodily disturbance that has been caused by a *"stimulus"*, acting upon a sense organ. The general consequence of the reaction is the removal of the stimulus. Sneezing, for example, usually expels whatever may be irritating the sensitive lining of the nose, and watering of the eyes flushes out foreign particles. The stimulus may come from within the body as well as from the outside world. One example of an internal stimulus is the reflex opening of the valve in the urinary bladder when the pressure on it reaches a certain intensity. The tension produced by the pressure is terminated by the emptying of the contents of the bladder through the open valve. If all the tensions, which occur in human beings, could be relieved by reflex action, there would be no need for any psychological development beyond that of the primitive system of reflexive relief. Many tensions occur for which there is no automatic remedy. For example, when hunger is felt by a baby, it cannot automatically produce food. Instead, it expresses restlessness and cries to attract attention. Without the help of an older person bringing food, the baby would die. When food in a suitable form is provided to the baby, the "hunger pains" in the stomach stop, leading to the baby quieting down. A baby, or a child, also has the instinctive fear of danger to his life. The slightest rough movement or sound, or the appearance of a stranger, causes it to sense danger. A baby would cry to bring his fear to the notice of adults and only the removal of the danger causes it to quiet down. As a child grows up, the sex instinct becomes active. The sight of a member of the opposite gender, a scene in a movie, or a dream may be the stimulus that causes the desire to have sex to arise. The process of finding a mate willing to have sex, however, is not as simple as finding an object to eat.

The Id has needs and desires that demand satisfaction. Since it is completely unconscious, it does not know about what is possible and what is not possible, what is real and what is not real. It only knows that its needs and desires have to be satisfied. We can say that a human child at birth has a mind that basically consists of the Id only. As a child develops into an adult, other systems of the mind, i.e., the Ego and the Super-ego, also develop.

Ego

To meet the goals of survival and reproduction, neither reflexes nor wishes can provide the hungry person with food nor the sexually motivated person with a mate. In fact, impulsive behavior may cause punishment by the environment. Adults have to seek and find food, a sex partner, and many other objects necessary for life. In order to accomplish these missions successfully between the person and the world, the necessary requirement is the formation of a new psychological system - the ego. In the well-adjusted person, the ego is the manager of his mind. It controls and governs the Id and the superego and maintains contact and interaction with the external world in the interest of the person and his wide-ranging needs. When the ego is performing its management functions properly, harmony and adjustment prevail. If the Ego surrenders too much of its power to the Id, to the superego, or to the external world, harmony and peaceful interaction with society disappears. The result is a maladjusted person in conflict.

Instead of the *"pleasure principle"* the ego is governed by what Freud called the *"reality principle"*. Reality refers to objects and relationships that exist in the real world, as opposed to the world of imagination. The aim of the reality principle is to postpone the satisfaction of a need until the actual object that will satisfy the need has been found. For example, a child has to learn not to put just anything into his mouth when he is hungry. He has to learn to recognize food, and to put off eating until he has located an edible object. Otherwise, he will have some painful experiences. The postponement of action means that the ego has to be able to tolerate tension until the tension can be relieved by an appropriate form of behavior. *The institution of the reality principle does not mean that the pleasure principle is forsaken. It is only temporarily suspended in the interest of reality.* Eventually, the reality principle leads to pleasure, although a person may have to endure some discomfort while he is searching for objects in the real world.

The reality testing is undertaken by a process which Freud called the *"secondary process"* because it is developed after and controls the primary process of the Id. In order to understand what is meant by the secondary process it is necessary to see just where the primary process gets the individual in the satisfaction of his needs. It gets him only to the point where he has a picture of the object that will satisfy the need. *The next step is to find or produce the object, that is, to bring it into existence.* This step is accomplished by means of the secondary process. *The secondary process consists of really producing the object that would satisfy a need, by means of a plan of action that causes action based on reason. This is* normally called problem solving or thinking. When a person puts a plan of action into effect in order to see whether it will work or not, he is said to be engaging in reality testing. If the test does not work, that is, if the desired object is not found, a new plan of action is thought out and tested. This continues until the correct solution, i.e., production of the real object desired, is found and the tension is relieved by a suitable act. In the case of hunger, the suitable action would consist of eating the real food that has been found.

Through this process of mental development, a person learns to distinguish the world of the mind, or subjective reality, from the objective world of physical reality. The system of perception develops deeper powers of understanding so that the external world is understood clearly. In this way a person learns to use information that has been stored in his memory system. This is how the psychological processes of cognition, memory, decision-making, and action are developed. Memory is improved by the formation of associations between images in memory and words, resulting in creation of a language. The result is that a person becomes capable of making and communicating decisions based on facts.

Another set of changes which take place in the development of the ego is the ability to control muscles and make complex movements, while being aware of the real world. Also, the ego develops the ability to produce fantasies and daydreams. Fantasies produced by the ego are playful and pleasurable imaginary objects and events. Although the ego is basically a product of an interaction with the environment, its development is specified by heredity. Thus, every person has inborn abilities of thinking and reasoning, which are developed by training, education and experience. Other instincts, like perception of danger and the suppression of the sex instinct with respect to close relatives, i.e., prohibition of incest, are also inherited by the ego. The information and details of behavior that the ego incorporates during its development are part of both the conscious and the unconscious mind. Thus, individuals are normally unaware of how they really make their decisions and how those decisions are based on the experiences of their ancestors.

Superego

The third major sub-system of the mind, or personality, is the superego. It represents the moral code developed in the mind of a person. It is, thus, a person's set of rules that determine what is right and what is wrong. It determines the person's identity - what is ideal for him and what he or she tries to perfect. It does not deal with reality or pleasure. The superego develops out of the ego as a result of the child's acceptance of his parents' standards regarding what is good and virtuous and what is bad and sinful. By absorbing the authority of his parents, the child replaces their authority with his own inner authority. The internalization of parental authority enables a child to control his behavior in line with their wishes, and by doing so to secure their approval and avoid their displeasure. In other words, the child learns that he not only has to obey the reality principle in order to obtain pleasure and avoid pain, but that he also has to try to behave according to the moral standards of his parents. The relatively long period, during which the child is dependent upon the parents, ensures that the superego of the child absorbs most of the parent's standards of morality and good behavior. Freud considered the superego to be made up of two subsystems, the *ego-ideal* and the *conscience*. The ego-ideal corresponds to the child's conceptions of his parents' morality. The parents convey their standards of morality, or set of virtues, to the child by rewarding him for conduct which is in line with these standards. For example, if he is consistently rewarded for being neat and tidy then neatness surely becomes one of his ideals. Conscience, on the other hand, corresponds to the child's conceptions of what his parents consider immoral. These conceptions are established through experiences with punishment. If he has been frequently punished for getting dirty, then dirtiness is considered to be something bad. Ego-ideal and Conscience are opposite sides of the same morality.

The rewards and punishments, by which parents control the formation of the child's superego, are of two kinds - physical and psychological. Physical rewards consist of things that are desired by the child. They are such things as food, toys, caresses, and sweets. Spanking and denial of things that a child wants, are physical punishments. The principal psychological reward is that of parental approval expressed either by word or facial expression. Approval stands for love. In the same way, withdrawal of love is the main form of psychological punishment. In the final analysis, rewards and punishments, whatever their form may be, are conditions that reduce or increase mental tension in a child. In order for the superego to have the same control over the child that the parents have, it is necessary for the superego to have the power to enforce its moral rules. Like the parents, the superego enforces its rules by rewards and punishments.

These rewards and punishments are given to the ego because the ego, by virtue of its control over the actions of the person, is held responsible for the occurrence of moral and immoral acts. If the action is in accordance with the ethics of the superego, the ego is rewarded. However, it is not necessary for the ego to produce a physical action to take place in order for it to be rewarded or punished by the superego; even thinking may be rewarded or punished. In other words, a thought is the same as a deed, which also makes no distinction between subjective and objective. This explains why a virtuous person may suffer pangs of conscience for wanting something that his own moral code forbids him. The rewards and punishments, that are available to the superego, may also be either physical or psychological. Psychological rewards and punishments employed by the superego are feelings of pride and feelings of guilt or inferiority, respectively. The ego becomes full of pride when it has behaved virtuously or thought virtuous thoughts, and it feels ashamed of itself when it has yielded to temptation. A person may be allowed to have a long rest, or sexual satisfaction, as a reward, or a person may have an accident - caused by the superego as a punishment.

The development of the superego corresponds to the internalization of parental authority and belief systems, including religion. However, parental authority may be related to much more than what the actual parents of a child believe in. They may teach a child what is good and what is bad, in terms of elementary behavior, but what they convey is much more. Pride in one's genetic heritage may be conveyed as something good, something to be valued and something to be proud of. The same may apply to the language, history and religious beliefs and traditions of a family, tribe, nation, or state that the child belongs to. The behavior of the tribe, or nation, in war or in peace-time, may be conveyed as something good, while the behavior of other tribes or nations may be conveyed as bad, to be looked down upon. This process does not stop with parents, but other "parents" may consist of teachers, religious figures, newspapers, books and television channels. The net result is that the superego incorporates the effect of all these influences.

In modern times, as the human species has developed, certain moral principles have come to be accepted by all societies and all groups of modern human beings. Some of these principles have developed due to long evolutionary human experience, like the value of human life, courtesy, compassion, sympathy and equality. Others have developed as part of, and due to, relatively very recent scientific discoveries and expansion of scientific knowledge - tolerance, value of diversity, respect for freedom of conscience, etc. Some or all of these principles may be conveyed to children and may be incorporated into their personalities, depending on the level of civilization of the society that the parents belong to.

The information that is incorporated into the superego is huge and requires a huge amount of memory, and the human mind does have a huge amount of memory. The superego, like the ego, has information that is retained by the conscious mind, but a lot of this information becomes part of the unconscious mind. Although it affects our decision-making, we may not be aware of its existence at a certain point in time. In Figure 16-01, the influences related to parents, family, religion, tribe, nation, state and scientific knowledge in general, are depicted as layers of the superego.

Defense Mechanisms of the Mind

The human mind has the ability to protect itself against painful and awkward situations. Rationalization is the most common way of justifying something that one finds painful to accept. Denial is another. Anxiety may be shifted to someone else as the source, in which case, it is referred to as "projection". Anxieties may cause other changes in behavior. Those changes in behavior, as defense mechanisms of the mind, are referred to as, "repression", "fixation", "regression" and "reaction formation", etc. Severe cases of these defense mechanisms in a normal mind need to be identified in political behavior, but, generally, they can be ignored safely, since they occur rarely in individuals involved in political activities and are easily noticed when they affect such behavior.

The Abnormal Mind

Our evolutionary experience has led us to have five senses - we can see, hear, touch, taste and smell. We have two eyes, which means that we can not only see objects but can also determine their distance and location to some extent. Similarly, we have two ears which enable us to sense vibrations in the air and to determine the direction of these vibrations or sounds. We can also sense the presence of objects by touching them and can make decisions about them based on past experiences. The same applies to the sense of taste and smell. All living beings are not born with these five senses. Some do not have eyes and they sense the environment in other ways. Some do not have ears but still can sense sounds with their bodies. Some living beings have sharper senses than us. Dogs have a keener sense of smell than human beings. Bees, with their five eyes, have a better sense of direction and location than we do. Thus, there are differences between us and other animals in terms of sensitivity and acuteness of our five senses. There are similar differences among human individuals. For example, some can see things in far sharper detail than others; some can hear sounds others cannot. However, virtually all of us fall within specific ranges of sight and hearing.

There are, however, people of exceptional ability. They may be able to see, hear and smell things in truly extra-ordinary detail. They are also able to retain memories of such sights and sounds for a very long time. Their sensory abilities may reach a point where they may begin to see images of objects that do not really exist, or hear sounds which are non-existent and which can be verified to be non-existent by other normal people. When things reach this point, then we say that those people are abnormal, that their minds or brains are going beyond reality. They are, thus, unable to distinguish between what is reality and what may only be imagination or a dream. Their minds are malfunctioning. This situation is generally believed to occur due to some kind of traumatic experience by those people, in early childhood, in most cases, when the mind has not yet fully developed. Generally, however, the causes of such mental disorders cannot be determined. In most such cases, where the disorder is not acute, the person may begin to read demeaning and degrading feelings in the words and actions of others. The person may begin to see in everyday events, some additional occurrences which did not happen. He may hear remarks which are not really made by people. When this condition occurs, then, we refer to the person as paranoid - meaning that the person can see and feel beyond reality. Paranoia can grow in such a way that it becomes a system of beliefs around a central theme, e.g., that someone or some organization is trying to harm him, or that someone has fallen in love with him, or that he is a superior human being and can work more than twenty persons, etc. These beliefs around a theme are called delusions and such a person is said to be delusional. A paranoid person can also develop obsessions related to people and organizations related to his, or her, delusions. Paranoid and delusional persons are generally aware of and in touch with reality, but in some ways and at certain times lose touch with reality. Superficially, they seem to be normal persons, and unless their behavior is studied closely, and in detail, they may remain unnoticed for long periods of time.

Manic mind is another type of abnormal mind. A manic person has an extra-ordinary amount of energy. His or her mania may develop in one or another direction, based on the individual's personality. He or she may be able to work very long hours and may be able to do a lot during those hours as compared to a normal person. Such a person may have extra-ordinary number of sexual partners. One common symptom of such a mind is the inability to sleep and rest. Hyper-verbalism is another. Such persons are generally leaders, managers and artists and are commonly considered people of extra-ordinary ability. Such persons may become very abusive depending upon the kind of mania they have developed. The more dangerous thing about such people is that they may have paranoia and other mental disorders associated with their condition - and those disorders may go unnoticed.

A person's ability to see and hear may become so abnormal that he may begin to see unreal people and events for long periods of time. This is an acute disorder referred to as psychosis and such a person is said to be psychotic. A psychotic person cannot distinguish between real and unreal objects, persons and events. He may see totally imaginary people or may hear unreal sounds. Such people, generally, cannot function normally and are quickly recognized as abnormal, since they live in some other reality, and some other world, and not the world most of mankind lives in. In some cases, personality of a person develops in such a way that he, or she, may have psychotic episodes of short durations. These are referred to as "transient psychotic episodes" and are a very dangerous phenomenon in a person with political power. These can easily go unnoticed by others.

The most common mental disorder is, however, depression. This condition occurs if a person faces extra-ordinary amount of stress over a long period of time, and cannot find a way out of the situation that is causing it. Such a person spends most of his, or her, mental energy dealing with stressful thoughts and, thus, the brain gets drained of a chemical it produces and uses during the thought process, known as serotonin. As the serotonin level goes down, the person is no longer able to function with normal energy. He may become lazy, sad and lose interest in many normal activities. Some depressed persons, especially men, can also become irritable, argumentative and violent in this condition, as the mind begins to malfunction, due to the lowered serotonin level.

Another mental disorder of importance, though it is not very common, is the manic-depressive disorder. A person with this kind of disorder may behave like a manic person for a period of time and then become extremely depressed for a period of time. His, or her, condition is thus bipolar, the mind switching from one to the other for some unknown reason.

Paranoid, delusional and obsessed persons have a tendency to become leaders since they spend a lot of their energies dealing with their fears and obsessions. Thus, these mental disorders are of great importance in those involved in politics and exclusion of such persons from positions of power is also of great importance. Hitler and Stalin were two such "great leaders", who managed to reach positions of great power over human beings with tragic results. Stalin was clearly paranoid, as can be concluded from accounts, by his close associates like Nikita Khrushchev, of his behavior when he was in power. Thousands of members of his own political party and hundreds of thousands of others were killed on his orders, because, in his mind, they were "conspiring to destroy him". Hitler was also paranoid and obsessed with race and superiority of the "Aryan", i.e., German race over all other races of mankind. His belief that the Jews were spoiling the great German race and how they had conspired to create a Marxist state, the Soviet Union, to destroy the Germans, brought death to perhaps six million Jews and twenty-six million soviet citizens, at the cost of about eighteen million German lives!

Summary and Conclusions

We are living beings and the state is also a living organism. To organize a state for the security and well-being of its population, it is crucial to know the needs and behaviors of the individuals, families, tribes and nations in the population. Thus, a general understanding of the human personality is essential for optimum organization of a state for the purposes that it is expected to serve. If we assume that a state has both primitive and modern segments in its population and that it has all forms of social groups and organizations that have existed in the last six thousand years of human civilization, then we can visualize the personality of its citizens in the form indicated in Figure 16-01. The Id is the inner self of a person and its needs and demands are mostly based on the genetic heritage of a person, but some newly developed demands are based on what are referred to as "derived instincts". The needs of all persons are basically the same but the modified needs may be demanded by some of the persons with varying degrees of urgency. In any case, the needs of all the members of a social group are basically in conflict with each other, especially in relation to food, clothing and shelter. These are objects of which there is always a limited supply and conflicts due to the needs of all individuals can become more acute if the supplies become short. Security and the sense of belonging are other such needs.

Thus, the basic needs of a population can be satisfied by equitable distribution of supplies. This action also leads to reduction of tensions and makes more time available to individuals to interact with each other in a more relaxed atmosphere. The attachments and perceived relationships to family, tribe, nation, state and humanity are, in general, more difficult to satisfy, since these are largely based on subjective reality. The "moral" demands and restrictions of these social groups effectively restrict the free exercise of the instinctive demands of the Id. Also, direct or indirect real and perceived threats to these social groups tend to act as stimuli requiring defensive action by an individual, or even the social group involved. Obviously, the formed and developed personality of an individual cannot be recreated or substantially changed by further education by a state. Thus, social change can only be brought about by *a very slow process*. The evolution of mankind has been a very slow process. Its study and the study of human personality lead us to the same conclusions regarding the wise approach to the future. We can, thus, come to the following conclusions relevant to the management of a state:

- Only a just society based on equal rights for all individuals can minimize the conflicts related to satisfaction of their basic needs. Such conflicts cannot be completely eliminated, but they can only be minimized in an atmosphere of social equality and justice.

- Individuals in a state must have autonomy in matters related to individual and family life. The state must not interfere in such matters unless there is some compelling reason involving the welfare of individuals therein, or the general population of the state. Freedom of conscience and belief, freedom of expression and freedom of association must be guaranteed to all individuals so that the individuals see the state as a free association of equals and not as a hostile entity suppressing their individual existence. Maximum freedom, to the extent it is feasible, must be guaranteed to individuals to engage in economic and social activities - including sexual interaction, sports and entertainment. Such guarantees tend to create state loyalties and attachments.

- All nations within a state must have equal national rights and their autonomy with respect to these rights must be guaranteed to them to avoid conflicts on subjective matters perceived by individuals. The relations between various nations in a state and between the state and all nations must be based on negotiated agreements with minimum of coercion and without any use of force if possible. Such relations would tend to help in fusion of nations in a state and would promote state loyalties.

- All tribes in each nation in a state must have equal rights and their autonomy with respect to these rights must be guaranteed to them to avoid conflicts related to subjective matters perceived by individuals of the tribes. The relations between a nation and its tribes, if any, must also be based on negotiated agreements with minimum of coercion and without any use of force if possible. Such relations would tend to help in the fusion of all tribes into their nation and would help in developing a national identity as opposed to narrow tribal identities. The worst thing that can be done by a nation or a state is to use military force in attempting to solve economic and political problems related to tribes. Use of military force creates a violent reaction to any perceived threat and destroys all possibilities of peaceful development of relations.

- Only education with equal access to global information can help individuals of tribes and nations in overcoming their tribal and national subjective fears. The process can only be very slow since evolutionary changes cannot be brought about in a short time. Global cultural and educational interchanges can help in minimizing fears of different nations and populations of different states. In this variety and diversity of cultures, only educational opportunities can promote peaceful development within states and relations among them.

Family life and the personal history of each political figure must be closely examined by the system of government of a state, so that persons with mental disorders are excluded from positions of leadership and power. Systems of examination of individuals must be in place for selection of individuals for each level of political office. Such examination must include sufficient exposure to the electorate concerned. For example, at this time, such a system of examination by the electorate is in operation in the United States. Each political party holds its primary elections to determine the support by members of the party for individuals seeking nomination for various offices. A general election is, then, held to determine the final nominee, for each office. There are severe shortcomings in this system, however. The Media are in the control of the ruling class. Thus, news, and exposure of the population to information, is controlled by the ruling class - in effect controlling the minds of the people. Also, the members of the ruling class are allowed, by laws made by themselves, to buy politicians to come up with deceptive programs to really serve the interests of the ruling class, while pretending to be meant for the well-being of the general population. These two factors have completely subverted what is, otherwise, a reasonable and good system of evaluation of individuals and political parties.

Sources of Conflict

As described in Chapter-01, human beings compete with each other as individuals and also as social groups - tribes, nations, labor unions, professional associations, corporations and other business organizations, political parties and states. They tend to get into conflicts with each other as individuals and as social groups within states. States also compete with each other and tend to get into conflicts. Thus, in human conflicts, both individual and group survival skills are involved. In this chapter, we would summarize what history teaches us about the nature of these conflicts, which cannot be easily eliminated because of the nature of markets and the nature of human beings who compete in them. The competition, in general, is over possession of resources and over division of benefits obtained from them by use of human effort.

Thus, uncontrolled freedom, like in a natural market or jungle, is the biggest source of conflict between individuals and groups. As indicated before, human beings, by nature, desire to be completely free and to do whatever they like. In other words, the "Id" (as defined by Sigmund Freud) of a human individual, demands satisfaction of its desires at any cost. It does not care, nor is it aware of, what harm the satisfaction of its demands would do to anything or anyone else. Conflict occurs when one individual's desire to do something deprives another of the freedom to act likewise. Such conflicts occur in all areas, and all "markets", of human activity - political, social and economic. Thus, for social existence of human beings, freedom has to be re-defined as the ability to do whatever an individual or group may desire to do, provided this activity does not deprive other individuals or groups an equal degree of freedom to act the same way. Thus, a society has to set limits to freedom of engagement in various activities - to prevent its abuse.

If limits of freedom could be specified for various human activities and those limits were observed by all individuals and groups, then conflict would disappear and the division between abusers and abused would not arise. In practice, however, this does not happen. There are always individuals and groups who, dominated by the demands of their Ids, refuse to abide by any rules of conduct. Thus, conflict is a norm. It can be reduced in practice but not completely eliminated.

In the following sections we describe the sources of our conflicts - both objective sources like land and natural resources and subjective sources like tribalism, nationalism and religion.

Natural Resources

Throughout our history, our conflicts have been over habitat, food and other resources. States are the most developed form of human social groups. Since we are the most advanced animal species on our planet, and the states have human populations associated with them, these states also behave like animals. Every state has its appetites. Every state gets angry and attacks, if provoked. Every state can get scared and attack back. A state can go into a state of grief, or it can become aggressive and attack others, because it cannot live with what it has and tries to grab what other states have.

Thus, the competition between states is basically over natural resources of this planet. We consider land, the sunshine, the air and the water on the planet as the natural resources we need to survive. There are mineral resources occurring on the surface of the earth's crust, or below it. We have learned to use them for our benefit. Similarly, the land, the lakes, rivers, mountains and other animal and plant species of the planet are freely available to us. We have learned to use and abuse them and they have become essential to our existence and, hence, are the source of conflicts between us, as individuals and as groups. States are constantly competing, sometimes fighting, over these resources. The individuals within these states are also constantly doing the same. As the human population has increased, the competition has become more and more intense. The use of these resources is accompanied by abuse and waste and we have begun to basically change the environment of the planet, giving rise to doubts about our continued existence. Conquests of land and its agricultural use or colonization was not the only motivation for wars over land. Panama is an example. It happened to be a good location for construction of a canal linking the Caribbean Sea with the Pacific Ocean. Britain constructed the Suez Canal in Egypt, its imperial possession, for trade and transportation.

At this time, mineral resources are the main "bones of contention", rather than agricultural land. There may be competition over land, within a state, between different nationalities, or different tribes. That continues to some extent. But among states, the main fights are over oil, gas, coal, iron, aluminum, and, to some extent, precious metals, like gold and silver. In recent times, fissionable materials, like Uranium or Plutonium, have become a source of competition and conflict, due to the discovery of nuclear power and, specially, nuclear weapons.

Oil is widely used for fueling transportation vehicles, for heating homes and for electric power generation. Global consumption of oil has steadily grown. Oil is a non-renewable resource and its discovered reserves are limited. The constraints on oil production were not realized till recently until the first shortages began to develop in the 1960's and early 1970's. So, the struggle has sharpened, and has ultimately led to the virtual occupation of the Arabian Peninsula, which had the largest reserves at that time. Gas and oil occur in the same kinds of rock formations of similar age and in the same areas, generally. They have developed due to the same phenomena which occurred millions of years ago. Marine animals and plants were buried underground by upheavals that occurred on the planet's surface. Some of these ended up under huge layers of rocks. Due to intense pressure of the rocks, under very high temperatures, with no oxygen being available for them to burn, they changed into hydrocarbons. So, the petroleum that we find today consists of remains of what were plants and animals before. Those reservoirs, that were formed earlier, have changed to lighter hydrocarbons, which occur in gaseous form.

The Persian Gulf area has most of the proved reserves of oil in the world. In this region, Saudi Arabia has the largest reserves. The Ghawar field alone contains more than 170 Billion barrels of oil in place, which means that about 100 billion barrels may be recoverable. However, about 70 billion barrels of oil have already been produced from this field. Iraq, Kuwait and Iran have roughly one hundred million barrels of proved reserves of oil each. Other states of the Persian Gulf area also have substantial proved reserves. The successor states of the Soviet Union have about the same size of reserves as Iran. Substantial reserves exist in the area between Nigeria and Angola, including the off-shore fields. In South America, Brazil and Mexico have substantial oil reserves, while the largest reserves in the world are in Venezuela. Indonesia, China and Canada also have considerable reserves. There are substantial oil reserves in the North Sea where fields are owned by British, Norwegian and Danish companies. Russia, Iran and Qatar have the largest reserves of natural gas. United States had large reserves of oil and gas. Most of its fields have been exhausted, but it currently produces more than half of its requirements, due to an increase in shale oil and shale gas production. Similarly, Russia's oil fields have been depleted to a great extent, but it has a huge potential for discovery of new reservoirs. It produces about 12 million barrels of oil per day. Saudi Arabia has about the same output and Iran, Iraq and Kuwait together produce about 10 million barrels per day. At this time, China is the biggest importer of oil and gas.

Demand for oil and gas is growing very fast, especially in the newly developing states like China and India. Till recently only about twenty percent of the proved reserves of oil could be recovered from an oil field. This technology has developed further now. For example, gas can be pumped into oil wells and this increases the amount of oil that can be pumped out. Electrical submersible pumps, or mechanical pumps powered by diesel engines, can be used for this purpose. With improved technology, now, the recovery rate has almost doubled. New technologies can be applied to oil reservoirs that have been exhausted, to recover more oil. There are improvements in the technologies for oil exploration also. Now deeper wells can be drilled and drilling can be undertaken in deeper, off-shore waters. Despite all these improvements in technology, new oil discoveries have lagged behind increases in consumption and it is difficult to imagine that the current reserves can meet the projected global demand for more than a hundred years, or so, in the future.

Oil-rich regions of the world have seen a lot of conflict, even before shortages developed. The Persian Gulf area has seen the most conflict and is, at present, mostly under occupation. All the states in this region have been converted into one or another form of subservient states, by the United States and Britain. Only Iran has become independent and remains so, at this time. The most recent wars over oil, however, have been the American invasions of Iraq in 1991 and 2003 and the coordinated NATO attacks on Libya. Similarly, Central Asia, Nigeria, Angola, Indonesia, Mexico and Venezuela have faced a lot of abuse at the hands of imperial powers. Venezuela has the largest proved reserves of oil at this time. It is currently facing sanctions imposed by the US. A few years ago, China signed a long-term strategic agreement with Kazakhstan to develop its oil fields.

In terms of gas, the situation is quite different. The territory of the former Soviet Union has about half of the natural gas of the world. Iran, Qatar, Mexico and Algeria also have substantial reserves of natural gas. Gas is, in a way, considered more valuable than oil, since it can be used for manufacture of plastics and fertilizers and for electric power generation, in addition to being used for residential heating and cooking. In some states, automobiles have been converted to use natural gas. Iran, Algeria and Mexico have been threatened and subjected to wars and imperial abuse in the past. As the demand for oil and gas increases, competition and conflicts over these resources are bound to increase.

United States, Russia and China have the largest deposits of coal and Iron ore. Also, Ukraine, Germany, Canada, India and Brazil have large deposits. In the early stages of industrialization, coal and iron ore were sources of power, because they were essential to industrialization of a state. Buildings, steam engines and other machinery required steel and iron and coal were used for its production. European states like Britain, Germany and France were the first to develop their steel industries and this helped them in becoming advanced industrial and imperial powers. Then Japan and the United States also developed their iron and steel industries. At the end of the Second World War, United States had become the biggest producer of iron and steel. In fact, it had almost three quarters of the world's production.

Things have changed dramatically, however. Today, China is the largest producer of iron and steel, producing about half of the total steel of the world. Europe and United States continue to produce large quantities of steel, especially specialized steel, however, they have lost their globally dominant position. This has basically happened due to policies of the Soviet Union. The Soviet Union focused on developing its heavy industry first, especially the iron and steel industry. Then it helped countries of Eastern Europe to develop theirs and then did the same for China, India, Pakistan, Turkey, Iran, Algeria, Egypt and Nigeria, etc. All these countries became competitors of the United States, instead of being dependent on it for imports. This dramatically undermined the monopoly of the United States, which was the objective of the Soviet Union because of its global conflict with the imperial powers including the United States.

Copper and Aluminum are used by electrical and electronic manufacturing industries, building construction, automobile, aircraft and aerospace industries and a variety of other areas. Big reserves of copper exist in the United States, Canada, Brazil, Australia, Russia, Congo-Kinshasa, Zambia and Chile. Aluminum deposits are in Canada, Russia, Brazil, Australia, India, United States, Ghana and China. Copper has been a source of conflict in recent times. Large deposits are found in Congo-Kinshasa and Zambia These states have been, and are, weak and underdeveloped. They were subjected to imperial abuse and were ruled by Belgium and Britain, respectively. Recently, China has invested in the copper mines of both these states.

United States was the first state to develop nuclear weapons. It had deposits of Uranium but these were not huge. Uranium had been discovered in the Congo, which was a Belgian imperial possession. United States had the whole deposit extracted and brought to its territory, during the Second World War. This Uranium was used to make the first atomic bombs, which were dropped on Japan. At this time, Russia, Kazakhstan, Uzbekistan, Australia and Canada have the largest deposits of Uranium. United States and, then, the Soviet Union developed nuclear weapons. Other states, which felt threatened by these powers, also developed nuclear weapons. As the technology for development of nuclear weapons spread, the then established nuclear powers formed a cartel to deny this technology to other states. The Nuclear Nonproliferation Treaty was signed by the major nuclear and non-nuclear states in 1970. Some states, like Brazil, Argentina and South Africa had their nuclear weapon programs well on their way to completion. United States and France had, in the meantime, helped Israel in developing its nuclear weapons, but continued to pressurize other states to sign the Treaty. Under pressure from the United States, Brazil and Argentina agreed to forgo development of nuclear weapons. India, Pakistan and North Korea refused to sign the Treaty and continued to develop and test such weapons. The Treaty required the major nuclear states to undertake reduction and elimination of their arsenals. However, they themselves continued to modernize their weapons instead of eliminating them. Some reductions of the arsenals were negotiated, however, and implemented. States, which felt threatened by the existing nuclear states, continued their efforts to acquire these weapons. Thus, possession of uranium and other fissionable material and development of this technology of war has continued to be a major source of conflict in the world.

Gold and silver have always been valued for ornamental reasons. Gold was used as money and it performed this function for quite a long time, due to the value attached to it and its unlimited life. It still has its ornamental value and serves as a hedge against inflation, along with silver and platinum and is being used by China, Russia and some other countries for maintenance of their foreign exchange reserves. China is using it to back its currency in purchases of oil and gas. New uses have been found for gold and silver in the computer, telecom and electronics industries in general. Currently, China, Russia and South Africa are the biggest producers of gold.

Exploitation of marine animals has reached its peak. Fish and other marine animals have been hunted for food. Japan, United States and Russia maintain large fishing fleets. Over-fishing has occurred in several oceans. In fact, some ocean species, like the great white whales, are threatened with extinction. Thus, what has happened on land is also happening to the seas. Many plant and animal species on land have gone extinct. Others are facing extinction, as human beings encroach on their habitat. The rate of extinction of species is accelerating.

If we examine the resources of the planet that each species uses or controls, we can see that we do not inhabit more than one to two percent of the surface of the earth. But the area controlled and used by us is many times larger. Two thirds of the surface of the earth consists of oceans which we do not inhabit, but we control and consume fish and other marine animals to a great extent. The species, which have not been studied in depth, or which pose problems of security, or are an economic threat, are the ones which are being destroyed. The habitat of animal species has decreased progressively as forests are cut down and more and more lands are brought under cultivation, leading to destruction and extinction of plant, animal and insect species at an increasingly fast pace.

Tribalism

Tribes have existed throughout our recorded history of about six thousand years and, as we have discussed, tribalism is based on genetic homogeneity and attachment to a piece of land. Every tribe has a unique belief system, a unique folklore. Every tribe feels superior to other tribes in some way and strongly rejects individuals who are not its members. Because of extreme attachment between members of a tribe and their attachments to their tribal land and customs, it is very difficult for modern human beings to understand tribal boundaries. As we have seen, tribalism is an essential characteristic of primitive feudalism. As a social system evolves, and feelings of insecurity diminish, tribes tend to fuse into other neighboring tribes that are close to them racially, linguistically and in terms of religion. The result is an advanced feudal society.

The kind of conflict that the Hutu and Tutsi tribes of Rwanda went through in 1994 seems totally irrational to outsiders. However, if we understand the tribal sensitivities related to "their" land and their customs, then, it is not difficult to understand that these sensitivities should not be trampled upon if peace is to be maintained. Tribes do not easily yield to threats of use of force. Everything must be negotiated with them, if some concessions are desired from them, even if these are only in the interest of the tribe itself. Forcible division of tribes, like those in Rwanda, or elsewhere in Africa, can only lead to violence. Tribal land boundaries must be respected in the interest of peace. Within a state, the tribal land should be clearly demarcated. This should be done even if a tribe is divided between two states, so that tribal identity is maintained and does not become a source of conflict and, if and when the two states decide to unite, the unification of the tribe would be automatic. Somalia is going through a high level of violence at this time, because tribal boundaries have not been respected and have become bones of contention, instead - within Somalia and between Somalia and Ethiopia. Tribes ultimately fuse into other tribes, or into a group of tribes, becoming a nation. And respect for tribal boundaries makes it possible for this process to occur without violence and furthering this evolutionary process should be the objective of a state.

Nationalism

We have defined nationalism with respect to four main contributing factors - land, genetic homogeneity, language and religion. These four factors make a social group into a nation. Historical experience, folklore, cultural activities may also be considered as factors that define a nation, but they are variables dependent on the first four independent variables mentioned before. A nation is formed when all tribes, having traditional land, racial characteristics, language and religion in common, merge into one social group and an advanced social group, the nation, results initially under the conditions of advanced feudalism. The larger size of the population of a nation, and its larger territory, gives the nation a higher level of feelings of security than a tribe, but the behavior of a nation is essentially the same as that of a tribe.

Thus, what is true about a tribe is also true about a nation. A nation resists any use of force against any part of its population. It develops its own ways of dealing with relationships between its individuals and the level of development of its group social skills continues with time and experience. The worst thing that can be done to a nation is to divide it by force. Behavior of the Vietnamese, the Koreans, the Kashmiris and the Germans has clearly demonstrated this. The Germans were divided into two states, as a result of the Second World War, but as soon as force, and the threat of use of force, was removed, the two states re-united into one German state. The same happened to Vietnam. The Koreans and the Kashmiris continue to struggle against American and Indian control, continue to die in this struggle and this counter-violence would continue till these nations are re-united, regardless of the state or the social system they become part of.

Europe has gone through a long period of intense violence due to the imperial ambitions of various states that developed on its territory. Several multi-national states have disappeared from its map and, essentially, a group of national states exists on its territory at this time. Britain, Switzerland and Bosnia are the main multi-national states which continue to exist. History of Europe over the last three hundred years, tells us that political management of a multi-national state is a very difficult and complicated matter and can easily lead to unnecessary violence. It is important that nations not be divided between states because of conflicts among those states. The occupation of the traditional national territory of a nation, by a neighboring state, is a recipe for never-ending violence. The boundaries of traditional national territories of nations, within multi-national states, must also be clearly demarcated, even if some such nations within a state are divided with neighboring states. Each nation, within a multi-national state, must be given some degree of autonomy to manage its national affairs. This is the only way to avoid unnecessary violence due to nationalism.

Religion

No one knows the answers to life's basic questions - Who are we? Where are we? Why are we here? What happens to us when we die? We feel insecure not knowing the answers. Throughout our evolutionary history we have tried to find the answers. In the absence of the needed explanations, we rationalize the meaning of our lives. We find explanations which may not really satisfy us, but they do help us to live on with some degree of comfort. Thus, we have developed belief systems, or religions. These religions encompass a lot of what we have learned during our evolutionary history, especially our group survival skills. These group survival skills, or social habits, like compassion, sympathy, courtesy, etc., form the foundation of all religious thought. We have developed myths to justify those feelings and rituals to reinforce them. Thus, every religious belief system has these three aspects - group survival skills, mythology and rituals.

Religious groups have spread as the original believers have spread, to new continents. Religious social groups go across boundaries of tribe, nation and state. Billions of people belong to Christianity, Buddhism, Hinduism and Islam. Thus, religious conflicts tend to be mainly sectarian conflicts within the boundaries of nations and states, although conflicts on a larger religious base also occur, like the Al Qaida and Islamic State related conflicts at this time. Religious conflicts are always about rituals and myths, since the foundation of all religious thought is the same. Reason has, generally, nothing to do with such conflicts and they cannot be resolved by reasoning.

The best way to deal with religious conflicts is to ignore them, unless they become violent conflicts. The worst thing that can be done is to use force against religious movements. A state's business is always about scientific matters. It is about security and well-being of its population. It is about economic matters, taxation, development programs for roads, telephone systems, airports, new technologies, social security, health-care, etc.; even military violence has become a scientific matter. Separation of state from religion removes the source of conflict to the extent that is possible. Feudal societies, however, tend to develop strong political parties which invariably want to dictate how society must operate, so that a conflict between the state and organized religion is natural, but it has to be managed to achieve the desired separation. Experience has shown that, ultimately, separation of state and religion does get accepted by society as it develops out of feudalism and into a modern society, making religion into a private, personal, or social matter outside the realm of a state's political system. The basic requirement for separation of state and religion is the availability of universal scientific education for all citizens. Thus, the people must have an understanding of the basics of scientific knowledge that they need for participation in a representative system of government. Knowledge of social and life sciences makes the minds of citizens of a modern state less susceptible to religious dogma, leading to a tolerant society. Thus, universal education is a desirable objective for the population of a state.

Gender

In the context of inter-gender relations in the political field, conflict has existed for a long time. Initially, women were not allowed to vote or to stand for political office even in Europe, North America and other regions and states of the world of *benevolent capitalism*. Women struggled for these political rights for a long time and have succeeded in this sector of the world. However, they continue to face difficulties due to the burden of child-rearing, which affects the full exercise of their rights. The other sectors of the world, i.e., the worlds of *intensive capitalism*, *advanced feudalism* and *primitive feudalism* have also tended to move in this direction, but, in these three sectors of the world, women have continued to face many economic and religious obstacles to full achievement of equal status in this field. This political inequality is related to the economic inequality in terms of rights of inheritance and inequality in the work-place. As the social systems evolve, women would ultimately achieve equality in this field. However, the issue remains a source of conflict throughout the world, at this time.

Class Structure of Society

As we have discussed in Chapter-1, since the beginning of our evolutionary history, and especially since the beginning of economic activities of social production and distribution, social groups have always tended to split up into two antagonistic classes, or sub-groups. In the age of savagery, a man dominated the family and the extended family. The chiefs or "elders" of a tribe formed the ruling class of a tribe. Nations and states also split up into two classes. Thus, society has always split up into two hostile camps - the abusers and the abused. This was the case in every feudal society. The division remained - in fact became more intense

- when feudalism evolved into intensive capitalism. It persisted even when a social system evolved into benevolent capitalism, although the antagonism became less intense.

Thus, society has always split up into two main classes, each of which has tended to split up further into many sub-classes based on the occupations of the individuals in each class. The sub-classes have tended to proliferate as society has developed and its economic system has become more advanced, requiring further specialization of skills. Thus, the antagonism between the ruling class and the people has persisted throughout our evolutionary history - spanning hundreds of thousands of years. This is the most fundamental of all sources of conflict in human society and its elimination is the prerequisite for the creation of a peaceful world.

Evolution of Political Parties

As society evolves, so do the political parties representing its different segments. At the stage of primitive feudalism in a national state, only conservative parties would dominate the state. This is the situation in Kenya, Uganda, Chad and Saudi Arabia ("Saudistan") today. Kenya is basically inhabited by one multi-tribal nation, although there is, also, a small minority nation of Somalis inhabiting its territory. A conservative party has ruled the state ever since Kenya was de-colonized by Britain and became an independent state. There are severe inter-tribal conflicts in its population and two conservative political parties have come to dominate the politics of this state. The situation in Uganda is similar. Only one political party rules. It does not tolerate any other political party, but opposition does exist to its rule consisting of conservative individuals with different tribal affiliations. The situation in Chad is similar to that of Uganda. In Saudi Arabia, which is a Bantustan controlled by the United States, the Saudi tribe rules. It uses the conservative Wahabi clergy to control the tribal society inhabiting its territory on the Arabian Peninsula. Any opposition is violently suppressed and formal political parties are not tolerated, but its society is slowly developing towards advanced feudalism and a multi-party political system is bound to emerge there when it evolves into a normal or Panamized state.

In a multi-national state at a primitive feudal stage of development, nationalist parties would also exist and wield considerable political power, because, at this stage, many economic conflicts are seen in terms of narrow national interests. This is the situation, at present, in Sudan and Afghanistan. As a society evolves into advanced feudalism, liberal parties emerge and begin to have some political power. At this stage, political parties representing the national interests of the smaller nations in a multi-national state, tend to align themselves with the liberal parties. They may even merge with such liberal parties, or lose some of their members to them. The same happens to religious minorities, who tend to support liberal political parties.

As society moves into the stage of intensive capitalism, economic conflicts become sharp and explosive. Nationalism and religion give way to class consciousness and political parties begin to merge into two groups, those representing the interests of the ruling class only, and those who take the interests of the people, also, into consideration. Broadly speaking, this is the present situation in India, Pakistan, Brazil and Russia. In India, the Indian Congress is the main liberal party. Other liberal parties and organizations representing small nations and minorities tend to coalesce around it to form coalition governments at all levels of government. Even the Communist Party of India (Marxist) tends to support the Congress Party, against the group of conservative and reactionary parties led by the Bhartiya Janata Party (BJP). In Pakistan, the Peoples Party is the main liberal organization and other liberal and nationalist parties, representing smaller nations of Pakistan, tend to support it. The Muslim League is the main conservative party and the religious reactionary parties and the Army (effectively an armed political party, since it has ruled Pakistan during most of its existence) tend to support it. In Russia, three groups of parties exist at this time, but the tendency is towards two political parties in the near future. This is so because Russia has recently separated from the Soviet Union, which was a one-party dictatorship for a long time. The situation in Brazil is similar, since Brazil has gone through a long period of domination by military dictatorships. Numerous political parties exist in Brazil. However, they are converging into two groups - liberal, or Social democratic, parties and conservative parties. In Britain, two main parties have emerged - the liberal Labor Party and the Conservative Party. However, due to the ill-treatment and extreme suppression of the minority nations in the past, nationalist parties still exist in Scotland, Wales and Northern Ireland, despite the fact that British society has reached the stage of *benevolent capitalism* since the middle of the twentieth century. The population of these nations and these parties, however, tend to support the Labor Party, which, naturally, has strong support among the minority nations of the United Kingdom.

As the social system develops further, towards benevolent capitalism, ultimately, all the conservative and reactionary parties tend to merge into one political party and all the liberal and radical parties merge into another opposing party - and a two-party system emerges. Both the parties, at this stage of development, represent the interests of the ruling class, but the liberals do make concessions in the direction of the interests of the people, in an effort to obtain political support from them. Thus, the conflict between the ruling class and the people intensifies during intensive capitalism and, then, becomes less severe as the political power shifts in favor of the people, as a result of a labor shortage. Once the benevolent capitalism has progressed to a degree that the economic relations of the previous stage have become obsolete, generally, only two camps remain, determined by the persistent economic divisions within society.

Evolution vs Revolution

The law of evolution is the basic biological law of development of all living beings. It is the fundamental behavior of all living beings to adapt to circumstances as they change and to compete with each other for their survival - groups, i.e., tribes, nations and states and their sub-groups, compete with other groups; and individuals in these groups compete with other individuals using all the individual and group survival skills that they have acquired during their evolutionary history. However, dramatic events have taken place, in recent human history, which are generally described as revolutions. The French revolution of 1789 is one such event. Similarly, the Russian revolution of 1917 was such an event. It led to similar revolutionary upheavals in China, Vietnam, Cuba, Ethiopia, Angola and Mozambique, etc. These events are referred to as revolutions because, as a result, a whole ruling class of a society was overthrown and was replaced by another class.

At the time of the French Revolution of 1789, European society had reached the stage of advanced feudalism. Kings and Aristocrats ruled most states of Europe and intolerance and violence reigned supreme all over Europe. The people were being subjected to economic and social abuses of all kinds and, despite large-scale emigration of the population to colonies on other continents, the population of Europe was being subjected to *Intensive Death Rate* on a large scale. Under these conditions, the working people of France rose in rebellion against the aristocratic authorities of the time. All institutions of the state were taken over by ordinary citizens. Workers took over farms and buildings belonging to the ruling Feudal ruling class. After a large-scale violence, a new class of individuals seized power. It is generally referred to as the "bourgeois" class. This class consisted of managers and owners of small workplaces, or workshops, skilled artisans, administrators, lawyers, accountants and technicians of different types. The king and his fellow aristocrats were deprived of their former level of power. This was a major upheaval in French society. It changed French society for ever. In fact, it was the beginning of a process - as aristocrats lost power in other European states, one after another, and were replaced by a new class. The new class became a class of factory and business owners and society changed into a modern industrial society, with a new factory-based mode of production, instead of a feudal society based on agriculture as the primary mode of production. Thus, the French revolution of 1789 was the beginning of the dominance of *intensive capitalism* in Europe.

After the French revolution, Europe remained in turmoil for more than a hundred years, as economic abuse, violence and war dominated its life. A revolution on a pan-European scale was expected by many intellectuals at the time. For revolutionaries, Germany seemed to offer the greatest hope. In 1905 a revolutionary change almost happened in Tsarist Russia, but failed to replace the ruling class. The aristocracy of the Tsarist Empire was ultimately overthrown in 1917, as the result of the Russian revolution. The rest of Europe, however, did not go through a revolutionary change. The change in Tsarist Russia was very dramatic and violent. For the first time in history, a political party representing the working class seized control of all institutions of power in a state and established itself as the ruling class. Virtually all agricultural land and industry was expropriated and taken over by the state and new organs of the state were created from the bottom up. A one-party state, the Soviet Union, headed by Vladimir Lenin, was set up on the territory of the Tsarist Empire. This was a momentous development of historical proportions. The Russian revolution led to revolutions in China, Vietnam, Cuba and several other states, leading to the overthrow of the ruling classes in those states and brought dramatic changes in the political landscape of the world.

Thus, the twentieth century saw several revolutions in the world, like the Chinese or Cuban revolution. However, there was a common thread to all these events in different states on different continents. In every state that went through a revolutionary change, society had become stagnant and immobile. All change was violently resisted by the ruling class, or a foreign power, and the ruling class succeeded in stopping the progress of society towards more power to the people. Thus, the people continued to progress socially and psychologically, but failed to make tangible physical improvement in their lives, in fact their lives became worse in all these states. In other words, evolutionary changes did take place in the mental and social lives of the people but these were *forcibly* stopped from being translated into improvements in the status of the working class and the people in general. The suffocation felt by the working class and the painful suffering and bloodshed faced by its members ultimately exploded with uncontrollable fury - and that is now referred to as a revolution.

Thus, all the revolutions described before were, in reality, expressions of pent-up evolution of society. As evolution was forcibly resisted by the ruling classes, a violent explosive change occurred instead. *We can, therefore, define a revolution as "pent-up evolution".* Pent-up evolution, however, may not necessarily lead to a revolution. The change may be a major and violent one, but it may not be a revolutionary change, i.e., it may not lead to the overthrow of the ruling class and its replacement by another class. The change may just cause the ruling class to stage a major retreat, making concessions to the working class in terms of economic relations, or in other aspects of social life. The civil rights movement in the United States in the 1960s, the movement for voting rights for women in the United States and elsewhere, the end of slavery in the United States, Brazil and elsewhere, and the recent demonstrations by the people in Arabia, leading to the overthrow of dictatorial regimes in several Arab states, would fall into this category of changes in society.

Evolutionary changes in society are very slow and they take place gradually and without a high level of violence if the political system provides the opportunity and the space for the people to express their desires and needs in peaceful ways. If social change is resisted by the ruling class of society, change still occurs, but it is accompanied by a level of violence in proportion to the violence with which the ruling class suppresses the people. There is a level of violence and suppression by the ruling class, which ultimately leads to a violent overthrow of the ruling class itself. When would that level of tension be reached by a social system, also depends on the level of development of the social system itself. If strong and sophisticated institutions of government exist, then a revolution may not occur for a long time. A system of representative government, which guarantees basic human rights, is the most important factor in providing channels for evolutionary social and political development, without the necessity of a violent revolution. This is what happened in Europe in the early twentieth century. All the European states, except the Tsarist Empire, were able to avoid a revolutionary change, although many thought Europe was "ripe" for such a change. The same has happened in North and South America, where only Cuba went through a revolutionary change. In Asia, the ruling classes in several states, like China, Vietnam and Laos, tried to use force to perpetuate themselves in power, even with the help of foreign imperial powers. Naturally, social explosions and the resulting revolutionary changes followed.

Thus, to avoid violent events resulting in revolutions, it is necessary for a state to adopt a system of representative government, no matter how primitive its society may be. This would ensure a relatively smooth evolution of all its economic, social and political relations.

Chapter-17

Human Rights

All human beings have a desire to control their lives and to live according to their own wishes. This desire to be free is deeply ingrained in our psyche and is the result of millions of years of evolutionary experience. At the same time, we have the desire to be connected with other human beings in various ways and feelings of sympathy, compassion and consideration are also deeply ingrained in our genetic heritage. Thus, for a society to function with least amount of tension, certain freedoms to act must be guaranteed to all individuals, without any distinction of any kind such as race, color, gender, language, religion, political or other opinion, national or social origin, property, birth or other status. Since all human beings are born free and equal in dignity and rights, there have to be limits to the exercise of these rights, or freedoms, to ensure that freedoms of some do not result in denial of the same freedoms to others.

What applies to individuals, also applies to social groups. All tribes must have the same rights if there is to be peaceful existence within a nation. Similarly, all nations within a state must have the same rights. Equal rights must also be ensured to all national states and all multi-national states of the world if we desire a peaceful world. The same applies to all political, social and economic groups within each state - i.e., political parties, labor unions, professional associations, business entities like corporations, etc. In addition, since limitations on exercise of the guaranteed rights would have to be imposed by a state in order to ensure the same rights to all individuals and social groups, mechanisms for smooth conflict resolution would have to be provided.

Rights of individuals have been defined in the Universal Declaration of Human Rights (UDHR) in a historic document that was adopted by the United Nations General Assembly at its third session on 10 December 1948 as Resolution 217. The Declaration is a document in the public domain. Although most of these rights are guaranteed by many states, few actually observe them and even less make an effort to create an economic environment that would make it possible to really guarantee these rights. The next section deals with rights of individuals.

. Fundamental rights of different kinds of social groups, i.e., tribes, nations and national and multi-national states, would differ according to the nature and functions of those groups. We shall describe these rights in the following three sections, after the description of individual rights.

Rights of Individuals

A brief summary of most of these rights is given in the following paragraphs. However, some details have been omitted. For more details, the reader should refer to the original document.

Right to Life and Liberty

Everyone must be guaranteed the right to life and security of person. This is the most important fundamental right. No one should be held in slavery or servitude; slavery and the slave trade are prohibited in all their forms. Everyone, as a member of society, has the right to social security and is entitled to realization, through national effort and international co-operation and in accordance with the organization and resources of each State, of the economic, social and cultural rights indispensable for his dignity and the free development of his personality. Everyone has the right to a standard of living adequate for the health and well-being of himself and of his family, including food, clothing, housing and medical care; and necessary social services; and the right to security in the event of unemployment, sickness, disability, widowhood, old age or other lack of livelihood in circumstances beyond his or her control.

Right to Citizenship

Everyone has the right to a citizenship and no one should be arbitrarily deprived of his citizenship nor denied the right to change it.

Freedom of Conscience

Everyone has the right to freedom of thought, conscience and religion; this right includes freedom to change his religion or belief, and freedom, either alone or in community with others and in public or private, to manifest his religion or belief in teaching, practice, worship and observance.

Freedom of Expression

Everyone should have the right to freedom of opinion and expression; this right includes freedom to hold opinions without interference and to seek, receive and impart information and ideas through any media and regardless of frontiers.

Freedom of Association

Everyone has the right to freedom of peaceful assembly and association and no one should be compelled to belong to an association.

Equality Before the Law

Everyone has the right to recognition everywhere as a person before the law. All are equal before the law and are entitled without any discrimination to equal protection of the law. All are entitled to equal protection against any discrimination. No one should be subjected to arbitrary interference with his privacy, family, home or correspondence; or to attacks upon his honor and reputation. Everyone has the right to the protection of the law against such interference or attacks.

Protection from Arbitrary Arrest, Detention, or Exile

No one should be subjected to torture or to cruel, inhuman, or degrading treatment or punishment. No one should be subjected to arbitrary arrest, detention, or exile. Everyone has the right to an effective remedy by the competent national tribunals for acts violating the fundamental rights granted him by the constitution or by law. -- Everyone is entitled in full equality to a fair and public hearing by an independent and impartial tribunal, in the determination of his rights and obligations and of any criminal charge against him. -- Everyone charged with a penal offence has the right to be presumed innocent until proven guilty according to law in a public trial at which he has had all the guarantees necessary for his defense. -- No one should be held guilty of any penal offence on account of any act or omission which did not constitute a penal offence, under national or international law, at the time when it was committed. Nor should a heavier penalty be imposed than the one that was applicable at the time the penal offence was committed.

Freedom of Movement

Everyone has the right to freedom of movement and residence within the borders of each state. -- Everyone has the right to leave any country, including his own, and to return to his country. -- Everyone has the right to seek and to enjoy in other countries asylum from persecution, except in cases of prosecutions genuinely arising from non-political crimes.

Right to Form a Family

Men and women of full age, without any limitation due to race, nationality or religion, have the right to marry and to found a family. They are entitled to equal rights as to marriage, during marriage and at its dissolution. -- Marriage should be entered into only with the free and full consent of the intending spouses. -- The family is the natural and fundamental group unit of society and is entitled to protection by society and the State. Motherhood and childhood are entitled to special care and assistance. All children, whether born in or out of wedlock, should enjoy the same social protection.

Property Ownership

Everyone has the right to own property alone as well as in association with others. No one should be arbitrarily deprived of his property. Everyone has the right to the protection of the moral and material interests resulting from any scientific, literary or artistic production of which he is the author.

Political Participation

Everyone has the right to take part in the government of his country, directly or through freely chosen representatives. -- Everyone has the right of equal access to public service in his country. – The will of the people should be the basis of the authority of government; this

"Will" should be expressed in periodic and genuine elections which should be by universal and equal suffrage and should be held by secret vote or by equivalent free voting procedures.

Right to Work

Everyone has the right to work, to free choice of employment, to just and favorable conditions of work and to protection against unemployment. -- Everyone, without any discrimination, has the right to equal pay for equal work. -- Everyone who works has the right to just and favorable remuneration ensuring for himself and his family an existence worthy of human dignity, and supplemented, if necessary, by other means of social protection. -- Everyone has the right to form and to join trade unions for the protection of his interests.

Leisure, Sports and Recreation

Everyone has the right to rest and leisure, including reasonable limitation of working hours and periodic holidays with pay. Everyone has the right freely to participate in the cultural life of the community, to enjoy the arts and to share in scientific advancement and its benefits.

Education

Everyone has the right to education. Education should be free, at least in the elementary and fundamental stages. Elementary education should be compulsory. Technical and professional education should be made generally available and higher education should be equally accessible to all on the basis of merit. -- Education should be directed to the full development of the human personality and to the strengthening of respect for human rights and fundamental freedoms. It should promote understanding, tolerance and friendship among all nations, racial or religious groups.

Reasonable Restrictions on Exercise of Rights

Everyone has duties to the community in which alone the free and full development of his personality is possible. In the exercise of his rights and freedoms, everyone should be subject only to such limitations as are determined by law solely for the purpose of securing due recognition and respect for the rights and freedoms of others and of meeting the just requirements of morality, public order and the general welfare in a democratic society.

Tribal Rights
Control of Traditional Territory

Every tribe should have the right to control all its traditional tribal territory. The "traditional" territory of a tribe is defined as that territory on which it has lived for a long period of time, perhaps thousands of years, and the tribe's claim is accepted by neighboring tribes and the nation of which they are a part.

Control of Natural Resources

Effectively, control of exploitation of all national resources on its territory, above ground as well as underground, should rest with a tribe. This follows from its right to control its traditional territory. In a nation consisting of several tribes, and, perhaps, some areas where its population has developed a dominant national identity as compared to tribal identities, the national government would have to negotiate an agreement with all its tribes to clearly define tribal rights vs. national rights. This would not be necessary in non-tribal and purely "national territories".

Control of Economic Activities

Every tribe should have control over movement of goods and people and consequent economic activities over its territory. This, naturally, follows from its right to control its traditional territory.

Concurrent Activities

The central government of a multi-national state, or a national government in such a state, have interests in many areas of activity over the entire territory of the nation. The same applies to a national state and the tribes within its national territory. Thus, in some areas, the rights and activities of a tribe and its national and central government may come into conflict. A nation and its central government would have to negotiate with the tribal authorities in order to engage in the activities that those governments desire to engage in.

Generally, the national government would want to build roads and railroads in a tribal territory, or extend its communication system to cover it. It may also want to set up an educational system and a judicial system that suites its purposes. It may want to set up projects to exploit natural resources of the tribal territory. The education system would have to include teaching of physical, social and life sciences in addition to what a tribe may desire to be taught. Similarly, the judicial system would have to serve the main purpose of ensuring fundamental rights of all citizens. For this purpose, procedures would have to be so modified that unwarranted search and arrest are minimized and those accused of crimes are not meted out cruel and unusual punishments, which is a natural tendency of all tribal and primitive societies. This could, for example, be ensured by mandating a right of appeal to higher national courts. These matters would have to be negotiated to arrive at agreements that are acceptable to the relevant tribal authorities. The result would be agreed guarantees of rights of all sides to engage in such activities concurrently. The main purpose of the national and central government in negotiating such agreements should be to clearly demarcate rights of all sides to ensure smooth and peaceful fusion of a tribe with other tribes, or its national population.

National Rights
Control of Traditional Territory

Every nation should have the right to control its traditional national territory. The "traditional" territory of a nation is defined as that territory on which it has resided for thousands of years and the nation's claim is recognized by neighboring nations.

Right of Self-determination

Every nation should have the right to decide which state it wants to be part of, and to set up a separate national state of its own, if it so desires. For this purpose, every nation in a multi-national state should have the right of secession from such a state, on terms negotiated by the parties, in each case.

Language and Culture

Every nation should have the right to promote its language and culture.

Ownership and Control of Natural Resources

The ownership of all national resources on its territory, above ground as well as underground, should rest with a nation. In a multinational state, the control, but not the ownership, of some critical natural resources may be transferred to the federal government, by a constitutional agreement among all the nations of the state.

Control of Economic Activities

Every nation should have control over movement of goods and people and other economic activities over its traditional territory, except the tribal territories within its control.

Concurrent Activities in a Multi-national State

The central government of a state has interests in many areas of activity over its entire territory. External and internal security, basic control of its currency and economy, all external relations with other states and smooth and efficient operation of its judicial and educational systems are its basic functions. Thus, in some areas of activity within a multi-national state, the rights and activities of a nation and its central government may come into conflict. The central government of such a state would have to negotiate with the national governments in order to engage in activities that are primarily controlled by them. Generally, the central government would want to build military and other bases for state security services, offices of the central bank and taxation departments, courts to deal with areas of central government jurisdiction, etc., throughout its territory.

In addition to engaging in primary central government functions, the central government may want to promote a state language, or languages, and may need to engage in certain activities in those matters that are primarily within national rights. These matters would have to be negotiated to arrive at agreements that are acceptable to both the central and national governments. The result would be agreed guarantees of rights of all sides, by means of constitutional provisions, to engage in such activities concurrently. The main purpose of the national and central government in negotiating such constitutional provisions should be to clearly demarcate rights of all sides to ensure peaceful coexistence of all nations in a multi-national state and their smooth and peaceful fusion into a population with strong state loyalties, as compared to national and tribal loyalties.

State Rights

Independence and Sovereignty

A state, by definition, is independent and has complete sovereignty over its territory, air space and its territorial waters and continental shelf, if any. It, thus, controls, and has the right to control, all economic, social and political activities within its borders. It also has the right to order its relations with other states according to its own interests.

Equality

All states have equal rights in all fields of inter-state relations. They have equal rights of navigation and commerce in oceans and waterways outside the control of any individual state. Similarly, they have equal rights in airspace over areas that are outside the control of any state. Further, they have equal rights of space exploration. These rights should be defined by collective agreements between the global community of states. Rights over rivers that originate in one or more states and flow through more than one state must be defined by global agreements between the community of states, so that riparian rights, and rights to natural resources in lakes cannot be denied, by force, to one or more states, by any individual state.

Individual and Collective Defense

All states have an equal right to defend themselves against aggression by any other state; by any means they deem fit, including, with the help of, and in collaboration with, other states.

Development of Social and Political Systems

All states should have an equal right to order their internal relations without interference by any other state, subject only to collective agreements among the global community of states. Obviously, inter-state institutions are needed for collective action regarding global issues related to the environment of the planet. Such institutions are also needed for dispute resolution in inter-state relations, regarding security; navigation in and exploration of oceans, waterways and space; and division of flows of water in river-basins shared by two or more states.

Summary & Conclusions

The most important rights of an individual are the right to life and liberty, as described before in this chapter. Other important rights, as described in this chapter, are: the right to freedom of expression and association, the right to education and the right to political participation – However, these rights can only be guaranteed if the population of a state is also guaranteed the right to comprehensive and objective information. All these rights require a just socio-economic system with equality of all individuals in a state. Further, the rights of tribes, nations and states, as indicated above are pre-requisites for guaranties of individual rights. If the social system is not just, then equal access to education and information cannot be ensured and this makes it impossible to ensure equality of association and self-expression.

Chapter-18

Representative Government

Democracy and Representation

Democracy has become an ideal for all of mankind. It is considered to mean the rule of the people of a state, i.e., a state is considered to be a democratic state if all its affairs are decided by the people, directly or indirectly. It implies equality of citizens of a state and equal rights for all citizens in all areas of human activity – political and socio-economic, including sports and entertainment. Democracy specially implies equal rights of all citizens to rule and to participate in political decision-making. Thus, it also implies freedom of the people to form their opinions on various issues affecting a state, equal access to information and equal rights of all to propagate their views and form political, economic and social organizations.

..... **"Four score and seven years ago our fathers brought forth upon this continent a new nation, conceived in liberty, and dedicated to the proposition that all men are created equal."**

..... **"We here highly resolve that the dead shall not have died in vain, that this nation, under God, shall have a new birth of freedom and that government of the people, by the people, and for the people, shall not perish from the earth."**

President Abraham Lincoln (1863), "*The Gettysburg Address*"

At present, no state guarantees, or can guarantee in practice, equal rights for all its citizens in all these fields, i.e., political, socio-economic and in the field of sports and entertainment. Constitutional and legal guarantees, however, do exist. Thus, democracy remains an ideal and a dream. Abraham Lincoln offered the best definition of democracy - *"government of the people, by the people and for the people."* What this means is that the people should rule themselves. They should rule in their own interest and this would make their government a government that belongs to them - a government of the people. But what we have right now, virtually all over the world, are *governments of exploiters, by the exploiters and for the exploiters - in the name of the people!* The ruling class owns the government and most of the means of production and exploits the people by means of the power it, thus, obtains. It forces the people to work for subsistence wages, or low wages in most states of the world. Thus, large numbers of the people are subjected to a controlled labor market, many as wage-slaves. The ruling class uses its resulting economic power to control all communication and education media, schools, colleges, universities, newspapers, television stations, radio stations, publishers, etc., and thus controls and manipulates all cultural activities and information relevant to a clear understanding of all social, political and economic issues. It thus controls the minds of the people and people are programmed like robots to express "their" opinions, i.e., the opinions of the capitalists controlling the mass media. Thus, free flow of information is controlled and distortions of the truth are spread whenever and wherever this process suites one or another segment of the ruling class. Thus, political debate in a society, effectively, becomes a debate between groups of capitalists of the ruling class. If a "representative government" exists in such a state, it represents the same groups of the ruling class, in reality.

"And then democracy comes into being after the poor have conquered their opponents, [killing] some and banishing some, while to the remainder they give an equal share of freedom and power; and this is the form of government in which the magistrates are commonly elected by lot.

Yes, he said, that is the nature of democracy, whether the revolution has been [achieved] by arms, or whether fear has caused the opposite party to withdraw.

And now what is their manner of life, and what sort of a government have they? for as the government is, such will be the man. - Clearly, he said.

In the first place, are they not free; and is not the city full of freedom and frankness - a man may say and do what he likes? - [It is] said so, he replied.

And where freedom is, the individual is clearly able to order for himself his own life as he pleases? - Clearly.

Then in this kind of State there will be the greatest variety of human natures? - There will.

This, then, seems likely to be the fairest of States, being like an embroidered robe which is spangled with every sort of flower. And just as women and children think a variety of colours to be of all things most charming, so there are many men to whom this State, which is spangled with the manners and characters of mankind, will appear to be the fairest of States. - Yes.

Yes, my good Sir, and there will be no better in which to look for a government. - Why?

Because of the liberty which reigns there - they have a complete assortment of constitutions; and he who has a mind to establish a State, as we have been doing, must go to a democracy as he would to a bazaar at which they sell them, and pick out the one that suits him; then, when he has made his choice, he may found his State. – He will be sure to have patterns enough."

Plato, *The Republic, Book VIII*. Translated by Benjamin Jewett. Urbana, Illinois: Project Gutenberg. Retrieved October 16, 2019, from www.gutenberg.org/ebooks/1497.

Note: Plato has written this book as a conversation between his teacher Socrates and two of his brothers. Here, in Chapter-8 (called Book-8), "Socrates" is talking to one of them, named Glaucon.

But this is only one aspect of distortion of the concept of representation, in many states of the world. Rules are normally set up in such a way that only persons with substantial capital in their possession can stand for election as representatives of the population. The result is that only members of the ruling class, or persons with substantial support of that class, get "elected". Similarly, rules are set up, by the ruling class, effectively making it a prerequisite, for political parties, to have large amounts of capital to take part in an election campaign. Thus, members of only the ruling class can form large political parties to wage an election campaign under a "representative government" controlled by the ruling class. Any other political parties, which may develop despite these discriminatory laws, are thus forced to operate in a severely unfavorable environment. They are, naturally, not favored by large corporations in their political donations. They cannot publicize their views and programs, because this publicity entails high costs. They cannot freely hold meetings because the meetings require payments of high rents which only organizations of the ruling class can afford. They cannot buy time on television stations, or space on newspapers for the same reasons. Thus, the economic exploitation that divides a society into two hostile camps – the ruling class and the people -, also results in domination of the political system by the ruling class. The political system, in this way, becomes another weapon, in the hands of the ruling class, to maximize the exploitation of the people. Thus, those who exploit the people also lay claim to representing their victims and this is referred to as "democracy"! In reality, such a system can only be accurately described as a *"looteracracy"*, or *"exploitocracy"*, *i.e. a government of exploiters, by the exploiters and for the exploiters*. And since the people are thus forced to choose between one or another exploiter-dominated political party, the people end up choosing those who would exploit them for the next four or five years. They really have no other choice. If such a system functions smoothly as designed, it effectively results in a *dictatorship of the ruling exploiter class*, which the people are helpless to change. Thus, the currently existing "democratic" governments have the "form" of democracy, but not the content. However, this system of government is sensitive to how the people feel and how they vote. It controls all information made available to the people and uses its control of the information media to "guide" the people how to feel and how and what to decide, when time for "elections" arrives.

We can see that the political system gets distorted because of the distorted and abusive nature of the existing economic system. The economic system has become central to our social and political existence. However, we know that economic inequalities have existed throughout the period of civilization and growing socialization of mankind and even before. Progress of mankind has been towards less and less abusive relationships between different layers of the ruling class and the people, as the social systems and social relationships have changed with time. We no longer have the likes of Genghis Khan, Hulagu Khan, or the Pharaohs, running around beheading people for sport, or subjecting people to forced labor. We do not have the old-fashioned slavery openly existing anywhere in the world of today. We do not often see whole tribes of human beings being subjected to mass murders anymore. We do not see arbitrary rule of kings in most parts of the world. In fact, we have seen heads of many kings roll in the last half century. The rest have been forced into becoming constitutional monarchs, or decoration pieces! The times of Kingdoms began to yield to rule by collections of oligarchs or mini-kings, or Rajas. The rule of the oligarchs ultimately changed into the rule of aristocrats under a much-diminished king. Aristocracies have disappeared from most parts of the world and constitutional representative governments have been set up instead, first in Europe and North America and more recently in South America, Africa and Asia, replacing despotic military regimes. So, the emergence of representative government is a welcome change, despite its severe shortcomings. It is certainly an achievement of the people of the world in their march towards the rule of the people, by the people and for the people. This system of government is now the only way human beings want to be governed even if it does not wholly satisfy their needs and their desires for happiness for all.

We can see that, despite all its shortcomings, the system of the so called representation does provide the people the opportunity to develop truly democratic institutions, like labor unions, professional unions and, even, political parties which could one day take the people to their desired goal - as the distortions in the economic system are forced to disappear by collective action. Thus, it is desirable that this system of representation be adopted and promoted in the societies which do not have it yet - especially in South and South-East Asia, Arabia and non-Arab Africa.

Historical Experience

The first parliament in a modern state came into being in England, which then was a monarchy with unlimited powers held by the king. After much violence, the powers of the king were taken away and a parliamentary government was set up with a monarch with virtually no powers. That parliament, ultimately, evolved into the parliament of "Great Britain" and later into the parliament of the "United Kingdom", as the state is now known. Because the British parliament was the first such parliament in history, it is sometimes referred to as the mother of all parliaments. The system of government in Britain has evolved with time. Four provinces exist at this time, including England, Scotland, Wales and Northern Ireland. Each province constitutes the national territory of majority of its population. Provincial parliaments were set up about twenty years ago and have very limited powers "devolved" to them. There is no uniform local government system in Britain and a mixture of different local government systems exists in the provinces. Political parties operate in all provinces and the majority party forms the government, which is formed, nominally, in the name of the King or Queen. At this time, the state is referred to as the "United Kingdom of Great Britain and Northern Ireland".

Since England has more than eighty percent of the total population of Britain, the English nation dominates the political system. This domination is further strengthened by the fact that taxation authority is virtually held only by the Central Government, which distributes the state revenue to the provinces according to its own priorities. The centralization of power is further enhanced because two minority nations, the Scots and the Welsh, do not have borders with neighboring states and these are not divided nations. Only the Irish are a divided nation and Northern Ireland has a border with the rest of Ireland. Ireland, being a former English colony, has struggled with Britain for its freedom for a long time. Severe tensions continue because Britain has divided Ireland, and its Irish nation, and has annexed its Northern part, with a Protestant majority, into the state. Thus, because of its Island territory, Britain has been able to get away with a strong centralized system of government and has been able to deny the minority nations some of their rights, for quite a long time. Since the majority political party, or coalition, forms the government, it does give some sense of participation to minority nations and provinces. This process of political and economic centralization, for a multi-national state, also forces the people of all nations to move all over the territory of the state, under economic pressures, thus, pushing the process of the formation of the segment of its population that has strong state loyalties, and for which the state-loyalty is stronger than nationalism.

Because of its desirable tendencies for a multi-national state, Britain's parliamentary system of government has been adopted by many other states. These include, to different degrees, former colonies and currently multinational states of Canada and Australia. Also, New Zealand, a former colony, and multi-national states like India, Pakistan and Malaysia, and other national states like Germany, Japan, Spain and Italy have adopted this system, because of considerations that are similar to Britain. The main conclusion that can be drawn from the experience of the states with a parliamentary form of government is that *this system is especially suitable for a multi-national state or a national state with a territory that, because of its vast extent, or natural hurdles of terrain, may tend to isolate communities within a nation.*

United States declared its independence on July 4, 1776. It adopted its constitution after a military conflict with Britain lasting several years and formed its first government in early 1789, almost at the time of the French Revolution. It adopted a Presidential form of government, which has served it well over more than two hundred years of its history. There are fifty "states", i.e., provinces in the state called the "United States of America". The name "state" used for the constituent regions of the United States, implies that these regions, or provinces, enjoy a substantial degree of autonomy. Thus, the United States is a federation with fifty "states", or provinces, at this time, and has a President as the head of its Federal Government. It is a former British colony, but because of its immigration policies has begun to look more and more like a global colony. People from all nations and states of the world are part of its population. Although, it is a multi-national state, no national group in its population can claim a traditional national territory within its borders, because the native population of its territories was wiped out during the Great Genocide. Thus, "states" of the United States have been established, during its historical development, only to give autonomy to its constituent regions. The population of the United States is referred to as a "nation" by its media, other power groups within its ruling class and the people of the United States. Thus, the term "nation", in this context, refers to the population of the state, implying rightly, that the population is a community with strong state loyalties.

The President of the United States is elected directly by the population, although an "electoral college" established by its constitution, giving votes to its states in proportion to their populations, and by its rules, ensures that smaller political parties have no say in the election of the President, thus virtually guaranteeing a clear winner for the Presidency, the President being the nominee of the larger political party in the regions, or states, dominated by the two largest political parties. The President, being the head of the Federal Government, has, on paper, vast powers. But these powers are held in check and are balanced by the powers of the bicameral legislature, called the Congress. The Congress, or the Parliament, consists of two houses, the House of Representatives and the Senate. The House of Representative consists of representatives elected by the population of the states. The number of seats for such representatives for a state depends on the share that the state has, of the total population of the United States. The Senate consists of two representatives, called Senators, from each state. All legislation proposed by the President, through members of his party in the Congress, has to be approved by majorities of both houses of Congress. This, legislation includes the budget and the money appropriation bills for all expenses of government. Thus, a constant political conflict between the President and the Congress is a deliberate feature of the constitution. However, the President does have overwhelming powers in waging wars against adversaries. Any serious conflict between the Executive Branch, headed by the President, and the Legislative Branch, consisting of the two houses of the Congress, is mediated by a semi-independent Judiciary, headed by the Supreme Court. The judges of the Supreme Court are proposed by the President, but their appointment depends on the approval of the Congress. Thus, all the three branches of government, the Executive, the Legislature and the Judiciary are quite Independent in their functions, but are dependent on each other for smooth functioning of the government.

The Presidential form of government has been adopted by most other former colonies, like Brazil, Mexico, Chile and Argentina, and by Indonesia, Philippines, Finland, Russia and other successor states of the Soviet Union, the successor states of the former Yugoslavia and Czechoslovakia, and several other European states. *The Presidential system of government has, thus, been found to be especially suitable for national states, especially smaller ones, and former colonies with substantially homogeneous populations.*

Variations of the two main systems of representative government are possible, and do exist, and they generally involve differences in the powers of the central government and the powers of governments of the constituent regions. These, differences, and the systems themselves, have evolved as part of the historical experience of various states. France, for example, is unique for having developed a "semi-presidential" system of government.

Requirements

What would, then, be the requirements for the establishment of any reasonable system of representative government, even if an economic system of inequality and exploitation still exists in a state? We would look at the basic requirements in the following sections. Also, to have a grasp of some of the details, we would have to assume that we are dealing with a federation with all the complexities expected in a multinational state. We should, then, be able to construct the political structure of such a state. The issues related to a national state can, then, be established by the simplification of the abstract structure of a multi-national state. Political parties are generally considered to be a necessary element of any state based on representative government and their existence has to be expected. Whether they function in a Presidential or Parliamentary form of government depends on whether we are dealing with a national state, or a multi-national state, since the form of government has to be decided by an actual state in existence, with its special requirements. For a federation, we have to assume a parliamentary form of government and this would be the basis on which we try to construct the fundamental political structure, omitting details which can only be determined on the basis of actual conditions of a specific state. However, we would point out the shortcomings of the present systems of representation, as compared to the system we describe here. The simplifications that are desirable for a national state would be described in the last section dealing with the federal government.

Civilian Supremacy

As long as a kingdom exists, or a people are ruled by arbitrary military power, the question of a representative government simply cannot arise. Holding of general elections and election of individuals requires that there be no compulsion on the population. The military mind only sees hierarchical relationships in which any layer of control has to say 'yes sir" to the upper layer of authority. The people, then, are expected to obediently and blindly follow whatever the lowest layer of authority dictates. That is military government, exactly opposite of what freedom of choice means, when the choice of the people determines, even if in theory only, each upper layer of authority - coupled with autonomy at every level of authority and checks and balances to prevent abuse of power in the exercise of that authority. Accountability to the people and the lower levels of government organizations is one of these checks on abuse of authority.

"197. As conquest may be called a foreign usurpation, so usurpation is a kind of domestic conquest; with this difference, that an usurper can never have right on his side, it being no usurpation but where one is got into the possession of what another has right to. This, so far as it is usurpation, is a change only of persons, but not of the forms and rules of the government; for if the usurper extend his power beyond what of right belonged to the lawful princes, or governors of the commonwealth, it is tyranny added to usurpation............

155. It may be demanded here, what if the executive power being possessed of the force of the commonwealth, shall make use of that force to hinder the meeting and acting of the legislative; when the original constitution, or the public exigencies require it: I say, using force upon the people without authority, and contrary to the trust put in him that does so, is a state of war with the people, who have a right to reinstate their legislative in the exercise of their power. For having erected a legislative, with an intent they should exercise the power of making laws, either at certain set times, or when there is need of it; when they are hindered by any force from what is so necessary to the society, and wherein the safety and preservation of the people consists, the people have a right to remove it by force. In all states and conditions, the true remedy of force without authority, is to oppose force to it. The use of force without authority, always puts him that uses it into a state of war, as the aggressor; and renders him liable to be treated accordingly."

John Locke (1871), *The Second Treatise of Government*. Urbana, Illinois: Project Gutenberg. Retrieved October 16, 2019, from www.gutenberg.org/ebooks/7370.

Thus, to have a representative government, rule by the military has to be excluded, in every form. Every state of today has a military organization, because it would not exist if it did not make arrangements for its defense. A military force is essential for the existence of a state in the world of today, as it was in the previous history of mankind. What this means is that to establish a representative system of government, not only must a state have a military force, but must organize it to defend the people and not try to rule them, instead.

More than two thousand years ago, Plato thought of this issue in the context of civilian rulers, or a civilian regime which was not expected to be a representative government. His conclusions are as valid today as they were during his lifetime. According to him we, as human beings, love other human beings and institutions and we demonstrate our love by treating others in ways that we think are best for us ourselves. We love institutions the same way as we would like to be loved. In other words, if loyalty, tenderness, respect, sympathy, civility, self-sacrifice by others, are the characteristics that make us feel loved by others, then we behave with the same feelings and passions towards those we love. Those men and women should be chosen by a state to be "guardians", or military men, who love each other the way they want to be loved, and those who would love the state, and its people, the same way. In other words, we must select those to be our guardians, who, as we test them, hold throughout their lives to the belief that it is right to pursue eagerly what they believe to be to the advantage of the state, and who are in no way willing to do what is not. The other requirement, according to him, is training of the soldiers thus selected. As he puts it, "a really good education would endow them with the greatest caution in this regard, in order to attain the greatest degree of gentleness toward each other and toward those whom they are protecting". He also warned against allowing soldiers to acquire wealth, because the greed to acquire riches would turn them into enemies of the people and the state - leading to ruin of the state.

> ... "To keep watch-dogs, who, from want of discipline or hunger, or some evil habit or other, would turn upon the sheep and worry them, and behave not like dogs but wolves, would be a foul and monstrous thing in a shepherd? - Truly monstrous, he said.

> ... And therefore every care must be taken that our [military men], being stronger than our citizens, may not grow to be too much for them and become savage tyrants instead of friends and allies? - Yes, great care should be taken.

> ... And would not a really good education furnish the best safeguard? - But they are well-educated already, he replied.

> ... they should agree to receive from the citizens a fixed rate of pay, enough to meet the expenses of the year and no more; and they will go to mess and live together like soldiers in a camp. Gold and silver we will tell them that they have from God; the diviner metal is within them...

> ... But should they ever acquire homes or lands or moneys of their own, they will become housekeepers and husbandmen instead of guardians, enemies and tyrants instead of allies of the other citizens; hating and being hated, plotting and being plotted against, they will pass their whole life in much greater terror of internal than of external enemies, and the hour of ruin, both to themselves and to the rest of the State, will be at hand"....

Plato, *The Republic, Book III.* Translated by Benjamin Jewett. Urbana, Illinois: Project Gutenberg. Retrieved October 16, 2019, from www.gutenberg.org/ebooks/1497.

After the Second World War, during the policy of containment of the Soviet Union and its allies, the United States supported the militaries of the front-line states in establishing dictatorial regimes. Such dictatorships were established in Turkey, Iran, Pakistan, Thailand, Taiwan and South Korea. Similarly, military dictatorships were established and encouraged behind the front lines – in Asia, Africa, Central America and South America. After the disappearance of the Soviet Union, most of those dictatorships have also disappeared. Direct dictatorships still exist in Syria and Egypt and Libya. The militaries in Pakistan, Myanmar and Thailand continue to rule by indirectly "electing" and manipulating civilian puppet regimes.

Separation of Religion and State

As we have pointed out before, any religious belief system can be divided into three parts. The feelings of sympathy, compassion, community and respect form the foundation of religious thought. These feelings are the expression of the group survival skills developed by mankind over its long evolutionary history. In the long period of development of feudalism, as mankind passed through various stages, an elaborate web of myths has developed in all religious thought. Two major belief systems exist. One is centered on Palestine and the Arabian Peninsula and includes Judaism, Christianity and Islam and the other is centered on India and includes Hinduism, Buddhism and Sikhism.

> **"In this age the quiet surface of routine is as often ruffled by attempts to resuscitate past evils, as to introduce new benefits. What is boasted of at the present time as the revival of religion, is always, in narrow and uncultivated minds, at least as much the revival of bigotry; and where there is the strong permanent leaven of intolerance in the feelings of a people, which at all times abides in the middle classes of this country, it needs but little to provoke them into actively persecuting those whom they have never ceased to think proper objects of persecution."**
>
> **John Stuart Mill (1859), *On Liberty*: Project Gutenberg. Retrieved October 15, 2019, from www.gutenberg.org/ebooks/34901.**

Both the major religious belief systems, developed during the last four or five thousand years - during the age of feudalism. In addition to the basic feelings and an elaborate mythology, each religion also has a set of rituals to reinforce the basic beliefs and the mythology. Thus, all religions display the feudal tendency of extreme intolerance. All religious thought is absolute and does not tolerate any new interpretation. Any expression of disagreement on any subject is automatically labeled as "heresy" and, thus, has to be suppressed. The concept of freedom of conscience, freedom of association and freedom of expression are not tolerated by religious thought. On the other hand, the very basis of representative government is the freedom of belief, free expression of thought and freedom of association. Thus, a state based on representative government cannot function if it is controlled by religious authorities, because extreme intolerance, violence and bloodshed results if this happens. This has happened again and again in the history of the development of the state.

"Congress shall make no law respecting an establishment of religion, or prohibiting the free exercise thereof; or abridging the freedom of speech, or of the press; or the right of the people peaceably to assemble, and to petition the government for a redress of grievances."

September 25, 1789, "*First Amendment to the Constitution of the United States*".

A solution has, thus, been found in the separation of state and religion. Thus, a normally functioning state, based on a representative government, does not have a religious preference. It tolerates all religions of its population, and all variations and sects, but does not take sides on religious matters. The business of the state is based on scientific knowledge and not on religious mythology or rituals. A state would, thus, deal with military personnel and equipment and military strategy and warfare and all these matters of modern warfare are grounded in science. It may deal with management of the state's economy, which is again a scientific matter. It may deal with roads, bridges, airports, sea ports, aircraft, ships, transport vehicles, buildings and canals, electrical power systems, development of energy and mineral resources and agriculture, etc., and all these activities are also based on scientific thinking. In short, all the functions of the state are based on scientific thinking. In the field of education, a state may have functions overlapping with religious institutions, and this conflict is resolved by separating religious education from the education provided by the state for its areas of activity. Thus, the people are free to believe in any religion they want to believe in, and religious education can be provided by religious institutions like churches, mosques and temples, religious schools and by families at home. The state, then, deals with non-religious matters and makes laws for its purposes.

... "If you work hard in cooperation, forgetting the past, burying the hatchet, you are bound to succeed. If you change your past and work together in a spirit that every one of you, no matter to what community he belongs, no matter what relations he had with you in the past, no matter what is his color, cast or creed, is first, second and last a citizen of this State with equal rights, privileges and obligations, there will be no end to the progress you will make.

... You are free; you are free to go to your temples. You are free to go to your mosques or to any other place of worship in this State of Pakistan. You may belong to any religion or caste or creed. That has nothing to do with the business of the state."

Mohammad Ali Jinnah (August 11, 1947), Presidential Address to the Constituent Assembly of Pakistan.

If this approach is adopted, people with all kinds of religious beliefs can be citizens of the state, the state gives them equal rights in all respects and representative government can function under these conditions. This approach has evolved in all areas of the world with civilized societies. In the Christian dominated states, all over the world, this approach has been adopted and virtually no dictatorial or religious regime exists in the Christian world at this time. The same applies to the Buddhist civilization. Except for the indirect military dictatorships in Burma (or Myanmar) and Thailand, no military or religious regime exists in the Buddhist world either. Both the states of the Hindu civilization, India and Nepal, are governed by representative governments, not military or religious regimes.

Only in the Islamic world, this phenomenon of religious regimes still exists. Religious political parties, in keeping with their religious desires, naturally claim that Islam cannot be separated from politics. It is claimed that there is such a thing as an "Islamic Republic". In other words, the claim is that a state based on representative government, or "democracy", can function under the control of Islamic laws, as interpreted by religious political parties and, thus, such a state must be an Islamic Republic. These claims are in line with claims of existence of "Islamic Economics", "Islamic Banking", "Islamic Military", "Islamic Courts", "Islamic Educational System", etc. Religious reactionaries have, thus, been claiming that Islam and all these aspects of the political system cannot be separated and that a political system must be dominated by religion and its "sole and rightful" interpreters, the religious mullahs, the religious groups and religious political parties.

"The political emancipation of the Jew or the Christian - of the religious man in general - is the emancipation of the state from Judaism, Christianity, and religion in general. The state emancipates itself from religion in its own particular way, in the mode which corresponds to its nature, by emancipating itself from the state religion; that is to say, by giving recognition to no religion and affirming itself purely and simply as a state. To be politically emancipated from religion is not to be finally and completely emancipated from religion, because political emancipation is not the final and absolute form of human emancipation."

Karl Marx, *On the Jewish Question*,
www.marxists.org/archive/marx/works/1844/jewish-question/index.htm, accessed October 16, 2019.

Despite the claims of the inseparability of state and religion, however, Muslim states have been able to establish functioning representative government systems. These include states like Indonesia, Malaysia, Bangladesh, Kirgizstan, Uzbekistan, Turkmenistan, Azerbaijan, Turkey, Bosnia and Albania. These states have been successful in separating religion from the state, to a great extent. Even Pakistan, which has a semi-representative government at this time, and the Arab state of Lebanon, with about half of its population consisting of Muslims, has achieved some degree of such a separation.

At this time, some states, especially Iran, Sudan and Pakistan claim to be "Islamic Republics". All these states have faced incredible violence at the hands of religious wise-men. Hundreds of thousands of the citizens of these states have lost their lives in the resulting blood-shed and unnecessary internal and external wars. It has to be realized that the feudal and tribal society, with its economic and social system characterized by slave ownership, multiple wives and unlimited numbers of concubines and savage warfare between tribes, including treatment of prisoners, males, females and children - as "property" and "spoils of war", at the time of the advent of Islam about fifteen hundred years ago, cannot be considered a society based on democracy and representative government, no matter how much facts are stretched to suite one's religious desire for such a characterization. Nor can religious punishments, on the lines adopted by the ancient pre-Islamic civilizations of

Mesopotamia - like beheadings, cutting of ears, hands, feet, stoning to death, etc., be renamed "Islamic" and imposed in today's societies that have moved beyond extremely primitive feudalism. They are more likely to be viewed as "cruel and unusual punishments". Separation of state and religion, thus, is absolutely necessary for the establishment of a peaceful representative government. The degree of separation of religion and state, however, has to be based on the level of development of a society. Of the three Islamic Republics referred to above, Pakistan, had adopted a constitution on these lines, in 1973, and had achieved substantial progress in this regard, but despotic inroads were made into its provisions by the religious-cum-military dictatorship of General Zia ul-Haq. At this point, most of his "Islamization" provisions are simply ignored by its government, while an active Islamic insurgency has been faced by the state, because of the violations, in the past, of this critical principle of separation of state and religion.

Federal, National and Provincial Governments

All rights of nations and tribes in a state, as described in the last chapter, should be recognized and the traditional national and tribal territories should be demarcated. Large national territories may be divided into provinces and provinces may be divided into counties and tribal territories. Depending on the size of a tribe, a tribal territory may consist of one or more counties.

In a large multi-national state, which may include some nations at the stage of primitive feudalism, it is necessary to have a bicameral federal parliament including a "House of Representatives" and a "House of Nations". Experience has shown that a reasonable tenure of a parliament is five years. The federal house of representatives should consist of members elected by each district, or nominated by the set of tribal counties of a district. A district, for this purpose, may consist of a fixed number of counties. The term of the members of this house could be five years.

The House of Nations should, similarly, consist of "Deputies", or "Senators", elected by a "zone" or mega-district consisting of one or more districts, or nominated by a tribal territory. To ensure continuity, half or one-third of its members could be elected after every five-year period.

Constitutional provisions should be made to allow inhabitants of a tribal territory to choose to convert their counties into regular counties subject to laws meant for other non-tribal counties. This could provide for a smooth transition for tribal populations, in terms of representation in provincial and federal houses of representatives.

The federal budget, taxes, development projects and all money appropriation bills should require the approval of a majority of the two Houses of Parliament. Matters involving multiple national territories may require approval of a simple majority of the House of Representatives and a two-third majority of the House of Nations, including a majority of members of each national territory concerned. Amendments to the constitution may require a two-thirds majority of both the houses of Parliament and a majority of the house of representatives of at least two-third, or another specified portion, of the number of national territories of the state. The federal government should have the exclusive control over the following subjects:

- State language of communication. The national language of the largest nation of the federation would, normally, be the official language of the state. What this means is that the citizens belonging to the largest nation in the state, would not have to learn a new language, thus saving on effort to learn a common language of communication between the citizens, and official institutions, of the state. However, if the state has several large nations included in it, more than one state language could be selected.

- Defense and foreign affairs, establishment of institutions anywhere on the territory of the state or in foreign states, including all financial dealings with foreign states.

- Currency and all financial institutions, including a central bank to manage money.

- State take-over of financial institutions and industries, or reversal of such takeovers. National territories and provinces may share in the resulting state ownership of industries or financial institutions.

- Federal taxes including income taxes on individuals and businesses and duties on imports and exports and fees on use of federal institutions etc.

- Control of such water and energy resources and mineral deposits as may be considered of essential strategic state interest, as determined by the federal legislature.

- Establishment and operation of federal institutions in areas of provincial jurisdiction, if considered of strategic state interest.

Obviously, the state structure developed above assumes that either two or more nations are included in the state, or the national territory is very large and has linguistic, or religious, variations requiring more than one province. In such a multi-national state, a national territory would have to be autonomous with respect to promotion and use of its language and all modes of expression of its cultural heritage. The official state language, as well as the national language may be taught in schools of each national territory. The state language, however, may be the medium of instruction in educational institutions to promote state unity and the business of the state in general. A province may levy personal and sales taxes on individuals and business income taxes on companies on its territory. All residual powers would need to be vested with the provinces.

In a multinational state, each national territory may be sub-divided into provinces, due to variations in national population, or due to its large size. Each national territory would need to have a "national assembly". to coordinate national issues among its provinces and between the nation and the federal government. The members of the national assembly may also be members of the Federal House of Nations. The "House of Representatives" of a province may consist of one representative from each county. A tribal county would be part of the traditional territory of a tribe. A process of negotiation would be necessary for smooth passage of legislation in provincial, national and federal houses of parliament of such a state.

If the state is a national state, with a homogeneous population, but its territory is very large, or tribal territories are present, then a federal system of government with a parliamentary form of government is still useful and suitable for such a state. The vast territory of that one nation would have to be divided into several provinces for this purpose. However, if the nation is quite homogeneous in terms of its genetic heritage, language and religion; and the national territory is also not very large, then multiple provinces are not needed for such a state. There would only be the state, or national, government, and in such a case, a presidential form of government with only one house of parliament would be suitable.

At this time, however, the territories of nations and tribes in most states are not demarcated and rights and powers of such social groups and federal governments of those states are, also, not defined – specially as related to tribes. This is specially a problem in states which were imperial possessions in the past. Even the state borders, as defined by imperial powers, have not been modified – leading to inter-state and intra-state conflicts.

Local Government and its Autonomy

Every village, town and city must have a local government to administer it. Small population centers may be called "villages", large villages with two or three times the average size of the population of a village may be called "towns" and those with a larger population may be called "cities". Big metropolitan areas would have to be divided into cities. Each population center must have a government elected by its residents. It should handle collection of real-estate and business property taxes, collection of fees for building and other licenses, construction and maintenance of streets, garbage disposal, water, electricity and gas supply, management of local police and courts, etc.

Each county needs to have its elected government which should administer all matters except those related to towns, villages and cities on its territory. Land rent, or tax, may be its main revenue source.

The administration of a tribal territory has to be left to the tribe to manage. However, procedures to ensure guarantees of basic human rights of individuals may be negotiated between the elders of the tribe and the provincial, or national government. The tribal representative of a tribal county may be selected by the tribal elders of each tribe as is the tribal norm. The provincial and national governments would need to negotiate economic and educational development of a tribal county with those tribal elders, to encourage fusion of the tribe with other tribes and its nation. This must be accomplished without the use of any compulsion or violence. Similarly, the right of appeal to national or provincial judiciary, for individuals convicted of crimes by a tribe under its own traditional system, would have to be negotiated.

At this time, most states at the stage of primitive or advanced feudalism, do not respect the rights of tribes. Use of military force and insensitivity towards tribal rights, has led to unnecessary violence. Some states, especially those that were former imperial possessions, do not even have local government institutions. This leads to neglect of local housing and other related problems and greatly increases corruption in their societies.

Social Organizations

A democratic state has to allow creation of social organizations like labor unions, professional associations and companies involved in production and distribution of goods and services. These organizations must be able to operate as democratic institutions and must be able to protect their professional rights regarding wages and working conditions and other activities.

Currently, in virtually all states of the first or the second world, labor unions are discouraged and workers are not encouraged to participate in the ownership or management of their companies. Some companies do assign small numbers of shares of their stock to some employees, but those shares are, generally, of negligible amounts. Employee participation in "management" is, also, generally, focused on improvement of production processes to increase efficiency and on quality assurance. Very few, worker-owned enterprises exist at this time. Corporations, thus, remain tyrannies dedicated to maximum profits for their owners in the ruling classes.

Individual Rights and Family Autonomy

Most rights of individuals were described in the last chapter. All those rights should be guaranteed by a democratic state. *At this time, most states do claim to respect human rights of their inhabitants, but ignore them in practice. Most states, especially, do not respect the economic rights of individuals.*

Regarding individual rights, family privacy and autonomy, some of the restrictions imposed by a socially backward, feudal, society of a state may have to be accepted for some time - in the interest of peace and tranquility. *At this time, most states, with societies at the stage of primitive or advanced feudalism, do not deal with tribes this way – the result, normally, is resistance and violence.*

To promote innovation, the right to study and research any phenomenon and develop and produce any products, alone or in a group, should, naturally, be guaranteed, including the right to patent processes and products, and receive some reward for, the tangible or intangible results of such activities.

Independent Information Media

For the proper functioning of democratic institutions, it is important that the people have reliable sources of information, because they cannot be expected to make correct decisions regarding current economic, social and political issues of their societies, unless they have the required education and training in educational institutions and are exposed to publishing, broadcasting and news media that provide accurate and reliable information. People need reliable information to make decisions based on their own needs and interests – especially when elections are held.

Currently, virtually all publishing, television and radio broadcasting and internet-based sources of information are owned and operated by corporations owned by the ruling classes of the pseudo-democratic states. Thus, the people of such states are unable to make decisions based on their own interests and are "guided" into serving the interests of the ruling classes, instead.

Independent Judiciary

An independent judiciary is a basic requirement for a representative government, *although such a judiciary cannot really exist in a society with a high level of economic inequality.* Judiciary is the third pillar of government, in addition to the Executive and Legislature. Without its function of resolving conflicts between the Executive and the legislative branches of government, a representative government, simply, cannot function. A relatively independent judiciary helps in the quick and smooth resolution of such conflicts and problems, without serious damage to the interests of the population.

Qualifications of candidates for appointment of judges in local government courts and higher courts may be specified in the constitution. Judges may be elected by the people of each population center and county at the time of general elections, subject to those qualifications. A hierarchical system of federal and national courts should be set up to deal with national and federal matters. Judges of higher levels of the Judiciary - corresponding to counties, provinces, national territories and the central government, should be elected by the relevant House of Representatives and should require approval of a high majority of representatives. The federal Supreme Court should be the final court of appeals in cases involving constitutional issues. Its judges could be proposed by the central government, out of the judges of the national, or provincial level courts. Such judges should be required to obtain the approval of a high majority of the federal, or central, House of Representatives and the House of Nations, if any.

To ensure the independence of the Judiciary, salary, benefits and other privileges of judges should be linked to the salaries, benefits and other privileges of the members of the relevant Parliament, by constitutional provisions. Each judge of the senior judiciary should decide about his own retirement. Judiciary itself should deal with removal of judges for misconduct. No other institution of the state should have this authority.

For a small national state, since there would only be the state, or national, government, the structure of the Judiciary would also be simplified, with local village, town, city, county, district courts and a State Supreme Court.

It is very difficult to be a judge – this requires a high-level education of social sciences, besides the laws of a state. It, also, requires a person to achieve an extra-ordinary ability to judge other individuals and institutions – without bias of any kind. It is, thus, very difficult for a state to create and maintain a judiciary relatively independent of political, national and class considerations – specially in a state with feudalism, or capitalism, as its social system. In addition, laws and adjudication systems in such states are generally weighed in favor of those with substantial capital in their possession. Also, states which were former imperial possessions have inherited laws and legal procedures designed by their imperial powers. Those laws and legal procedures were designed to maximize the loot and plunder of the people of those states. Thus, it is not surprising that courts in most states, including former imperial possession, deliver "justice" only if it can be bought, or it is in accordance with the wishes of authorities of those states. Civil cases specially related to property issues in former imperial possessions, like India or Pakistan, can take decades to decide – thus favoring those who have the needed financial ability.

Political Organizations

As human evolution has proceeded, conflict has always developed within each social group. This conflict has been primarily between individuals, but these individuals have tended to create sub-groups, within a group, based on real or perceived common interests. Within a tribe, sub-groups may form around common genetic heritage or around some issue of intra-tribal conflict like possession of some piece of land over a hill or near a river. Within tribes, genetic heritage tends to be the focus of close relationships between individuals, but in modern states, conflicts tend to be primarily economic, resulting in formation of sub-groups. Labor unions form to promote, and fight for, higher wages and social benefits of workers. Professional associations promote the economic interests of carpenters, engineers, fishermen, physicians and teachers etc. On broader issues of interest to particular nations, religious groups, agricultural land owners and industrialists, etc., political parties may emerge within a state. Such organizations, in general, wield considerable political power and enter into alliances with other political parties to influence the functioning of a state and may try to capture political power and form the government of the state to directly control state policies to promote their agenda. The kind of political parties that emerge in a state depends on the level of social development of the state.

Political Spectrum

Since the beginning of recorded history, societies have bifurcated into the ruling class as the class of exploiters, and the people as the class of exploited. The balance of power between the two classes has changed as a society has progressed. The balance of power has tended to shift in favor of the people in recent times. This trend is likely to continue. In general, four segments of the population can be identified in a society at any time:

- Those who are happy with the society as it is, and do not want any change. They may resist the changes that society is going through. Such individual may be referred to as "Conservatives", i.e., those who wish to conserve the present status of society, or the status quo.

- Those who are not happy with the present status of society, but wish to restore those social relationships which have become weak, or have disappeared completely. They wish to take society back to the actual past, or a past of their imagination. Thus, ultra-religious individuals may wish to take society back to the "golden" past dominated by their religious practices. Others may wish to restore the more abusive economic relationships of the past, of which they were the beneficiaries. These individuals are referred to as "Reactionaries".

- Those who are willing to go along with the changes that society is going through and may welcome them. Such persons are referred to as "Liberals".

- Those who are not happy with the pace of evolution of society and wish to push it forward. Such persons are referred to as "Radicals".

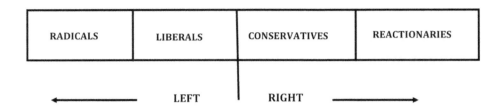

Figure 18-01 The Political Spectrum

Conservatives and *Reactionaries* tend to be members of the ruling class, since they are the beneficiaries of the current status of society, or the *status quo*. Reactionaries have the desire to make the present abusive society even more abusive and, thus, derive greater real or imaginary benefits from it, in terms of economic wellbeing, or "moral" authority. Churchill, Nixon and Reagan were some well-known conservatives. Similarly, Abul A'la Maududi of Pakistan, Hitler, Modi of India, Osama Bin Laden of Arabia, General Zia ul-Haq of Pakistan are some well-known reactionaries, who have done immense damage to their societies by trying to restore the "glory" of the past to their states and societies.

As a society develops, some members of its ruling class may begin to see their privileges as unnecessary. They may wish to let society change in the direction it is moving. Such persons, referred to as *Liberals*, emerge in the ruling class, as a result of education and acquisition of scientific knowledge. President Carter of the United States, Prime Minister Harold Wilson of Britain, Chancellor Willy Brandt of Germany, Prime Minister Benazir Bhutto of Pakistan and President Julius Nyerere of Tanzania are some well-known liberals.

Radicals, in general, are not members of the ruled class. They, in general come out of the working class. They wish to push society in the direction of less abuse and greater freedom for the people. It does happen, though rarely, that some members of the ruling class would develop radical tendencies. Such individuals tend to have highly developed group social skills, i.e., they feel sympathy for the sick, the dispossessed, the handicapped and the poor and they only emerge from the ruling class because of acquisition of scientific knowledge, becoming aware of their unjustifiable power and privileges and the sufferings of the people due to the economic and social excesses of their own ruling class. Karl Marx, Lenin, Rosa Luxemburg, Mao Zedong, Ho Chi Minh, Che Guerra and President Castro of Cuba are some well-known radicals. Zulfiqar Ali Bhutto of Pakistan also falls in this category, since, despite his origins in the feudal class in Pakistan, he tried to bring radical changes to Pakistani society and paid for this "treason" to his class with his life. The reactionaries of the ruling class and religious reactionaries celebrated his murder by Zia-ul-Haq's reactionary military regime.

In general, we can say that, in traditional societies at any of the four stages of evolution, the "left", consisting of the liberals and radicals, tries to move society forward towards greater power to the people and the "right" tries to block society from moving forward, or tries to move it back towards more power to the ruling class. Thus, we can describe the "political spectrum" of a society by a simple diagram, as indicated in Figure 18-01.

Future of Representative Government

Representative government has developed roots in most of the world of today. Europe has representative governments all over the continent. Since most of the multi-national states have broken down into national states, many of these states have adopted the Presidential system of government. Europe is also developing into a continental super-state. Most European states have surrendered some of their sovereignty and have developed a common economic space. Most of those states are members of the European Union. Most members of the European Union have a common currency, the Euro, controlled by a European Reserve Bank, known as the European Central Bank (ECB). A European Parliament has also come into existence, though with limited powers.

The colonial states of North and South America have, mostly, chosen the Presidential system, because of their initially homogeneous populations. Because of their colonial origins, the Presidential form of government still suits them, even though their voting populations have not remained homogeneous because of the empowerment of native tribes and African slaves, who, formerly, had no say in the governments of their states. South American states are also developing deeper economic relations and are moving towards emergence of a super-state in South America. A customs union has already been established.

Canada and Australia, both former British colonies, have adopted the parliamentary form of government and have retained the British King, or Queen, as their titular head of state. Both these states have large territories, but relatively small populations. The parliamentary system is especially suitable for Canada, since it has a large French population concentrated in the province of Quebec. Canada has also adopted two state languages, English and French, because of the same reason. The United States, also, has a unique Presidential form of government, which guarantees a considerable degree of autonomy to its constituent provinces, which are called "states". North America has also developed a common economic space binding Canada, United States and Mexico together in the NAFTA Treaty. This is another emerging continental super-state.

The Central Eurasian region of the world is also developing representative governments as it also develops into a super-state. China, Vietnam, Laos, Cuba and North Korea are one-party states. All of them, except North Korea, have been able to reform the Soviet model that they had adopted. China and countries of South East Asia are also developing deep economic links and are moving towards becoming another super-state. Most of these states have developed representative governments. The South Asian region is developing some economic links and is likely to become another super-state, though it may take a long time. In non-Arab

Africa, most of the states have developed representative governments but their systems have not become firmly established, since the problem of divided nations and tribes, inherited from colonial times, remains a big hurdle for establishment of smooth-running governments. Also, since most of non-Arab Africa has societies in the primitive feudal state, characterized by a high degree of tribalism, and recent experience with military dictatorships, it has not found a suitable system of representative government.

Monarchies and military dictatorships are continuing to rule many Arab states of the Middle East. Most are Bantustans or Panamized states. Only Lebanon may be referred to as a state with a representative government. In the Muslim world, however, several states have emerged with representative governments. These are spread over South and South-East Asia, Central Asia and Europe.

We can, thus, conclude that the representative form of government is being adopted by the whole world, although advance of different regions of the world varies in this respect - depending on their socio-economic evolution. The level of democratic tendencies in such representative governments also varies from state to state

Chapter-19

Some Fundamental Societal Issues

Crime and Punishment

The relationship of a human individual to society was described in Chapter-3 in detail. How society and the environment determine and influence the development of an individual and how an individual affects society was explored there. Similarly, the development and structure of human personality was described in Chapter-16, in some detail. To summarize, we can say that there are two processes that ultimately create a mature human being - the process that led to the genetic heritage that a person was born with as a child and what the environment did to it.

Our physical bodies, including our brains, are what we are born with. The gene-pool passed on to us determines how we would grow and what form and physical shape we would take. The intelligence our minds are capable of developing and our other innate abilities are predetermined for us at birth. Even our future growth and performance potential is partially predetermined for us at the time of birth. Every child is born with certain genetically determined instincts. The instinct of self-preservation and the sexual instinct are two such basic instincts, which govern the behavior of a human being.

As we have discussed before, the growth of an individual and his personality development are the result of the interaction of the environment and the individual at every stage of his development. There is a big difference between a human child and the child of any other animal. In all species, parents teach their children the skills they need to survive and impart all the knowledge they need for survival. This takes only a few months in some species and may take somewhat longer in others. But a human child takes a very long time to acquire the knowledge and skills needed to compete and survive. This is so because the amount of knowledge to be imparted is so huge that it is literally impossible to transfer all of it to the child. What is actually conveyed to the child is what the parents, teachers and others in the child's social group, besides the family, can convey - in short what the society, as a whole, has acquired in terms of scientific knowledge, etc. Further, whatever knowledge and skills are conveyed are constrained by the availability of sufficient time in contact with the child. The result, after many years of education and training, is a person with a body that has been impacted by the physical environment in which a child has grown up, i.e., the houses, the schools, the air, the water, the sunshine etc. Further, the child develops a personality, or mind, that has been shaped by the living environment – i.e., plants, animals and other organism around it and parents, teachers, tribal elders, thinkers, writers, poets, political leaders, etc. At every stage of its development, the child uses what it has acquired from the environment, to acquire more knowledge and skills. Thus, the society completely creates the individual and his behavioral pattern and skills. The individual may do harm to other individuals he comes in contact with, or he may do good. This would depend on what education and training the individual received from society since birth and what he, or she, was born with. In other words, the so called "free will" does not exist.

Now, if a person breaks the rules of behavior that society has developed, he or she is said to have committed a crime. For causes of such crimes, one has to look at the society in question. Normally, the question, then, is - why did the person commit that crime and what punishment should be meted out to him. The questions that should be posed are - How did society educate and train that person to become a criminal? What was wrong with the training that the individual received before? What crime did society commit through that person? What punishment should society receive for committing the crime of creating a criminal? What could be corrected in society so that it does not create such criminals? Obviously, a child has no knowledge of where it is and what rules society has, which he has to follow. No child is born a criminal. A child does not know what murder, theft and fraud, etc., are. Only the society, in which the child grew up and developed, has to answer for "his" crimes.

Historically, people have been hanged, decapitated, crucified and shot to death for killing other human beings, or robbing them. Societies have cut the hands and feet of persons found to be stealing from others. People have been executed for distributing harmful drugs or for embezzling money. They have been lashed and tortured in other exotic ways for committing other relatively minor offenses. The tendency has been to mete out more and more harsh punishments for breaking society's rules, in the conviction that this is justice and that it would deter and prevent crime in the future. Driven by the desire to take revenge, these practices have continued, and even intensified, over time. *But crimes have continued to occur and have even increased in some societies. Thus, by this approach to crime prevention, injustice is being perpetuated. An eye for an eye, a tooth for a tooth, only creates twice as many blind and toothless people. It, also, does not protect the innocent from criminals of the future.*

A society, which puts criminals into jails as punishment, protects its members from harm by isolating itself from those criminals, but this doesn't stop the process which creates criminals. All criminals can be killed or put into jails, but even larger numbers may be created by society itself, if the processes responsible for their creation remain unchanged. For example, millions are in jail today in the United States, but the crime rate remains among the highest in the world. It should be obvious that the problem cannot be fixed by "fixing" the criminals. This is a problem of American society and the United States. The society of the United States needs to be fixed!

There is another aspect of punishment for crimes that needs to be looked at carefully. If a person kills another without any justification, we refer to this act as murder. The person who committed the murder may be caught. In most states of today, laws require that the criminal be executed one way or another. If we study this issue carefully, we realize that executing a criminal does not bring the murdered person back to life. If the murderer was free for some time and the process of his apprehension and trial took several months, then, we cannot claim that justice was done, on any grounds, since the offender was able to live for some time after committing his crime. Also, executing him did not undo the crime, i.e., it did not bring the dead person back to life nor did it stop the pain and suffering of those who were closely related to the victim. It did, however, fulfill the desire, for revenge, of the victim's family and friends. Most importantly, it did not stop new murderers from emerging in society, as experience of thousands of years has shown that the so-called "deterrence" does not work.

It is possible that those committing a crime are not caught, but innocent people are mistakenly caught and convicted. This would double the injustice and it happens quite often, because no system of investigation and prosecution is fool-proof. These flaws in the administration of justice become even more serious when innocent persons are put to death for murder and other crimes, because killing of innocent people is a big crime by definition, even if it is committed by a state. In general capital punishment for people, as punishment for any crime, needs to be abolished as it serves no good purpose and always results in murder, by the state, of some innocent individuals along with the guilty. *Thus, the only way to ensure that injustice is not done to innocent people in society is to prevent crime from occurring.*

The only way to reduce crime in society is to make the society into a just society. It is well known by now that Sweden and Denmark have a low crime rate. The same is true of all European states, Australia and Canada, as compared to the United States or Brazil. European societies, specially the Swedish and Danish societies, have a high level of social protection for their citizens and the disparities of income are much lower as compared to other societies named before. This reduction of social tensions clearly has an effect of lowering the crime rate. Another, factor is free and universal education and health-care for all children. The social protections to be provided by a state should include universal and free health-care, unemployment insurance, minimum housing for all and adequate care of the elderly and handicapped. *History and experience have shown that this is the only correct approach to crime prevention and promotion of justice in society.*

It is natural that the ruling class in a Lootera-dominated state would want harsh punishment for those who commit crimes like rape, theft, robbery and murder. Most of such criminals are members of the dispossessed working class, or the people, in general. Also, naturally, a lenient view is taken of "white collar" crimes involving deception, theft and embezzlement of money and other monetary instruments, as virtually all such criminals belong to the ruling class itself. The ruling class, in general, has been able to get away with this deceptive "self-righteousness" regarding punishment of criminals, so far. It has been able to hide the basic economic crime committed against the people. This crime of organized loot and plunder of the fruits of labor of the working class is the basic source of all injustice in a society, as it distorts relationships in all other areas of human activity - economic, social and political. *This fundamental crime in society, which has not yet been widely recognized as such by most societies, has to be eliminated - if production of criminals in all other areas is to be brought to an effective end.* The reduction of social tensions by universal minimal housing, health-care, free education, un-employment insurance, etc., would decrease the crime rate in proportion to the degree of reduction of such social tensions, but complete and effective end to the fundamental crime is what society needs for creation of, effectively, a crime-free society.

The Mating Game & the Sexual Markets

We have discussed the causes of bifurcation of genders and the whole spectrum of inter-gender relations in human beings in Chapter-1. The two genders are inter-dependent in many ways and engage in competition in all aspects of human activity - economic, social and political. The core issue between the two genders is, however, the issue of sexual relations and reproduction, which is the fundamental reason for the emergence of the two genders. Activities of the two genders, in pursuit of successful mating and reproduction, are heavily impacted, and are constrained in many ways by, existing social, economic and political relations in society.

Men compete with each other in wooing and attracting females to achieve mating as a temporary relationship, or a more lasting one. Traditionally, marriage has been considered the only socially acceptable form of sexual relationship for procreation during the last six thousand years of existence of civilized human society. Marriage, as an institution, has also evolved from one in which one man has a harem of wives and concubines, to simple polygamy, to the serial monogamy of the present time. During the course of its evolution, it has further become inter-twined with religious belief systems and is, now, looked at as a more or less religious institution. The modern state has divested the institution from sexual relations *per se*, and has converted it into a contract between two individuals for living together as a family and having and raising children. Thus, the modern marriage contract has provisions for property rights of the two parties, and their children, and responsibilities to each other and to the children to be raised. As all responsibilities are ultimately translated into financial terms, marriage has basically become a financial contract and the laws of modern states do not deal with sexual responsibilities and conflicts. In case of the break-up of a marriage, a no-fault divorce is becoming the norm under the laws of these states.

In all states that have reached the stage of *intensive capitalism or benevolent capitalism*, the marriage contract as stipulated by the laws of those states is increasingly being discarded by their citizens, who prefer to live without a formal contract. Thus, in their societies, there are no restrictions on intimate contact between members of either gender. Dating is considered the normal procedure for finding a mate. For those couples who become reasonably confident that they are likely to become life partners, informal sexual relations and even co-habitation is considered normal. Many couples tend to retain life-long informal co-habitation. Some do end up going through a formal marriage as recognized by the state. This is the situation in Europe, North and South America, Australia and Central Eurasia, etc.

In East and South East Asia, the situation regarding marriage is basically the same. However, there is much less tendency of couples to co-habit without a formal marriage. In non-Arab Africa, marriage remains a primitive institution with multiple wives and lack of formal laws governing it. South Asia and Arabia are the only two conservative regions of the world as regards separation of the two genders. The Muslim world is the most conservative. Polygamy has begun to disappear in this religious community, but the separation of the two genders in all areas of activity, is still in force, to a great extent, in most Muslim states. In these states, marriage is regarded as purely a religious institution, with very few rights for women and a tendency to regard women as property - as a continuation of the historic traditions of humanity.

We have discussed the origin of homosexuality in Chapter-1. This is a natural phenomenon which follows from the fact that the human species has emerged after a very long evolutionary experience. At some stage, in the very distant past, human beings also were unisexual beings, which reproduced themselves without the presence of two genders. Then the species evolved into a two-gender species. As the law of evolution indicates, both the genders inherited some characteristics from their unisexual ancestors. Obviously, even today, men and women have many common characteristics. Characteristics related to reproduction are, however, different in the two genders. It is not surprising that some human beings exhibit homo-sexual preferences due to the fact that genetic inheritance always involves a range of variation of characteristics. In fact, we can say *that all human beings are really bisexual.* Both the genders have sexual feelings toward their own gender as well as the opposite gender. In most individuals, heterosexual feelings are predominant, but the homosexual feelings also exist, although to a very low degree. The balance is different from individual to individual and may tilt toward one's own gender in some cases. Thus, homosexuality has existed as long as the two genders have existed. We can see this in other animal species also.

In the last six thousand years of civilization, homosexuality, being a minority preference, has been consistently treated with hostility. All religious thought shows a prejudice against it. Because of this hostility, those who had homosexual feelings have always tended to engage in their sexual practices in secret, away from public knowledge, to the extent possible. In recent times, homosexuals have tended to make their situation public in modern, relatively tolerant, societies and their demands for equal rights, and removal of all discrimination due to their sexual preferences, has begun to be accepted. Some states have even developed laws covering marriages between individuals of the same gender, despite hostility of the heterosexual communities. Homosexuals in feudal societies still face high levels of discrimination. This can only change as those societies evolve into more tolerant societies. Some effort, however, is being made in this direction, even at this time, in these sectors of the world, with some success.

The hostility and discrimination faced by homosexuals is also faced by those individuals who do not fully develop into males or females ("transsexuals", referred to, in South Asia, as "Khawaja Sarai's"), but show characteristics of both genders in their genetic heritage and behaviors. Even modern capitalist societies have not found a solution to this problem to the satisfaction of this community and have failed in developing legal frameworks to cover relationships of such individuals with each other and with the homosexual and the heterosexual communities. Some progress has, however, been made and the issue of discrimination is being addressed to some extent, but a lot more needs to be done.

There is a need to modify the marriage contract, as covered by family laws of states, so that it covers individuals of all sexual orientations. The debate, as to whether a homosexual "marriage" should be allowed by state law, or not, seems to be misplaced. What has happened is that the original and traditional "marriage" institution has been reformulated as a secular "marriage" under a state's family laws. The ceremony may be performed in a church or in a government office, but it is still referred to as a "marriage". However, a state's family laws do not deal with sexual issues and, in case of a later divorce, sexual behavior is not considered the business of the state and a "no fault divorce" is what the state deals with. This is a good approach in the sense of separation of state and religion, but causes unnecessary confusion and raises religious and political tensions. A further step could have removed the confusion.

A marriage could take place under any religious institution and with any religious ceremonies, and could be called a religious marriage, but when it was, optionally, registered under a state's family laws, it could, also, be called a "civil union". A state's family laws could, then, apply to all civil unions, whether they were unions of heterosexual, homosexual or transsexual individuals. Obviously, the responsibilities of the members of these civil unions would be the same - to each other and to any children born to the couple, or adopted. The property issues, including inheritance, would also be the same. In case of later break-up of a civil union the divorce issues would also be the same. Thus, one set of laws would be sufficient to cover any of the three kinds of civil unions, or any permutations and combinations of the same, if they arise in the distant future. Obviously, in this arrangement, only unions between heterosexual couples could be called "marriages", which is the traditional name of the religious institution. Homosexual and transsexual civil unions would not have a religious ceremony associated with them, but that would not make any difference to the responsibilities and rights of the individuals involved. Any rules of sexual behavior would remain in the private domain of the couple and would not be dealt with by the state, as is the practice at present in most states of the worlds of intensive or benevolent capitalism. However, currently, the trend, in societies of both of these stages of capitalism, is towards retention of the old terminology. Thus, all such unions are being called "marriages". This has given rise to continuing resistance by conservatives and religious reactionaries. The resistance of such groups in primitive or advanced feudal societies is much stronger, as should be expected.

In addition to the marriage market, the two genders have continued to engage in pursuit of other kinds of sexual relationships. The market for sexual favors to men by women, in return for financial or other favors, has existed since time immemorial. This has been described as the "oldest profession" or "prostitution", which is a derogatory term traditionally used for such activities. Recently, the prostitutes have begun to be described as "sex workers". In this market, the sex workers have always been looked at with hostility by the female segment of society and the religious community, since it poses a threat to both. Women feel the competition from the "low and vile" sex workers and the religious authorities sense that the very existence of the informal sex market is the violation of the principles of control they have propagated in the name of morality, partly to impose and consolidate their own power over society and partly to consolidate the "property rights" of the ruling class over the female segment of society.

For the causes of the existence and persistence of this market, one has to look at the evolutionary experience of mankind, as discussed in chapter-1. Men have always had a tendency to mate with multiple women. The origin of this tendency can be found in the fact that women always know that the child they are carrying is their child, but men have never known, especially in the age of savagery and even primitive feudalism, that the child that their wives or other sexual partners are carrying is their child. Thus, it is natural for men to want to mate with as many women as they can, in competition with other men, in an effort to maximize the chances of passing their genes to the next generation. Further, since women have a short child-bearing age, they tend to lose their attraction for their mates, while men are still capable of impregnating females. The result is that men do not stick to their wives, or sexual partners, but tend to try to find young women of childbearing age, throughout their years of sexual potency. Similarly, some women find their formal relationships, with their men, unsatisfactory or frustrating in different ways and tend to seek sexual satisfaction, or supplement their income and improve their financial status by resorting to informal sexual relations - becoming call girls, prostitutes, strippers, "mistresses", or "keeps". Others lose the support of their parents and families, due to sexual abuse within their families, or outside. Some face accidents and financial hardships due to lack of, or insufficient, family support, or insufficient societal support in general. *Thus, the existence of an informal, heterosexual, sex market is a natural phenomenon imbedded in our genes because of a very long evolutionary experience. It is furthered by economic and social pressures of unjust societies. It cannot be eliminated by any means, but can be reformed.*

Although, the informal sexual market has existed throughout the six thousand years of civilization, and has been recognized as a natural phenomenon by many intellectuals and other leaders in many societies, it has not been tolerated in most primitive societies. The result has been that those female sex workers, who are driven into their profession by economic pressures and abuses, are further abused by their middle-men and "business owners". Some such business owners even resort to abductions and forcible recruitment of girls, to meet the demands of their businesses and clients, effectively making women into sex slaves. Under these conditions sexually transmitted diseases tend to spread among the sex workers and the general population. The result can be a major disaster if highly contagious, or fatal diseases like the HIV infection, are allowed to spread this way. In the First World and some states of the Second World, reason has finally begun to prevail. In some states of these worlds, sex workers have been given legal status and the market has been de-criminalized. Sex workers are supervised by the state and are required to go through periodic medical examinations to control the spread of sexually transmitted diseases. Those sex workers who wish to leave their profession, which may have been adopted under difficult and intolerable personal circumstances, are helped to get themselves educated and retrained in other professions of their choice.

There is a need, in all societies, for the state to manage this market, while extending social protections to the whole population of the state, so that this market may function smoothly and without fear, tensions and abuses. The reduction of social tensions by universal minimal housing, health-care, free education, un-employment insurance, and support for the elderly and the handicapped would create an atmosphere in which forcible recruitment of sex workers would also become difficult, or even impossible, if the level of such protections is high enough.

Like the existence of the informal sexual market, in which female sex workers provide sexual services, markets have also emerged in which male sex workers provide sexual services to female clients. This phenomenon is less wide-spread than the phenomenon of female sex workers. Females find themselves less often in situations where the demand for paid sex with a male would arise. However, this does happen. Markets, naturally, also exist for homosexual men and women and for transsexual individuals, where individuals of one or another predominant sexual makeup may provide services, to other transsexual individuals or persons of other sexual orientations. All these markets exist because there is a need for them, and should be treated just like the main informal sexual market.

Inter-Gender Relations & the Work Place

The fundamental fact is that the relationship of the two genders is complimentary and asymmetrical. It is complimentary in the sense that the two genders have to cooperate to ensure successful procreation. Also, the basic characteristics of the two genders are complementary. Men are, in general, strong and agile. Thus, women are protected by men from dangers. Women are gentle and patient in their behavior, making it possible for them to raise children and perform repetitive activities. Men can keep their emotions in control while facing tragedies of life and at times of danger and can protect their families under such conditions. The relationship of the two genders is also asymmetrical as far as rearing of children is concerned. Women not only carry the burden of child-bearing but also perform most of the functions of raising children when the children are small and young. When children grow up, they need to learn survival skills that, generally, only fathers can teach. Thus, the burden of raising a child is shifted to the father, as a child grows up. So, both parents contribute to this effort, but at different times and in different ways.

In the work place, in general, men have a natural advantage over women in jobs requiring physical strength. Similarly, women have an advantage in jobs requiring gentleness and patience. So, these relative advantages in one type of work or another would tend to equalize the two genders in terms of competition in all fields together. But, since women carry the burden of child-bearing, a level "playing field" cannot be ensured for the two genders, if the burdens of reproduction are ignored and the principle of "equal pay for equal work" is applied to all workers regardless of their gender. The same applies to all other social relationships and activities in which the two genders have a competitive position - political activities, marriage and sexual relations, divorce and property ownership and inheritance, sports and entertainment, etc.

Thus, it is not enough to talk about, and allow, equal rights for the two genders. What needs to be realized is that, by nature, reproduction is the most important function of all human beings, and other living beings. To bear children is to perform the highest form of "socially necessary" work for society as a whole. Whole society must bear this burden - not just the female segment of society. *In short, the cost of child-bearing and, to some extent, child rearing, should be borne by society as a whole.* Thus, what is needed is to tilt the playing field in favor of the female gender as far as reproduction is concerned. Only then, equality may be said to exist in the work place and other fields of activity - cultural, social and political.

In practical terms, what it means is that women should be allowed reasonable paid pre-natal care and maternity leave during pregnancy, sufficient paid leave to take care of the new-born and on the job retraining when they return to work. There must be monthly days off, or "reproductive days off", related to their reproductive biological functions. All day-care and special health-care costs related to reproduction, or its complications, must be borne by society. Since, employers would discriminate between the genders if they have to bear them, these costs must be borne by the state and society as a whole. In addition to these provisions, equality of rights in all fields of human activity must be ensured, including all economic, social and political activities, especially equal rights and responsibilities must be ensured in terms of property relations, during marriage and in case of divorce.

Chapter-20

The Ideal State

The State & its Functions

When we compare the families, extended families, tribes and collections of tribes and nations to modern states of today, we may define a state as a highly organized, sovereign social group on a clearly-defined territory. Human social groups consist of living human beings and, thus, behave like living organisms. Like a living being, the political organization, or structure, of a modern state consists of a number of "organs" dealing with different functions – e.g., an armed force to protect the state from external threats, a police force to protect the state from internal dangers, i.e., dangers to individuals from other individuals in society and dangers to social sub-groups of society from other such groups. Also, states generally have an organ dealing with relations with other states, a taxation authority, an internal conflict resolution institution to provide judicial functions, and other institutions for management of other internal functions. About 200, national states and multi-national states exist today. The state has existed in its less-developed form, even during the early hunter-gatherer stage of human evolution. Tribes and nations developed later as more or less organized states.

> **"If one enquires into precisely wherein the greatest good of all consists, which should be the purpose of every system of legislation, one will find that it boils down to the two principle objectives, liberty and equality."**
>
> **Jean-Jacques Rousseau (1920),** *The Social Contract & Discourses*. **Urbana, Illinois: Project Gutenberg. Retrieved October 14, 2019, from www.gutenberg.org/ebooks/46333.**

The main functions of a state are, generally, described as "the security and well-being of its population", but they should ideally include its peaceful co-existence in the global community of states, to ensure well-being of the global population. By "well-being" we mean safe and happy existence of individuals and social sub-groups of its society in terms of a wide spectrum of human activities – social, economic and political. Maintenance of the health of its population by organizing preventive public health and health care programs, enforcing food and environmental standards, providing education and training, organizing property protection rules and standards, and ensuring fairness in all economic, political and other social activities by individuals and businesses, are some of the responsibilities which ensure the "common good" or well-being of the population of a state.

The Evil Within Us

As we look at the evolutionary history of mankind, it is quite clear that our history has been a history of violence, cruelty and economic and sexual abuse. Although most of us have developed high levels of group survival skills, i.e., sympathy, consideration and respect for the rights of others, injustice has been the basis of the past behavior of many of us as individuals and as members of social groups. The social violence and cruelty of those individuals has been driven by their blind and insatiable greed. In our societies, some individuals develop social skills and learn to respect the rights of others, while others do not develop those social skills to the same extent. This causes some individuals to engage in violence against others or to take advantage of others by trying to rob others of what they possess, or steal from others what they desire themselves but cannot obtain by their own efforts and skills. This is what causes pain and suffering to the victims of such crimes.

Some individuals in our societies become clever enough to develop skills which enable them to form sub-groups to abuse others. Thus, all our societies and states of the past and present have been driven by greed of their ruling classes at the cost of those who were less knowledgeable and, hence, less powerful. The working people were forced by one scheme or another to work hard, while the fruits of their labor were expropriated by the ruling classes. This insatiable greed of the ruling classes resulted in greater and greater inequality in our societies and states. It also led, in the distant and recent past, to endless wars. mass murders of whole families and tribes and genocide on an ever-growing scale. The ruling classes have always consisted of individuals with similar outlook in their lives – a desire for accumulation of wealth, a desire to engage in sexual exploitation of others and the desire of personal aggrandizement – with no sympathy and little or no respect for the rights of others.

The abusive behaviors of individuals affecting individuals and of groups affecting individuals and other groups, have to be controlled and the behaviors based on sympathy, courtesy, mutual respect have to be developed further, so that the human species can have a peaceful existence and our social groups and states can really ensure security and wellbeing for all of us. Consequently, the main questions that arise are – what are the forces that cause crimes of physical and sexual abuse among individuals? What motivates groups of individuals to use and abuse other groups? How can we control such behavior among individuals and social groups? How can we promote sympathy and mutual respect in all individuals in a state? How can we ensure peaceful existence of all states without any violence, or exploitation? All these questions boil down to just one question – how can we develop goodness and suppress evil in our societies and states and promote the greatest happiness for all?

The "Id", as part of the system of our minds, is what we are born with. It is similar for all animals – based, as it is, on the genetic heritage of an individual and his or her instinctive, or hereditary needs and demands – basically, food and the feeling of safety and comfort. As described in Chapter-16, the Id is totally unconscious. Another way to say this is that it is "blind". It demands satisfaction of its desires at any cost and with no scruples, i.e., it is totally blind to the needs and rights of others. Some of its desires are legitimate while others may not be. In humans, the "Id" is tamed by the development of the superego, as described in Chapter-16. The Id grows in its needs and desires as an individual grows into an adult with new needs and desires, e.g., the desire to explore the world, the sexual desires and sex-related needs.

As human societies go through further civilization and as group survival skills develop further, the superego of individuals would also develop further and become more and more dominant. The ego negotiates and mediates between the id and the super-ego, to control the Id in terms of what is acceptable behavior to society in general. In Chapter-3, we discussed how society changes and how it changes the individuals in it and how the changed individuals change the society and the rules of society around them. What it means is that as states and societies evolve, the individuals within also keep on changing. Since, our social groups have developed their group social skills further, the individuals therein are also becoming more sympathetic and considerate toward each other and are developing greater respect for each other's rights. Thus, "goodness" within us and our societies is growing and the "evil" is getting under control – and this is the reason behind the decrease in violence and savagery in our societies over our evolutionary history. Our dreams about an ideal state and our tendency towards this goal are a continuation of this process. The rudimentary ideal states which have come into existence are reinforcing our determination to speed up our march towards further perfection of this ideal. Crime would always exist in our societies to some extent, because we, like all other animals, are accustomed to life without any controls - like in a jungle. The level of crime in our societies is, however, likely to diminish and may virtually vanish with time.

In chapters 1, 2, 6 and 8, we have described how ruling classes developed and behaved in the times of savagery, feudalism, intensive capitalism and benevolent capitalism. The behavior of the ruling elite, in all states of this long period, was the same – exploitation and abuse of the population and abusive violence against other states, motivated by the blind desire for accumulation of capital. This remains true in the present semi-global empire. However, the methods of exploitation have become more sophisticated as described in chapter-13. At the same time, the opposition to such behavior has also developed globally. Looking at the development of rules and institutions of social and political behavior, as described in chapters 17, 18 and 19, it can be reasonably expected that the abusive behavior of states would also continue to diminish with time as civilization develops further and socialization of mankind, along with further development of super-egos of individuals, becomes a more and more powerful factor in human existence.

Visions of an Ideal Society

"Liberty, Equality and Fraternity" was the slogan of the French Revolution and it expressed what mankind had developed as a vision of social life at that time, after its very long passage through its evolutionary history from the age of savagery, to different stages and forms of feudalism. French society was a feudal society at that time, ruled by kings and aristocrats. People of France demanded equality where rampant inequality and abuse existed in all fields of human activity. They demanded freedom, or liberty, where only the aristocrats were free to use and abuse the people. They dreamed of fraternity, where total disregard for rights of the people was the norm, where the ruling class considered the people as lower beings whose job it was to work as peasants, serfs, or servants, and whose lives were worth nothing and could be taken at the whims of "superior", or "noble" aristocrats. The slogan of *"liberty, equality and fraternity"* summed up what mankind had learnt as a result of development of its group survival skills till that time. Not much more than two hundred years have passed since. These ideals of mankind define, even now, the vision of how it wishes to live. However, a deeper

understanding of the vision has developed because of the historic experience of the last, momentous, two hundred years.

Human beings have always wanted to live freely, like other living beings. It is the basic characteristic of life to want to satisfy one's needs no matter what the cost to others. Basically, for living beings, there are no rules which would hinder satisfaction of their basic desires. Plants and animals do whatever they can to satisfy their needs. They eat other living beings and fight to eat whoever they want to eat. Plants compete with each other for nutrition and sunlight. Animals eat plants and other animals and compete with each other in grabbing what they want to grab. Human beings have also behaved this way in the initial stages of their evolution.

.........." The creed which accepts as the foundation of morals, Utility, or the Greatest Happiness Principle, holds that actions are right in proportion as they tend to promote happiness, wrong as they tend to produce the reverse of happiness. By happiness is intended pleasure, and the absence of pain; by unhappiness, pain and the privation of pleasure. To give a clear view of the moral standard set up by the theory, much more requires to be said; in particular what things it includes in the ideas of pain and pleasure; and to what extent this is left an open question. But these supplementary explanations do not affect the theory of life on which this theory of morality is grounded - namely, that pleasure, and freedom from pain are the only things desirable as ends; and that all desirable things (which are as numerous in the utilitarian as in any other scheme) are desirable either for the pleasure inherent in themselves, or as means to the promotion of pleasure and the prevention of pain....

.... This firm foundation is that of the social feelings of mankind; the desire to be in unity with our fellow creatures, which is already a powerful principle in human nature, and happily one of those which tend to become stronger, even without express inculcation, from the influences of advancing civilization. The social state is at once so natural, so necessary, and so habitual to man that except in some unusual circumstances, or by an effort of voluntary abstraction, he never conceives himself otherwise than as a member of a body; and this association is riveted more and more, as mankind are further removed from a state of savage independence. Any condition, therefore, which is essential to a state of society, becomes more and more an inseparable part of every person's conception of the state of things which he is born into, and which is the destiny of a human being. Now society between human beings, except in the relation of master and slave, is manifestly impossible on any other footing than that the interests of all are to be consulted. Society between equals can only exist on the understanding that the interests of all are to be regarded equally."

John Stuart Mill (1897), *Utilitarianism*: Project Gutenberg. Retrieved October 15, 2019, from www.gutenberg.org/ebooks/11224.

In those species which discover the power inherent in forming a group, individuals learn to respect each other, feel compassion and have regard for the needs of others. We, with our highly developed group survival skills, have learned to respect the needs of others and have learned to be considerate and compassionate. Freedom and liberty no longer mean complete freedom to do anything one likes to do - but that one has to respect the freedom of others, also. Freedom means doing whatever one wishes, without violating the freedom of others, in each area of human activity. No human community allows an individual to kill others, because this would violate the equal right of others to live. Similarly, violation of the rights of others to hold and enjoy what is their property, by stealing their property, or by robbery, is not considered freedom.

.... "Consequently the smallest germs of the feeling are laid hold and nourished by the contagion of sympathy and the influences of education; and a complete web of corroborative association is woven around it, by the powerful agency of external sanctions. This mode of conceiving ourselves and human life, as civilization goes on, is felt to be more and more natural. Every step in political improvement renders it more so, by removing the sources of opposition of interest, and leveling those inequalities of legal privilege between individuals or classes, owing to which there are large portions of mankind whose happiness it is still practicable to disregard. In an improving state of the human mind, the influences are constantly on the increase, which tend to generate in each individual a feeling of unity with all the rest; which feeling, if perfect would make him never think of, or desire, any beneficial condition for himself, in the benefits of which they are not included. If we now suppose this feeling of unity to be taught as a religion, and the whole force of education, of institutions, and of opinion, directed, as it once was in the case of religion, to make every person grow up from infancy surrounded on all sides both by the profession and the practice of it, I think that no one, who can realize this conception, will feel any misgiving about the sufficiency? of the ultimate sanction for the Happiness morality....

.... Neither is it necessary to the feeling which constitutes the binding force of the utilitarian morality on those who recognize it, to wait for those social influences which would make its obligation felt by mankind at large. In the comparatively early state of human advancement in which we now live, a person cannot indeed feel that entireness of sympathy with all others, which would make any real discordance in the general direction of their conduct in life impossible; but already a person whom the social feeling is at all developed, cannot bring himself to think of the rest of his fellow creatures as struggling rivals with him for the means of happiness, whom he must desire to see defeated in their object in order that he may succeed in his. The deeply-rooted conception which every individual even now has of himself as a social being, tends to make him feel it one of his natural wants that there should be harmony between his feelings and aims and those of his fellow creatures."

John Stuart Mill (1897), *Utilitarianism*: Project Gutenberg. Retrieved October 15, 2019, from www.gutenberg.org/ebooks/11224.

Certain human activities are considered crimes universally. They include rape, murder, robbery, theft and torture, etc. Certain forms of economic abuse are also considered crimes, *but abuse of human beings by setting up factories and other businesses and using the market mechanism to deny them the fruits of their labor and, effectively, loot and plunder of their labor, is not yet universally recognized for what it is - a fundamental crime that leads to many forms of abuse in other fields of human activity.* People do kill others even now and this happens everywhere - a minority still does not respect the morality and principles that mankind has adopted. We can find human beings who would even eat other human beings, but this is a very rare phenomenon and has become history. Selfishness in dealing with other human beings, especially in economic matters, without regards to whether this *"exercise of freedom"* causes pain and suffering to others, still persists.

The other visions of how society should be organized and how we should deal with each other, however, have progressively become much more powerful. Even capitalism, whether of the *intensive* or the *benevolent* kind, has become much less abusive in some regions of the world, like North America, Europe, or Central Eurasia, where it reigned supreme until recently. The ruling classes of most states of the world do continue to sing praises of the "daring initiatives", "innovative ideas" and "entrepreneurship" of those who continue to exploit and abuse the rest of mankind, but this is what can only be expected. After all, the existence of the ruling classes is based on such behavior and any reduction of the rate of capital accumulation means reduction of the degree of their lavish living styles, based on exploitation. Sections of these parasitic classes wish to drag mankind towards the *"good old days"* of *"Pure Capitalism"* and greater exploitation and abuse of the past, when they had even more power over the people and, thus, could afford even more luxurious living at the cost of industrial workers and peasants.

"We hold these truths to be self-evident, that all men are created equal, that they are endowed by their Creator with certain unalienable Rights, that among these are Life, Liberty and the Pursuit of Happiness. – That to secure these rights, Governments are instituted among Men, deriving their just powers from the consent of the governed."

Thomas Jefferson, July 4, 1776, "United States Declaration of Independence"

Freedom for all, in all fields of human activity, basically means equal rights, or equality of all human beings. Because of the mass abuses of the past - slavery, mass murders of whole tribes, destruction of whole cities, and land-grab on a tribal and national level, the psyche of humanity has developed a loathing for such behavior. This has given rise to the desire for all to have the same rights in all areas - to be free from the abuses of those who claim rights that they deny to others. Most of humanity desires a world in which individuals could compete in the political, economic and other social fields as equals - being paid an equal wage for an equal amount of work, speaking on any subject as equals, standing for election as a representative of a group, or having an equal right to pursue any other competitive social activity - asking for a man's or a woman's hand in marriage, playing and competing in some sport, taking part in a dancing or singing competition, etc.

The Ideal State

The desire for equality has given rise to the demand for democracy. Treating all individuals of the population of a state as equals, gives them the equal right to stand for election as representatives of segments of the people and giving them the right to vote, as equals, for a representative of their liking. The *vision of equality* was formalized, for the first time in constitutional history, in the declaration of Independence by the United States. This desire has not only given rise to demands for equality in the political field, but also in the economic field. Thus, the concept of socialism, i.e., social living as equals, rose in the nineteenth century. Different versions of the concept developed as a result of political debates and continue to be promoted by political parties all over the world, at this time. As indicated before, *socialization and equality are basic to the concept of socialism. Socialism means the desire to live together and to solve social problems together as a social group. It presupposes equality and democracy in all spheres of human activity – social, economic and political. Democracy, ultimately, is nothing but the creation of a government of the people, by the people and for the people.*

Every human being belongs to a family. No matter how primitive we were, we always belonged to a family. We have learned to treat other members of a social group, i.e., a family, tribe, nation or fellow citizens of a state, with courtesy, consideration and compassion and to respect the life, liberty and rights of others. Family was always the ideal, even today; family is idolized, because of what it is - a relationship based on selflessness. A man and a woman get together and have children to form a family. When children are small, parents work as hard as necessary to meet their needs and the needs of their children and protect their children, even at the risk of their lives. When the children grow up into adults, the parents become old and need protection and help, the children, in an ideal family, then take care of their parents. This is how an ideal family works. Each individual works to the best of his ability to meet the needs of all the members of the family, without any kind of accounting. Thus, a human family functions on the basis of the principle - *"From each according to his ability, to each according to his needs".*

This principle is based on courtesy, consideration, love and sympathy - in short, compassion, and is devoid of any form of selfishness. This ideal has grown to a stage that human beings have begun to think of tribes as their family, or even huge nations, and states, as their family. Tribal, national, racial, linguistic and religious barriers, based on fear, have broken down. There are human beings who think of all human beings as a family and want to treat them this way, some are even beginning to think of all living beings as a family and want to treat other living beings with respect, to the extent possible. All living beings are genetically related and it is a scientific fact that we are, in reality, branches of a huge family. So, it is not surprising that many of us are sensitive to the needs of other forms of life. We have also won the struggle for survival against all other species. We are now so powerful as compared to any other species that we are willing to tolerate, and in fact encourage many other species to survive and flourish - not only those whom we find useful, but even those which do not provide any tangible benefit to us, directly. We, of course, cannot tolerate those species, like viruses and bacteria, which blindly attack us and cause us disease and death and which we cannot easily protect ourselves from.

For a state, what compassion means is this - that a state be based on the principle that all members of its population have a guarantee not only to be protected from physical attack or death, but of a minimum standard of living, no matter what they contribute to society - just like members of a family. Thus, those who are in need of protection are protected. Those who are children, no matter whose children they are, and whether their parents are alive or not, have a right to be raised, educated and trained to become useful adult citizens of the state. Those who are old, or sick, have a right to health care and a minimum standard of living, regardless of what they earned during their productive years. The same guarantees apply to those who are handicapped, or become disabled during their lifetime. Only the able-bodied individuals of society may be expected to work, thus producing goods and services for themselves, their families and the other members of the state population. Thus, the whole population of a state should function as a family.

Karl Marx felt that such a state would have to be initially organized just like the city-state of Paris was organized in 1871, during the upheaval that was termed the *Paris Commune*. At that time there were many political parties espousing socialism, and were called *"Social Democrats"*. To distinguish the thinking and program of his followers from others, Marx had named his concept as *Communism*, meaning communal living, or community living, implying living as a family. This vision of treating the whole population of a state as one's family is a powerful concept. It is the highest moral ground that a human being can aspire to, and motivates the hearts of all of humanity to some degree, at present. No wonder, these ideas have totally transformed the world, as we have described before in previous chapters.

To summarize, we can say that humanity has developed three ideals, or visions, of how society and states of the world should be organized:

- *A world based on freedom to do whatever a person wishes, including the freedom to use and abuse other human beings, by means of the "free market".*

- *A world based on equality and democracy, where government of the people, by the people and for the people is a reality*

- *A world based on courtesy, consideration, sympathy, compassion and selflessness, like in a family, based on the principle of "from each according to his ability, to each according to his needs"*

The Ideal State

It should be obvious that human beings want to live with some level of freedom, as their nature demands, and the competition between individuals and between social groups cannot come to an end. But, based on the desires of all of humanity, rules can be devised so that human beings can live in accordance with their desires for *compassion, equality and freedom*. Thus, competition between individuals can be made much less abusive. In fact, abuse can be effectively eliminated from society. Our evolutionary experience has taught us that some activities are undesirable, since they cause pain and suffering to other individuals of our communities. Murder, robbery, theft, rape, torture, enslavement and deception with the aim of depriving another human being of his or her property, or any other rights, are some such activities. All of humanity is in total agreement on this, as far as relationships within a state are concerned. However, most states still encourage, or tolerate, such acts in dealing with other states, although opposition to such behavior is steadily increasing.

States compete with each other on the same principles as individuals - *adaptation and competition*. As knowledge advances, new technologies are discovered and put to use for improvement of community life and also to develop new means of destruction - new weapons for survival of the state in competition with others. The competition is over natural resources - land, minerals and control and possession of plant and animal life on land and in the oceans, lakes and rivers. Competition between states has resulted in expansion of some states, or shrinking of territories under their control, because of expansion of other states at their cost. The battles over control of the hearts and minds of tribes and nations continue, but behind these battles is the ancient and ages-old desire to possess and control *habitat and resources*.

A cursory examination of history, of states which have successfully survived and grown in size and power, reveals that these states have encouraged development of society in keeping with the human ideals of freedom, equality and compassion, *to some extent*. Sweden, Britain, China and the United States are examples of four such states, to varying degrees. The opposite is true of the states which have not tried to adopt all these principles to the same level as their competitors. They have, thus, failed to compete successfully, have become smaller or have vanished completely. The Soviet Union, Yugoslavia, Czechoslovakia, South Yemen, South Vietnam, East Germany, the Japanese and other European empires, Somalia and Pakistan are examples of such states.

The question, thus, arises as to how the three principles of freedom, equality and compassion should be implemented by an ideal state, what constitutional and structural arrangements should be made by such a state and what economic and political policies should be adopted by it, so that it is successful in ensuring the security and well-being of its population, succeeds in surviving in competition with other states and, ultimately, settles down to a peaceful co-existence with those states, with no imperial desire to abuse them economically or otherwise. We would investigate these issues in this chapter. We, of course, are aware of the *"four worlds"*, or the four stages of evolution of mankind and the states into which it is organized at present, as described in chapter 15. We are also aware of *eight super-states that are emerging* and the *virtual semi-global empire that exists in the world of today*, as described in chapters 13 and 15.

To arrive at a political and policy structure of a state that remains relevant for some time, we would focus on a multi-national "super-state". An ideal state is not likely to come into existence, to its full extent, as long as global social systems continue to be as they are at this time – divided into four levels of socio-economic development. All the four worlds have to progress towards the level of benevolent capitalism for this to become practicable, exactly. However, humanity would continue to progress and some states of the world of benevolent capitalism of the present and the future, and the world of managed transformation, would continue to make social, economic and political progress, resulting in further socialization of their systems of production, distribution and consumption – in short socialization and socialism would continue to move forward and the social systems of all intensive capitalist, feudal and tribal states would continue to develop toward higher levels, resulting in reductions in economic inequality. However, this socio-economic progress, naturally, would continue to be opposed by the ruling classes of the world of benevolent capitalism– based on their economic power, especially that derived from exploitation of the second, third and fourth worlds that exist at present. Only when the rest of humanity reaches the stage of benevolent capitalism, and the ruling classes experience a drastic reduction in their powers of exploitation, can socialism make really dramatic progress towards creation of ideal states.

The structure that we arrive at should be applicable to any national, multi-national, or super-state of the future, or to a global state if and when it comes into being.

Political Organization of the Ideal State
Representative Government

Multi-party system of representation is the highest form of government that humanity has developed after many years of political experience., as described in chapter-18. It would have to be adopted for an ideal state, A one-party system of representation can also work for some time, depending on the level of social development of a state's population. It can work well for longer period of time if its structure is such that *it ensures a government of the people, by the people and for the people.* Till now, such a system has only been adopted after a revolution in a state. A multi-party system would have to be of the parliamentary form for a multinational ideal state. If an ideal national state emerges, then, it could adopt the presidential form. The organizational structure of such a state has been described in chapter-18. However, no tribes would exist in such an advanced state. The state would have a civilian government as required by the principle of civilian supremacy and would ensure complete separation of the state and religion.

Federal, National and Provincial Governments

Federal, national and provincial governments would be organized, and would have the same powers, as described in chapter-18. The federal and national houses of representative would also have the same functions. Only tribal territories would not exist and all legislative and executive positions would be filled by elections.

Taxation and Fiscal Policies

Sufficient taxes would have to be collected by the federal, provincial and local governments to implement all their socially necessary programs related to support for the non-working population and for proper functioning of all institutions in various sectors of the economy. Progressive tax rates would have to be designed to ensure that profits do not become windfall profits in any sector, and a reasonable rate of return is ensured to corporations, so that the economy may grow with time. Both federal and provincial governments should have the authority to impose income and capital gains taxes on individuals and businesses. The federal government, in addition, must have the authority to impose inheritance taxes and taxes on imports and exports.

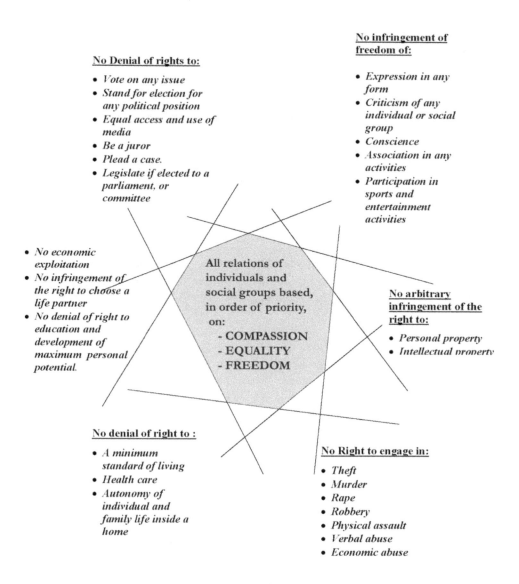

No Denial of rights to:

- *Vote on any issue*
- *Stand for election for any political position*
- *Equal access and use of media*
- *Be a juror*
- *Plead a case.*
- *Legislate if elected to a parliament, or committee*

No infringement of freedom of:

- *Expression in any form*
- *Criticism of any individual or social group*
- *Conscience*
- *Association in any activities*
- *Participation in sports and entertainment activities*

- *No economic exploitation*
- *No infringement of the right to choose a life partner*
- *No denial of right to education and development of maximum personal potential.*

All relations of individuals and social groups based, in order of priority, on:
- COMPASSION
- EQUALITY
- FREEDOM

No arbitrary infringement of the right to:

- *Personal property*
- *Intellectual property*

No denial of right to :

- *A minimum standard of living*
- *Health care*
- *Autonomy of individual and family life inside a home*

No Right to engage in:

- *Theft*
- *Murder*
- *Rape*
- *Robbery*
- *Physical assault*
- *Verbal abuse*
- *Economic abuse*

Figure 20-01: Limited Region of Allowed Activities in an Ideal State

Individual and National Rights

National territories would have to be demarcated in such a national or multi-national state. All individual and national rights, as described in Chapter-17, should be ensured for individual citizens and nations of the state. Individual rights would include economic rights, as described under "Right to life and liberty" in chapter-17. The right to have access to accurate and objective information would also be included in the rights of individuals. Ownership and management of news and entertainment media and information related to all sciences would be regulated by the state to ensure accuracy and objectivity.

Freedom from Economic Abuse and Exploitation

All citizens of the state must be given an undertaking that they would not have to face economic exploitation and abuse of the kind inherent in a system of intensive or benevolent capitalism. The state would have to guarantee this right to all citizens, through provisions of its constitution and other laws. The state should regulate and control all markets for this purpose and should, through its taxation and economic management policies in general, aim at effective elimination of all economic abuse and exploitation

Inter-Gender Relations

Laws of a state must guarantee equality and maximum freedom to both genders in pursuit of their sexual needs, as described in the last chapter. The marriage and other sex-related markets of all kinds must be regulated by the state, as described before, in order to provide protections to those who need them, to minimize abuses and to protect families and children.

Crime and Punishment

The stress should be on prevention rather than punishment, realizing that punishment cannot really serve as a deterrent and justice cannot really be done after a crime has been committed, as described in the last chapter. The death penalty must be abolished for all crimes. A system of criminal reform needs to be established for all criminals, so that they can become normal human being again, if possible. All individuals must be protected against arbitrary arrest, detention, or exile. No one should be subjected to torture or to cruel, inhuman, or degrading treatment or punishment.

Social Organizations

A democratic state has to allow creation of social organizations like labor unions and professional associations. These organizations must be able to operate as democratic institutions and must be able to protect their professional rights regarding wages and working conditions. Workers of all corporations should have the right to elect the managers of their economic institutions and remove them if required

There must be regulations governing safety on the job, and functions of labor unions. The main objective of these policies should be to ensure that maximum freedom is allowed to individuals and organizations to engage in economic activity, while ensuring that abuse of such freedom does not take place.

Maximum Freedom

All individuals should be allowed the maximum freedom possible in all areas of human activity - political, economic, and in the fields of sports and entertainment. *This freedom should only be constrained by the need to avoid crimes and by the requirements of fraternity and equality as outlined in this chapter. In other words, the ideal state has to give the highest priority to fraternity, then to equality and then to freedom.* Thus, the freedom to engage in activities desired by an individual would be constrained, as graphically depicted in Figure 20-01. The constraints on the freedom are indicated by straight lines representing the rights of individuals and social groups and prohibitions of certain possible activities. The area limited by those constraints is the *"feasible region"* to which the freedom of individuals and social groups would be confined. This *"feasible region"* shows the *"feasible"* activities, i.e., the activities that are permitted by a society. It indicates the environment that makes a state into an ideal state and leads to peace and general happiness in its society - which promotes state loyalties and innovation leading to further prosperity and strengthening of the state. Unnecessary constraints on freedom of individuals and social groups lead to an atmosphere of oppression and resulting unhappiness. The larger the feasible region, the more the state would be closer to an ideal state. The feasible region should be constrained only by those requirements that are absolutely necessary for maintenance of general peace, happiness and well-being of the population.

Judicial System

Even in an ideal society, some Individuals and groups are likely to break its laws, although the state would tend to create an environment in which crimes are minimized. Still, conflicts between individuals are likely to arise. The same would happen to various social, economic and political organizations and the executive, legislative and judicial branches of government. Thus, creation of a high-functioning judicial system is absolutely necessary for conflict resolution and adjudication, including conflicts and unlawful behavior within the judiciary.

Economic Management of the Ideal State
Fundamentals

Economy of an ideal state cannot be run by *command and control* alone, as the history of the Soviet Union and other East European states have amply demonstrated. Also, the evolution of states under feudalism and capitalism has demonstrated that if exploitation of the people continues in a state, purely under the conditions created by the *law of supply and demand*, then the state would have nothing in common with the ideals of mankind. It would become more and more hostile towards its people, leading to inequality, political and economic abuse and violence. Thus, the economy of an ideal state cannot be run on the basis of a totally free and abusive market, either. Therefore, the management of the economy would have to have three objectives simultaneously, based on the three visions of humanity, in terms of priority - *Compassion, Equality and Freedom.*

An ideal state would have to have exclusive ownership of all natural-resources of its territory, most financial institutions and major industries. It would, also, have to establish wholly-owned hospitals, educational institutions, child-care centers, long-term elderly care centers, and social institutions for supporting handicapped and disabled individuals. The economic market, the labor market and other markets would be closely monitored and strictly regulated to prevent abuse of workers and consumers.

Ownership of Natural Resources

Land and other natural resources, like rivers, lakes, mountains, forests, mineral deposits, should be owned by the national governments of a super-state. The Federal government, however, should have the right to take over land and other resources, to the extent considered necessary to promote the interests of the federation. Energy resources, like coal, oil, natural gas and deposits of fissionable material could be included in this category. Land should be leased by individuals, families, wholly or partly state-owned organizations and other social organizations, like professional associations, within a national or federal territory. Only, licensing and leasing of very limited number of mineral and energy resources of strategic importance and those within the boundaries of federal territories, including the continental shelf, should be managed by the Federal Government. The ownership of the continental shelf of a super-state should rest with the Federal Government.

According to a global, inter-state agreement, security boundary of a state extends to certain extent into the sea from its coastline. Economic zone of the continental shelf, belonging to a state, extends even further into the sea. This can be expanded, but whatever is in the middle of the oceans, and the continent of Antarctica, should be controlled by the *Interstate Relations Organization (ISRO)*, by an agreement between states and super-states in existence around the world. The present agreements on this issue have been arrived at under the control of imperial powers and are, naturally, not fair to the smaller and less powerful states. At present, satellites and space vehicles of any state can go through space, without regard to political boundaries since they move around a hundred miles above the surface of the earth. Air space above a certain height (100 Km, or so) is considered space, not air-space. Resources discovered on other planets are, at present, brought to earth by the state that discovers them and no inter-state organization monitors such activity. In the future this could be the responsibility of the *ISRO*.

Financial Institutions

At present, generally, every state has a state-owned, or state-controlled, central bank that manages monetary policy, as the Keynesian scheme of the management of the business cycle has been universally adopted by all states. Banks owned by individuals or groups of individuals of the ruling class are no longer allowed to create their own currency, since this practice had led to economic disasters in the past. In normal times, the central bank of a state creates money as the GDP of the economy increases. When a period of "boom" arrives, it tries to increase the cost of borrowing by gradually increasing the interest rates. Under such circumstances it also reduces the money supply by issuing and selling treasury bonds, so that the economy, driven by out-of-control greed, cools down. It may also increase the reserve ratio of banks, to reduce their ability to issue loans. Under conditions of a recession, or economic "bust", the central bank takes the opposite measures - it reduces the prime rate, buys back its bonds and may also reduce the reserve requirements for banks. These measures tend to "stimulate" the economy, by re-igniting greed among business owners. Banks, in general, have an enormous freedom to issue loans, far in excess of their deposits. They, thus, tend to collect immense amounts of money as interest on loans they issue. They also tend to maintain a large "spread" between what they pay depositors as interest and what they charge borrowers. This enormous power, without any productive work, cannot be justified for individuals. *All financial institutions, thus, should be wholly owned by the federal, provincial and local governments and other social institutions of the state. The financial institutions include banks, insurance companies and investment companies which manage investments in stock, bond and money markets. Companies dealing with commodity markets may delegate some of their operations to institutions that are not wholly state-owned.*

Ownership of Companies

Based on a full understanding of property ownership and its evolution, it is quite clear that most of the present ownership is based on violent and despotic actions of the past. It needs to be ended, so that society may begin afresh in determining the real owners of capital. This need not be done by despotic inroads into the current ownership of capital. Instead, market-oriented state organizations may be established to trade in individually owned, or company owned real estate, stocks, bonds, mines, forests, etc. Social property should be encouraged so that society as a whole may derive benefits from it, instead of individuals with dubious claims of ownership. Workers may become shareholders of their companies. Economic organizations, including corporations, partnerships and sole proprietorships, may be owned and established by individuals, professional associations and labor unions. All levels of government should have the right to set up production centers for major industries, organizations for transport and communication services and other service institutions for banking, insurance, education, health care and research. Companies, including those engaged in research and development, may be categorized, in terms of ownership, as follows:

1. Those involved in military and strategic industries may be limited to ownership by the federal government, professional associations and employees, or by other organizations of this category only.

2. Major industries may be owned mainly by the government, professional associations and the employees of those industries, the proportion depending on the degree the state has reached in its ideal status. They may also be owned by other organizations of category 1 and 2.

3. Ownership of the rest of the organizations may be limited by their size, the smaller organizations having higher proportion of ownership by employees and other individuals. These organizations may also be owned by other organizations of category 1, 2, or 3.

Economic Markets

All economic markets would have to be tightly regulated to minimize abusive practices. Laws would have to be enacted to ensure that in collective bargaining, the workers of enterprises are not handicapped by rules that tilt the playing field against them. An environment would have to be created by means of such laws so that bargaining and changes in wages and benefits take place with full knowledge of the performance and financial standing of an enterprise and in a peaceful atmosphere and strikes are avoided to the extent possible. Laws on work safety and environmental impact of all industries would have to be promulgated and enforced, to protect the workers and the environment. Workers rendered surplus due to changes in factories, farms and mines, would have to be provided financial support during unemployment and retraining for new jobs. Organizations would have to be set up to monitor and certify the quality and safety of products. All transactions would have to be recorded for later analysis by government departments to ensure conformance to the relevant state laws. All advertising would have to be regulated and processed through local and provincial government departments to ensure accuracy and truthfulness.

Economic institutions would, thus, operate in a restricted economic market. They would be able to make a socially necessary profit, or a reasonable return on invested capital, but their ability to exploit workers would be restricted by means of laws regulating the economic markets. There would be controls, but those would be legal or economic controls by fiscal policies or monetary policies of the state - not administrative controls or commands from a state organization, but regulations, instead, governing different markets, tax policies related to economic activity within the state and exports and imports by it.

Capital Markets

All capital markets, i.e., stock markets, commodity markets, bond and money markets should be regulated by the federal government to avoid abuses. The stock market may have the variety of stocks, exchange traded funds (ETFs), managed mutual funds, etc., that are traded in these markets at this time. The bond markets would include markets for bonds issued by federal, national, provincial and local governments and companies operating in all sectors of the economy. The list of commodities is very long and present practices in their trading may be continued. However, in all these capital markets, the players involved would be very different, since most of the capital would be owned by social institutions.

Labor Market

A level playing field would have to be created for competition among able-bodied adults. The state should encourage competition between able-bodied workers, whether male or female, on an equal footing. Competition would thus exist for those seeking employment, those who want to maintain their employment status and those who want to progress and seek rewards for better performance on the job. Thus, a labor market would exist, but the guarantees of a minimum standard of living would have to be such that extreme class difference cannot arise.

Inheritance

It is the natural desire of parents to want their children to progress in every possible way. Parents normally want their children to be healthy and to have the best education possible. They naturally prepare their children for competition with others, as they become adults. If possession of large amounts of capital is considered an advantage, parents would want their children to have all that they have accumulated in their lives - regardless of whether it was accumulated by fair or unfair means. In an ideal state, large scale accumulation of capital would not be possible. Initially, the only large-scale capital accumulated by families would be the capital passed on from the days of feudalism and later stages. This capital, accumulated by abuse of human beings by human beings, would have to be eliminated from ownership by families. This elimination may not occur suddenly in one step. Once this has been achieved, then large amounts of capital would not be accumulated by individuals, in general. Inheritance laws would have to be passed in an ideal state with the sole objective of not allowing inherited capital to be used for abuse and exploitation of citizens. Then, basically, inheritance would be limited to a reasonable share of a person's property and personal items of special sentimental value. Under normal circumstances, a reasonable share would be an amount of capital that can be consumed by the inheritors during their lifetime. This share would naturally vary according to the circumstances of the inheritors. Children who are physically or mentally handicapped, or incapacitated by accidents and disease, could be allowed a larger share than physically fit working persons.

The desire to pass on legitimately earned and accumulated capital to one's children is a very human sentiment, and that must be respected. A person may also want to donate part of his or her capital to a professional association or charity. An engineer could contribute to an engineering association, or a doctor could contribute to the research related to or treatment of a specific disease, or some organization dedicated to treatment of patients with specific medical problems or diseases. Inheritance taxes should be set up so that only property that is not donated by the owner to an association or charity is inherited by the next generation. What is not donated or inherited should be collected by the state as a tax. Under these circumstances, the greatest asset of inheritance that a person could pass on to his or her children would be knowledge and training. That is something that can give them the greatest competitive advantage, even though subsidized education is provided by the state. The purpose would be to create an environment in which all individuals would be free to compete with each other without any unjust advantage.

Innovation

Security and well-being of the whole population is the only legitimate purpose of the existence of a state. Basic well-being of the population can only be attained by the creation of an environment of tolerance, sympathy, civility and peace. In such a peaceful environment, citizens tend to focus on constructive and productive activities, instead of conflicts and destruction. Innovation is naturally encouraged by such an environment. An ideal state would naturally create such an environment. Progress of a society depends on innovation and development of new ideas and new products to meet its needs and solve its problems. What could be done to promote innovation? Service of mankind is a very great and honorable motive. It has motivated many in the past and continues to motivate many. Love of family is another very great motive. But financial reward is also very important for many individuals, not only for themselves but also because it helps in fulfillment of their desires to serve their families and mankind in general. So, it should not be excluded. *Scientists would have to be the highest paid and honored individuals in the state.* How to reward intellectual production to promote innovation? Much has been done in this direction in modern European and North American states. However, the focus is on companies and universities conducting research and development in various areas of science. These organizations, in turn, encourage those whom they employ for these purposes. Companies and individuals can patent new discoveries. Companies are guaranteed long periods of marketing of new innovative products they develop, at prices that are basically set by themselves - thus ensuring that the companies can cover all the costs incurred in research and development, and more, before other companies may produce the same products, e.g., new medicines. Thus, consumers are the ultimate judges of the usefulness of new inventions and discoveries. The consumers may be the military institutions of the state. In an ideal state, the focus would have to be on rewarding the individuals who invent or innovate. New discoveries and inventions may continue to be rewarded with lump sum financial rewards and a share in the resulting profits over a certain period of time. Individuals, however, do not require huge rewards, like companies working for maximum capital accumulation would. The rewards, of course, should be subject to the taxation policies of the state.

Information and Entertainment Media

The political system of a state can only function properly, if its people have access to accurate information and this information cannot be controlled, or manipulated, by government agencies, economic institutions, or companies owned by groups active in various sectors of the state's economy. To ensure that accurate information is available to the people so that they can make wise political decisions when elections are held, or otherwise, all information and entertainment media would have to be owned and managed by their participants. Thus, news agencies, film studios and companies publishing books, magazines and newspapers would have to be owned by their workers and professional associations. The same would have to be ensured for all publishing activities on the internet and for all organizations managing sports. All such companies would have to be made financially independent by legal provisions.

Health Care

A comprehensive and universal health care system would have to be created. All aspects of care would have to be dealt with by such a system, including dental, vision and long-term care. All hospitals and health-related institutions should be owned by federal, provincial and local governments, with the higher levels of government establishing the more advanced levels of hospitals and research centers.

Education

Education system of an ideal state would have to be wholly owned by the state. Its management may be delegated to other social institutions. Making educational institutions into businesses, simply destroys the education system. So, this must be avoided. Education should be highly subsidized for all citizens of a state, up to a level only limited by an individual's ability or desire to acquire knowledge. All children should be encouraged to develop to their full potential. Education would have to be free and compulsory up to high school level, as it is now in many states of the world. Higher education institution should be owned, mainly, by local and provincial governments and it should be subsidized by the state, based on the development status of its socio-economic system. Children should be educated to become useful not only to themselves and their families but also to society at large. Service to humanity and acquisition of basic knowledge of social, physical and life sciences should be stressed along with issues of morality, specially respect for the rights of others and sympathy for those who are not fortunate to be physically and mentally healthy and young.

Culture

Each nation should be free to engage in its cultural activities. Its national language should be taught in its schools and this language should be the language of communication in as many areas as possible, subject to the needs of communication with other nations of a federation. Dancing, singing, painting, writing and all other art forms of every nation in a federation, should be encouraged throughout the federation to encourage the processes of fusion of nations, which would be underway anyway, under the influence of economic forces and activities.

Minimum Standard of Living

All able-bodied men and women should have the right to work, to choose their professions, or change them, to engage in any activities - economic, political, and other social activities included in the fields of sports or entertainment and permitted under the rights guaranteed by the state. All individuals should have the right to move anywhere within the boundaries of the state and live there, or move to another state, if they so wish. All members of the population of the state should be guaranteed a minimum standard of living. The minimum standard of living would depend on the economic development of the state. It would rise as the state develops economically and would consist of guarantees of minimum housing, education, health care from birth till death, access to institutions for social activities and a minimum income regardless of what a person is capable of producing, or produced in his or her productive years of life.

Special Provisions for Children, the Aged and the Disabled

All special needs of those who are retired or disabled during their working lives, should be met beyond guarantees of a minimum standard of living, based on their performance during their working lives. Similarly, all children should have economic and social rights to be raised with a minimum guaranteed standard of living, in a loving family atmosphere, with full access to all institutions of physical, educational, social, economic and political development, so that they may be able to reach their full potential without any social or economic constraint.

Special Provisions for Women

All women should be guaranteed special rights related to their reproductive functions, as described in the last chapter. Women should have additional guarantees of income, from the state, when they are going through one or another stage of life related to their reproductive functions. This means paid monthly days off, maternity leave during pregnancy, especially some weeks before delivery, some reasonable period after delivery and for any required job retraining before restarting regular work. Thus, all men and women should be able to compete with each other in all fields, subject to the constraints that the state and society impose on them as requirements of fraternity.

Defense and Internal Security

When a state has become a super-state and has reached the "Ideal State" status, it may not need to organize much in terms of the means of defense from external threats, if other states of the world have also reached the same status. But in the process of achieving this status, it would require defense forces for protection against external threats. The objective, at any stage of this process, should be defense on a sufficiency basis. As the ruling class of the state starts losing its power over its people, its collective greed and the desire to exploit other states would also decrease progressively. This should lead to a stress on economic and social development rather than acquisition of armaments for foreign adventures. For internal peace and harmony among its people, a developing ideal state would also require police forces to deal with crimes between individuals and social groups, but the need for such forces would decrease progressively, with its social, economic and political development.

Summary

We can summarize the requirements of an ideal state by stating that such a state must create an environment for maximum exercise of freedom for its citizens, by ensuring the following:

- Protection against, individual and group crimes.

- Ensuring equality of political rights for all, without any distinction – racial, national, or gender.

- Creating an environment of equality of economic activities for all citizens of working age, by effectively removing all economic exploitation.

- Ensuring a minimum standard of living for all its citizens, by state-supported programs.

- All those who are retired or disabled during their working lives, should be provided benefits beyond guarantees of a minimum standard of living, based on their performance during their working lives.

- Those who become temporarily unemployed because of layoffs by their companies, should be guaranteed reasonable income and retraining during their period of such unemployment.

- Promoting the development of all individuals to their full potential, by providing universal health care and education for all.

- Imposing limitations on inheritance, protecting individuals requiring support, while discouraging misuse of inherited economic power.

- Promoting creativeness and innovation, in arts and sciences and promoting equality of participation in sports and entertainment activities

- Removing the additional economic burden of child bearing and child rearing from the female segment of society - making these natural and essential activities into state-supported activities.

Consequences

Peace and progress should be the result of the creation of the ideal state as described in the previous sections of this chapter. Such a state would promote the well-being of its population like no other state ever has or could. As class differences become virtually non-existent, the conflicts in society would tend to decrease and even disappear as a state becomes more like an ideal state. Economic conflicts would be minimized. The labor market would not remain abusive and other economic and social markets would also become virtually devoid of any abuse. As an ideal state progresses, its political system would tend to become more and more democratic. In fact, one can imagine a parliament consisting of the whole adult population of the state, voting on all legislation formulated by its deputies. Even in our present age of automation, computerization and the internet, such a parliament is not a dream. It can be an actual functioning parliament. It would, however, have to elect and create the executive and judicial branches of government.

Compassion, equality, freedom would reign in such a state, leading to peaceful cultural activities, scientific progress and innovation in all fields of human activity. The only internal danger such a state would face is the danger of the reemergence of a parasitic class, resulting in a re-division of society into two hostile camps. The politically conscious people, through exercise of their political rights, would have to ensure that such a situation does not arise.

In such an ideal state, real unemployment would not exist. Since all workers would have an equal opportunity to educate themselves, and to stay healthy and physically fit, they would not feel the need for a job just to survive. The bargaining power of all citizens would thus increase, because, even if they are not healthy, they would not be afraid for their lives or for survival of their loved ones. Children, their elderly or non-working parents, would always be cared for, no matter what happens.

Once such individual and family security is established, then, employers would lose the power that they have over the working class, under the system of intensive or benevolent capitalism. Workers would still compete, and the companies could offer low or high wages, but in this kind of situation, a non-oppressive labor market would come into existence. There may be slight differences among sub-classes of the people. Thus, theoretically, class differences would exist to some extent and an economic space would come into being where competition takes place, production increases, efficiency is maintained and individuals are free to engage in activities that they naturally desire, in any area, subject to the constraints of compassion and equality. However, the state would ensure, through its taxation and spending policies, that capital does not accumulate in a few hands.

Due to the absence of the extreme exploitation of the working population, the economy of the ideal state may not accumulate capital as fast as it does under intensive capitalism, or even benevolent capitalism. But the long-term growth rates may not be very different, since the cycles of boom and bust accompanied by huge wastage of resources would come to an end. New technologies like robotics, artificial intelligence and further automation of production would lead to shorter working hours, allowing more leisure time for the working families. The systemic high death rate of the population, by starvation, malnutrition and lack of health care, would be impossible under these conditions. Because of the increase in the well-being of the population, the population of the state may start to grow faster, but this process could be easily managed by the state and its people. With the end of severe business cycles, events like the great depression of 1931, or the Great Recession of 2008, would not occur because of the government and other social ownership of most financial institutions. Even severe recessions would not occur. This would happen because of several factors. First, the social ownership of economic enterprises would drastically reduce greed as the sole motivation behind economic activity. Also, the economic management of the state would have several effects on the business cycle. First, the social security and welfare benefits of the non-working, or retired population and unemployed workers would be so complete that the government would be spending a substantial part of its budget on expenses related to these "automatic fiscal stabilizers". We can also assume highly progressive income and business taxes.

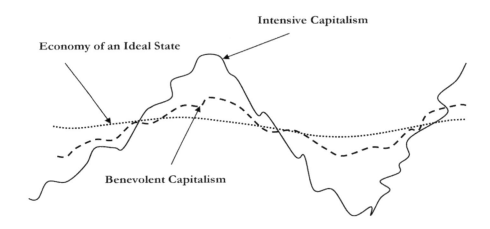

Figure 20-02: Recessions of the Future

As new sectors of economy develop, businesses would naturally tend to shift to those sectors, from the sectors that are no longer highly profitable. Such events could lead to mild recessionary trends in the economy. The resulting dislocation of workers and families would increase the social security spending and move the federal budget towards a deficit. The decrease in family incomes and company profits would also reduce taxes collected by governments. Due to the large share of the state sector in the economy, the effects of the automatic stabilizers would have a much greater impact. The economy would, thus, tend to stay stable. It may not require any drastic action by the central bank or the government in this regard. But, of course, the government could take the standard actions prescribed by the

Keynesian system of management of recessionary tendencies, by both fiscal and monetary measures, especially when the state is evolving towards an ideal one.

The steady rate of economic growth, and absence of boom and bust cycles, would be welcomed by the population, since it would mean working under less stressful conditions, more leisure time and more peaceful and enjoyable working conditions and leisure - leading to lives of much higher satisfaction for the citizens. *In short, the jungle would no longer be a dangerous wilderness and would be transformed into a wooded park where people can roam around in safety and peace.*

Due to the absence of economic exploitation and abuse in an ideal state, a parasitic capitalist class would ultimately cease to exist and its greed would not be able to create internal economic instability, as indicated by the extreme boom and bust cycles, and would not be able to motivate the state to engage in abusing other states - leading to an end to imperial abuse of states by states. Thus, to the extent that the world is dominated by such ideal states, imperialism would tend to die.

Examples

We have mentioned the names of four states as examples of those states which have moved on the path to becoming like the ideal state - Sweden, Britain, United States and China. Of course, none of these states is exactly like an ideal state, but all have some characteristics in common with such a state, to some extent at this time. Let us look at them one by one. China and Sweden are the closest to this concept.

Sweden became a separate and independent state in 1814. It managed to avoid getting actively involved in European conflicts and steadily developed its economy. In the twentieth century, it basically remained neutral in the two world wars. Under the leadership of the Social Democratic party, it set up a comprehensive social security system for its citizens, because of which it is still referred to as a "welfare state". Its constitution guarantees most basic human rights to its citizens. It is a national state, so that the question of internal national boundaries does not arise. It has a modern society which is in an advanced stage of benevolent capitalism. It has fulfilled the requirements of compassion and equality to a great extent. However, it has a political system that allows exploitation of the working class by its companies owned by its ruling class. Thus, it has not completely guaranteed the right of its citizens to be protected from economic exploitation. Still, its liberal ruling class is not driven by blind greed and unquenchable appetite for capital accumulation. The result, because of the small size of its economy, is that the state has not developed the desire to exploit other nations and states directly. However, it does take advantage of the controlled market created by the US, to trade with and exploit developing states. It has been a peaceful state for about two hundred years, without any colonies or imperial possessions and has not openly participated in any wars during this long period. Its people are highly educated and politically conscious with the result that all efforts by political parties, other than the Social Democratic Party, to dismantle the system of social protection for the non-working population have failed. Social and economic inequality persists, however, and the state needs to move in the direction suggested by Figure 20-01.

The Ideal State

China has become a good example of a state following the path towards ideal state status. Although it is a one-party state, yet it is in many ways more democratic than European or North American states. Its rulers are ordinary people of China, although they are highly educated and trained – with administrative experience at different levels of government. The government is not the government of a capitalist class, since capitalists cannot be full members of the CPC. It works only in the interest of its people, although a large business community has developed in its economy. However, China has restrictions regarding rights of free expression and association, because of which it is frequently criticized by the capitalist-controlled press and media of European and North American states. There is an obvious contradiction created by the policies of the Chinese government - while a capitalist class has been allowed to emerge in the Chinese economy, it is not allowed to acquire political power to govern. The capitalist class uses the capital so acquired not only to increase its standard of living, but also to try to maximize its exploitation of the people. It can do so in two ways – it can bribe the government officials to obtain concessions and privileges to increase the profits of its companies and it can obtain the support of the people under its control by confusing them about what is happening to them. Its members, naturally, desire to acquire political power to increase their exploitation of the people and tend to engage in, and encourage, propaganda to promote the global cultural hegemony of the ruling classes of capitalist states, aimed at maligning the government - hence the restrictions on "freedom" of such "speech". This situation can be rectified by enhancing these basic freedoms of all citizens and engaging in their political education. In a speech at Davos, President Xi Jinping had said that China is pursuing *"development of the people, by the people and for the people"*. Looking at the *"big picture"* of its political and economic system, it is quite clear that its government is *the government of its people, by its people and for its people and not for its capitalist class*. In this sense, China is one of the most democratic states of the world.

China is likely to stay on the path to becoming an ideal state, unless its development is disrupted by a political upheaval - which is, however, highly unlikely. At this time, all members of its minority nations of Mongols, Tibetans and Uyghurs, among many other smaller minorities, show unhappiness with conditions of their lives in China, although the territories of all those nations have seen dramatic economic and social development. They are not exactly happy, as naturally happens with minorities, with their status in China, because of their feelings of being so small and insignificant in the existing system! The one-party system of government has this difficulty. Ultimately, when the global economic, military, social and political environment permits, an inclusive system of government may be needed in such areas to change the perception of those minorities. Several steps in this direction have been taken, including the political and vocational training of the adult working-age populations. Further steps are possible and desirable. China may achieve the ideal state status, if it continues to liberalize its political system and continues to move towards social ownership and worker-controlled management of its companies and farms, while removing the deficiencies in its socio-political system as compared to an ideal state.

Britain's English nation developed group social skills at a very early stage. It became a tradition for the English men and women to follow the law and to be courteous and considerate in their dealings with each other. This gave power to their nation. Their ruling class used this power to abuse the Scottish and Welsh nations on their island state. Even sexual abuse of the minority nations was considered a "right", during feudal times. Thus, due to their developed group social skills, the English became powerful and conquered the whole island and then moved on to conquer Ireland, which remained a British imperial possession for a long time. As intensive capitalism developed in Britain, it intensified the appetite for colonies and imperial possessions in its ruling class, as the "tribute" of the domestic population was not sufficient to satisfy its hunger. The British Empire continued to expand, till it had subjugated about one-fourth of mankind, over about one-fourth of global territory. Due to wars with other imperial powers and the effort needed to keep other nations subjugated, Britain became weaker with time. The growth of the US also led to its relative decline. This ultimately led to the disintegration of the British Empire and its imperial hold over other nations decreased with time. Its imperial activities are now mostly indirect through its corporations and, also, through state relations with its Bantustans, which are primarily under US control now. Britain has also developed a system of social protections for its citizens, but there are still no effective controls on its capitalist ruling class which has continued to push it towards maintaining its relations of imperial exploitation towards the rest of mankind.

Three hundred years ago, the United States did not even exist as a state. It was part of the British Empire. After winning its independence from Britain, it grew from a small European set of colonies into the most powerful state of the world of today. Its constitution talked about equality of all men, although, it sanctioned slavery. It boasted of protecting the basic human rights of its citizens, while it intensified the genocide of the native peoples of North America begun by the British and other European imperial powers occupying lands inhabited by the natives. The blood-shed of the natives continued while the US expanded westwards and became a continental-size state. The practice of slavery was abolished after a civil war and women were ultimately given the right to vote, although racial and gender discrimination continued. During the times of violence and cruelty, during the expansion of the US territory, the Eastern states (i.e., provinces) of the US, became centers of peaceful existence relative to the violent atmosphere of European states engaged in never-ending wars. Innovation developed and many new discoveries and inventions were the result. Also, the population of the US developed increasing religious and national tolerance giving rise to growing loyalty to the state. This gave further impetus to the increasing power of the state, but the state remained a predatory state because it had, from the beginning, promoted freedom of economic exploitation and abuse as "freedom". The appetite of its ruling class continued to grow without any limit and has, now, reached a point where it must subjugate the whole human species to meet its demands. Without a global hegemony over the whole planet, the ruling class of the US does not feel "safe"!

As we have discussed before, the secret of the power and prosperity of Britain and the US lies, above all else, in the degree to which these states have adopted the principles of compassion, equality and freedom within their societies. After the Second World War, the ruling classes of both the US and Britain made big concessions to their people - under the threat of Soviet influence, and because of the shortage of young workers. Health care and social security systems were set up. The resulting well-being of their populations, relative to other states, was based on the increasing stress on these objectives. This led to increased power of these states. That this power was abused by the ruling classes of these states, is another aspect of their behavior that is closely related to the economic *"freedom of exploitation"* retained by them. Both Britain and the US became predatory states because they were dominated by classes of feudal lords and capitalists and not by those who would want to encourage development of greater equality and compassion. The development of their societies was very limited in these directions till after the Second World War. Britain's power and influence continues to exist, to some degree even now, despite the apparent dissolution of its empire.

Thus, the creation of the environment of tolerance and consideration in their societies at an early stage, though it was certainly not ideal, was the key to the success of the United States and Britain. By the beginning of the twentieth century, the United States had created an economy of a size that did not exist in Europe and the rest of the world. So, it naturally grew and became a cohesive and militarily powerful state, despite the racial and religious differences among its people and the violent and tragic history of slavery and denial of equal rights even to women. By that time, most of those problems of oppression had been resolved to some extent. *It became prosperous and strong because most, if not all, its people wanted to belong to it and felt comfortable with each other. This was the foundation on which the development of its vast natural resources occurred.* The presence of those natural resources was, to a great extent, accidental. We can consider it the good fortune of those who "cleared" the land on which the US was established.

At this time, all the states of the First World have developed their societies in the direction that leads to an ideal state. Since the 1980s, however, the people of both North America and Europe are being pushed towards a lower standard of living as their ruling classes do not have the Soviet Union to frighten them. It is doubtful, however, that they would succeed in pushing their people back to their pre-Second World War status. Their power is based on investment of capital in countries where wages have remained relatively low. This ability to exploit low wage structure of working people in the world of intensive capitalism or feudalism is, however, bound to come to an end with time, as the rest of the world progresses towards benevolent capitalism.

Chapter-21

The Future

As described in Chapter-14, two existential threats to our species have emerged in the twentieth century. They would continue to threaten our continued existence. The dangers are real and immense. Also, many inter-state conflicts continue to exist, making nuclear conflicts more likely. The following is a list of some of those conflicts, which endanger peaceful existence of mankind at this time:

- **Conflicts between the US and Russia, and between the US and China:** In 2002, the US had pulled out of the Anti-ballistic Missile Treaty with Russia. It pulled out of the Intermediate Range Ballistic Missile Treaty with Russia, in early 2019. It seems the Trump administration is, also, planning to pull the US out of the Strategic Arms Reduction Treaty with Russia, in 2021. If it does, then, both the US and Russia would be free to develop and deploy all kinds of nuclear weapons and the danger of a catastrophic nuclear war would grow further. This increase in nuclear forces seems to be aimed at both Russia and China. The objective seems to be, mainly, to force China to reduce or stop its economic progress by devoting more of its resources to its defense. The installation of intermediate range nuclear missiles in the Panamized states, close to China's coast, is expected to help in achievement of this objective. The US has also been imposing tariffs on imports of goods from China and other states with the same objective, instead of improving its own economic performance. The "trade war" is in progress at this time.

- **Israeli Expansionism:** Israel, known by the Biblical and Quranic name of a tribe of Jews, i.e., "Israel", has developed into an "apartheid" state like South Africa under the control of its former racist regime. At present it controls Palestinian, Syrian and Egyptian territories and has developed nuclear weapons with the help of the US and France. Its violence against the Arab population within its control, and against other Arab states, continues with the result that Arab and other Muslim states feel the need to acquire nuclear weapons to defend themselves. Since Israel's creation and expansion is based on support of the US and other imperial powers, no limit to its expansion, or subjugation of Palestinians and other Arabs, seems possible at this time. Pakistan has developed nuclear weapons and Iran has, also, begun to acquire nuclear technology. Thus, the risks of future wars and the use of nuclear weapons are growing with time.

- **Panamization of Afghanistan:** During the Obama administration, drone attacks were greatly increased over Afghan and Pakistani tribal territories. After protests by the Pakistani government over bombing of its territory by its "ally" - the US, the intensity of the bombing campaign was reduced. President Trump increased the number of US military personnel in Afghanistan, because the Taliban had taken over about forty percent of Afghan territory and, according to US generals, the US was not winning the war – forty years after initiating its terrorist campaign against the government and people of Afghanistan, in 1978. President Trump started putting the full blame for this situation on Pakistan and demanded that it do more to help the US forces. Since, Pakistan was, also, facing terrorism from the Taliban controlled territory within Afghanistan, it started building a fence along the Pak-Afghan border, instead, to control the infiltration of the terrorists. The terrorists had changed their name to "Taliban" from "Mujahideen (Jihadists)" in the 1990s. Some of their groups changed their name to "Islamic State" after 2015. They have continued their activities in both Pakistan and Afghanistan. However, the situation in Pakistan is getting under control and terrorist attacks have decreased there in frequency and effectiveness. Efforts have been made to end this conflict by negotiations between the US and Taliban, but have not been successful till now. The sticking point is that Taliban refuse to deal with the government installed by the US. They, naturally, consider it to be a puppet regime. An agreement was signed between the US administration and the Taliban in early 2020. It provided for a "decrease fire" between the US forces and the Taliban, releases of about 5000 Taliban prisoners from Afghan jails and direct negotiations between the Taliban and all other organizations in Afghanistan, including the puppet regime, after about six months. However, the agreement seemed to have broken down soon after its signing. Russia has, also, been involved in this peace effort, but no end to this war is visible at this time and it has become the longest external war that the US has ever been involved in.

- **Panmization of Iraq and Syria:** When the Panamization of Iraq had, apparently, reached quite an advanced stage, a new challenge developed in its northern territory after 2006. An extremist organization calling itself the "Islamic State of Iraq and Levant (ISIL)" captured large areas of North-Western Iraq and North-Eastern part of Syria. It captured several cities and towns, including Mosul, the second-largest city of Iraq. The organization was surprisingly well armed and it captured a large amount of weaponry from the Iraqi army organized by the US. It carried out many large-scale massacres and numerous beheadings of its opponents. It, also, abducted and raped many girls belonging to a Christian community called the "Yazidi" tribe. The main reason for the military success of ISIL was that a large number of the officers and men of the old Iraqi army, disbanded by the US, had joined it. A civil war started in Syria during the "Arab Spring" - when demonstrations occurred in Tunisia, Egypt, Libya, Syria, Bahrain and some other Arab states. The US, used this opportunity to launch a war of Panamization in Syria also, by arming opposition members and promoting attacks on government installations. The Islamic State also got involved on both territories of this conflict, i.e., in Iraq and Syria. The US had to put its

412

project of Panamization of Syria on hold, for the time being, after Russia intervened in the conflict. All territories and cities captured by ISIL were taken back by the Iraqi and Syrian armies after large-scale destruction. ISIL has since moved back into terrorist bombing activities. Kurdish territory in Syria, captured by Kurdish Peshmerga fighters, has remained in their control.

- **Panamization of Libya:** A civil war started in Benghazi, Libya in 2011, after the people rose up against the Gaddafi regime. US and the other NATO "intelligence" agencies got involved in it also, supplying arms to the rebels. Ultimately, NATO started bombing Libyan territory and the US "complemented" this effort with attacks by its drone aircraft. A convoy in which Muammar Gaddafi was travelling, was hit by missiles from drone aircraft. He was caught and savagely murdered by the NATO-supported forces. US Secretary of State, Hillary Clinton, was overjoyed after hearing of Gaddafi's murder. Libya, now, has two organizations claiming to be its legitimate "governments". The Panamization project has bogged down for now. Gaddafi had done a lot of good to his country, but he had one defect in common with all Arab rulers – when they get into power, they want to become kings, i.e., they want to rule for the rest of their lives and they want their sons to become kings after their own deaths. They don't seem to realize that the times of kings and kingdoms are gone!

- **Conflicts Among China, India and Pakistan:** Near the end of the nineteenth century, Czarist Russia had conquered territories in Central Asia and was advancing towards Afghanistan. The British Raj in India became alarmed that Russia might try to grab "its India" from Britain. Two expeditions were sent to Afghanistan and after much violence a buffer state of Afghanistan was created with borders agreed to by both imperial powers, while Britain retained control of Afghan foreign affairs for some time. After this achievement, rulers of Britain were still afraid that Russia might capture Tibet and then attack "their" India from the North. To preempt this possible attack, British forces in India launched an attack on Tibet in 1906 and conquered territory both in the South-East and West of Tibet. Neither the Tibet province of China nor the Chinese central government accepted this territorial expansion. The conquered territory of Western Tibet was added to a Bantustan called the "Princely state of Jammu and Kashmir" under British control. This Bantustan was one of many previously set up by Britain in South Asia. It was and is a Muslim majority territory. However, the "sovereignty" of this princely state was sold by the British to a Hindu landlord, who was charged a large sum of money in this favorable transaction. At the time of partition of South Asia, Britain decided that the nominal rulers of Bantustans in their control would have to determine whether their people wanted to join India or Pakistan. As the "Raja" of Kashmir hesitated in making a decision, the Gilgiti and Balti nations of the Bantustan rebelled and wiped out his military forces on their territories. Also, people belonging to some Muslim tribes in Pakistan entered Kashmir to help the Kashmiri nation. Indian forces invaded Kashmir, after the Raja of the "state" ran away to Delhi and, without determining the wishes of the people of Kashmir, signed a piece of paper saying he wanted "his state" to join India. India occupied about two-third of the territory of this Bantustan. This dispute remains unresolved till now. Similarly, India refused to return Tibetan territory, conquered by the British Raj, to China. This includes Northeastern part of Kashmir, called Ladakh, and a territory in the North-East of India, which the British had named the "North-East Frontier Agency (NEFA)" and is, now, called "Arunachal Pradesh", by India. Wars between India and Pakistan have occurred in 1948, 1965 and 1970. A war between India and China occurred in 1962, when China, in frustration, attacked NEFA and then returned all the captured territory to India, to teach a lesson to the Indian rulers, so that they would peacefully resolve this dispute. The Indian rulers refused to learn any lesson! They continue to claim the Bantustan of Kashmir and parts of Tibet as their imperial inheritance from Britain. Thus, India has transformed itself from an imperial possession to an imperial power and the "border disputes" remain unresolved. All three states, i.e., China, India and Pakistan have become nuclear powers and the danger of a nuclear conflict, especially between India and Pakistan, has increased with time since, in case of another large-scale invasion by India, Pakistan would be forced to use, at least, short-range tactical nuclear weapons for its defense.

- **Division of Korea:** Division of Korea occurred after the Second World War, as described in chapter-8. After the Korean war of 1950-53, a truce was arranged at the end of military operations but no peace treaty has been signed between North Korea and the US, to formally end the war. During the war, General MacArthur had wanted to attack China, Russia and North Korea with nuclear weapons. The result is that all three states have developed nuclear weapons to deter such an attack! North Korea has developed long-range missiles which can, probably, reach the US mainland. It wants the removal of US military forces from South Korea as a precondition for limitations on its nuclear program, while the US President, Donald Trump, has threatened it with *"complete destruction"*! No resolution of this dispute seems possible at this time.

- **Russia-Ukraine Border Dispute:** Ukraine was part of the Czarist Empire and, later, the Soviet Union. To mollify the nationalism of the Ukrainians, Lenin transferred a large part of Russian inhabited territory to the Ukrainian "Republic", i.e., the province of Ukraine. Later on, Stalin transferred to it some Hungarian inhabited territory in its West. In the 1960s, Nikita Khrushchev transferred the Crimean Peninsula to it. Because of the abrupt dissolution of the Soviet Union in 1991, a large part of Russian territory became "Ukrainian territory" – thanks to a power-hungry and perpetually drunk Yeltsin! After fighting started in Eastern Ukraine, as a result of a coup-like take-over of its government by extremist Ukrainians, Russia took the Crimean Peninsula back from Ukraine. In the Western mainstream press this is referred to as an annexation. Although Russian and Ukrainian nations are quite close culturally, they have resentments anchored in Czarist times and the Nazi invasion of the Soviet Union, when many Ukrainians sided with Hitler's army. The US has been trying to take advantage of these resentments, but a stand-off has ensued. Any rash action by the Ukrainian super-nationalists can initiate a large-scale war.

- **Russia-Japan Dispute over Islands:** Four islands in the North of the Japanese island of Hokkaido are in dispute between Russia and Japan. Considering that large parts of Eastern Cyberia and the Kamchatka were conquered by the Czarist empire in the not too distant past, it should be possible to hand over these islands to Japan, in return for satisfactory compensation to their residents and the resettlement of the residents in other parts of Russia, in accordance with their wishes. This unnecessary irritant in Russian-Japanese relations needs to be removed. Russia has no shortage of territory like Japan! A little humane treatment would go a long way in creating friendly relations between the two states.

- **China-Japan Dispute over Islands in the East China Sea:** Some Chinese islands were taken over by the US, during the Second World War. They were, then, handed over to Japan instead of China. China is asking for their return.

- **North and South Sudan Border Dispute:** South Sudan has become the newest state to come into existence, but it, still, claims some territory that has not been transferred to it by Sudan. Besides, a high level of inter-tribal violence is continuing in South Sudan. This tribal state needs external help in finding internal peace.

- **Russia-Chechnya Dispute:** Territories of Chechnya, Ingushetia and Dagestan, etc., were conquered by the Czarist Empire. When the Soviet Union was created and national territories were demarcated, these territories were included in the Russian Republic. These tribes/nations have always been rebellious, not only because of their religion but also because of their tribal and national identities. They have been badly treated over more than a century. Repeated military campaigns have only solidified their resentment of Russian subjugation. Their problems with Russia need to be solved on the basis of tolerance and tribal autonomy. If such efforts fail, then only self-determination can be a solution of their problems – Such a solution would require long-term plans of economic and educational development coupled with development of legal protections of religious and other cultural rights – followed by referendums, to determine their wishes for the future. Development of mutual trust between these tribes/nations and the Russian nation can only occur under such conditions, leading to long-term peaceful coexistence.

- **South China Sea Islands Dispute:** China claims some rocks and coral reefs in the South China Sea as part of its territory, based on its control of these islands about two thousand years ago. In recent years, some surrounding states in South East Asia have also started laying claims to these islands, probably because of the speculation that their territorial waters may contain some mineral resources. These are uninhabited rock formations which China seems to be developing for defensive purposes. Although the US cannot have any grounds to claim these islands, it has started to send its naval ships on "freedom of navigation" patrols through their territorial waters, although no hindrance to navigation has ever occurred. It seems this dispute would be resolved between the neighboring states concerned, with the passage of time.

The conflicts between the US and China, US and Russia, China and India, India and Pakistan, Israel and Iran, and North Korea and the US involve states possessing nuclear weapons and are, thus, much more dangerous than the other disputes.

Assuming that mankind would be able to control and eliminate the existential threats posed by global warming and nuclear weapons, by reducing its inter-state conflicts, what can be expected in the immediate and distant future? In historical context, we can divide further expected development of mankind into two stages as described in the following sections.

Global, Socio-Economic Convergence

At present, the surplus capital in Europe and North America, accumulated during the previous centuries of exploitation of the people of the whole world, is being heavily invested in East and South East Asia, Central Eurasia and South Asia, leading to high rates of economic growth in these regions. High accumulation of capital is also occurring in China. It is mostly occurring in the state sector and in terms of improvement of living standards of the people and development of skilled labor, i.e., in terms of human capital. The economies of East, South and South-East Asia are growing, as are the economies of South America, while those of Europe and North America are stagnating, due to diversion of investment to other regions of the world.

A global culture is also emerging as the reach of the internet and other media has become virtually global. This would result in a basic change in the balance of economic and political power. As the emerging economies and regions grow, and their populations stop growing or begin to decline, the standard of living of their populations is bound to rise, under twin processes of economic growth and slowdown of population growth, or its reversal. For the East Asian, Central Eurasian and South American super-states, or regions, this process would be similar to what happened in Europe in the second half of the nineteenth century and the twentieth century. China, and the region around it, would also go through similar developments. The rates of economic growth of the world, excluding the world of benevolent capitalism, would become progressively faster as fewer and fewer regions with low wages and low-skilled labor remain under-developed. Populations of Bantustans and Panamized states have grown in numbers and in terms of education and political consciousness. The demands of those peoples are already becoming irresistible for their ruling classes. The balance of power in such states would continue to shift in favor of their people.

The Future

The basic requirements for emergence and consolidation of super-states have been met by North America and Europe. These two virtual super-states have emerged in the present-day world of benevolent capitalism. The same would happen to the present worlds of intensive capitalism and advanced feudalism. As time passes, super-states would be consolidated in these two worlds also. From Europe's history we can see that the transformation of European advanced feudal societies into capitalist societies started in the eighteenth century and was completed in the twentieth century - *a period of about two hundred and fifty years*. This was the period in which Europe evolved from advanced feudalism to the end of intensive capitalism. Since the second and the third worlds are already going through advanced stages of intensive capitalism and feudalism respectively and are unlikely to get involved in large scale conflicts like Europe in the nineteenth and twentieth centuries, they should take a much shorter time, *maybe one and a half century*, to go through their respective transformations into benevolent capitalism and intensive capitalism respectively. States going through managed transformation would take much shorter time to transform their economies and societies into forms similar to benevolent capitalism, *although extreme inequality is unlikely to develop in such states*.

Thus, in the first period of about one and a half century, the Central Eurasian and South American super states should become cohesive entities – becoming part of the world of benevolent capitalism. China and other states of managed transformation close to it would form the East Asian super-state. Socio-economic development in this super-state is likely to be faster due to the guided nature of its development, which would result in higher levels of social justice and accumulation of human capital. Other East Asian and Southeast Asian states may also reach some economic accommodation with this super state. *During the initial period of one hundred years or so, China's economy is likely to become bigger than that of North America and Europe combined. It is expected to be one-third of global economy at that time! It is also likely to develop a society with a higher level of social justice than the world of benevolent capitalism today.* Southeast Asian states may reach the stage of intensive capitalism during this period. So would South Asia and Arabia. The populations of all Bantustans are already increasing and demanding better educational, employment and living conditions. These pressures are like to grow with time and all Bantustans are likely to disappear during this period. Some may get converted into Panamized states, however.

In the second period of about one hundred years, South Asia, Southeast Asia and Arabia would be expected to reach the stage of benevolent capitalism. India, Central Eurasia and South-East Asia are, however, likely to retain a mixed economy during this period with considerable levels of state-ownership. At the end of this period, only two worlds would come into existence. The First World would then, probably, consist of seven closely-knit virtual super-states - North America, South America, Europe, Central Eurasia, East and Southeast Asia, South Asia and Arabia. The Second World would consist of non-Arab Africa only. The First World would, also, have become greatly homogeneous in terms of economic mode of production, standards of living, diversity of population and social security.

In the third period, which may last less than a hundred years, most of the huge capital generated by economic activity in the First World would be invested, mainly, into non-Arab Africa which should develop very fast, as a result, to catch up with the rest of the "global village". By this time, the existence of Panamized states, and any other states based on religious or racial segregation, would become untenable.

Thus, it seems that, unless, some unforeseen disaster like global warming, or a major nuclear war, hits humanity in this period of three centuries or so, the whole human species would have reached a more or less uniform condition of existence. The population of the world would have gone down from its peak and would have become either stable, or would have started decreasing, with a large proportion of elderly population. A high degree of diversity is to be expected throughout the planet, at that time.

In today's world, we have the US, which is the most powerful state in human history - with global military reach. No state ever had the power to control the whole planet. Many people dreamed of it, people like Hitler and Genghis Khan and others dreamed about a European or a Global Empire, but it was never really achieved. Even the British Empire, although it was spread all over the planet, did not cover the whole world. States have grown as described and it seems to follow from the considerations of competition and adaptation, that one state would outstrip the others in adaptability, and may dominate others in terms of economics. If this happens, the planet could, then, have the greatest conflict of human existence - conflict over control over the whole human species.

The global empire, if it comes into existence, would tend to create a global state, or a closely integrated set of states, with two classes - the global ruling class and the global class of industrial and intellectual workers. Boundaries of all states may be demolished in the pursuit of more "tribute". First, labor, incorporated in products, would flow across state boundaries, as is increasingly happening at the current time. The flow of living labor to areas of labor shortage, due to a much higher development of robotics and automation, would do the same. This would happen because large areas of labor shortage are likely to develop in different regions of the world. Further, our psychological and social evolution would have progressed with time - we are already at a point where we don't need to stay within states where we are born, and be afraid of other human beings. The labor shortages on the economically more developed territories controlled by the semi-global empire would increasingly attract immigration into those territories. Boundaries of nationalism and religion would be increasingly ruptured and demolished.

However, it seems that a global empire is not likely to emerge. The semi-global empire would be continuously opposed by the virtual central Eurasian, East Asian, South Asian and South American super-states. Those states which have been able to maintain and safeguard their independence during the twenty-first century till now, would certainly continue to oppose its hegemony. No matter how this conflict is resolved, momentum would continue to be towards formation of one group, as a species, on a global scale - a tightly knit set of virtual super-states or a virtual global state! Under the resulting conditions, the standard of living of the whole species would also tend to converge with time. During this period of convergence towards one level of socio-economic development, the creation of an *Interstate Relations Organization (ISRO)* would be necessary. The current United Nations Organization might be modified to truly represent all super-states of the world, or a new organization might be developed to represent all states on the democratic basis of their populations.

418

Global Post-Convergence Development

Evil has existed in our species, like it has existed in all living beings. It exists in our states, our nations, our tribes, our families and even in our individual selves – so does goodness. Evolutionary history of mankind is testimony to the fact that, slowly but surely, goodness has won over this evil of blind greed and insensitivity towards our fellow human beings. This is why we are the dominant species on this planet. This is why, ultimately, evil would be brought under control and banished to the unreachable depths of our minds and to the very small and insignificant sections of our societies.

Let us consider what type of global state, or community of closely linked super-states, would have developed in three or four more centuries. Sympathy, compassion and respect for each other are evolutionary survival skills. Looking at the evolutionary history of mankind, it appears that these group social skills are the driving force behind the growth of social systems and states – from families, to tribes to nations to multi-national states, to super-states. These social skills are bound to become *irresistibly* powerful and the desire to live in peace and harmony with each other and with most of the other species on this planet is likely to become dominant in human societies. The eight super-states would become highly interdependent economically, socially and culturally, so that military conflict among them would become increasingly difficult to contemplate, especially since some of these super-states would remain, or would have come, under the hegemony of the North American super-state, ruled by the US ruling class – while the other super-states become increasingly independent and powerful with nuclear weapons and high-speed missile technology. The presence of highly developed industry, with high intellectual inputs, combined with a labor shortage due to slow or negative growth of global population, would tend to increase the level of scientific knowledge and political consciousness among the working people and would give a high level of power to the global working classes in terms of political and economic decision-making in work places and at various levels of political management. Thus, maintenance of all institutions that create social inequality and economic abuse would become increasingly untenable.

This has already happened in China, Tsarist Russia and some other states, as a result of violent confrontations between the people and their ruling classes. These social explosions have resulted in establishment of one-party governing systems controlled by their people. Gains of the working classes in the successor states of the Soviet Union, have been reversed to some extent in economic terms, but have become irreversible in other aspects. This has also happened in states like Norway, Sweden and Denmark, where a peaceful transition to a socially just economy has occurred and their people have successfully resisted any reversal of their social progress. Similar processes are at work in all other states of the First World, where the drive towards universal health care, social security, workplace safety, care of children, free education for all children in schools, elderly care and care of disabled has become more and more irresistible and irreversible.

The models of socio-economic development, as developed by China, Norway, Sweden, or Denmark, are highly likely to be adapted by other super-states, in accordance with their prevailing socio-economic conditions, leading to stable and sustained growth in the well-being of the people. Dramatic increases in automation and artificial intelligence would result in greater leisure time for working people. What path those states would take would depend on their specific economic, social and political status. Under the resulting conditions, the working people would be able to confront and successfully resist their ruling classes. The global, or semi-global, empire and the ruling classes of exploiters in the virtual super-states would tend to lose their power of exploitation and would, ultimately, cease to exist as exploiters under the conditions of social and political consciousness and the resulting organizational abilities of working people. Workplaces would become democratic, with worker ownership increasing with time. The need for large stocks of weaponry would tend to disappear, especially, if and when a global state develops - the need, for nuclear weapons, bombers and fighter aircraft in the air, sub-marines, aircraft-carriers, destroyers and cruisers on the oceans and tanks, armored personnel carriers and howitzers on the ground, *would disappear*. In short, the super-states would tend to *"wither away"* in military terms and in terms of institutions of economic exploitation and abuse. The world would, thus, inevitably become an ideal global state, or a set of ideal super-states - *of the people, ruled by the people themselves, for themselves*.

Notes

Chapter 1
Life on Earth

1. Audesirk, Gerald, Teresa Audesirk and Bruce Byers, *Biology, Life on Earth with Physiology, Chapter-1: An Introduction to Life on Earth – What is Life?* (Pearson.com, Pearson Education, Limited, 2016[2013]), p. 2.

2. Ibid., *Chapter-4: Cell Structure and Function –What is Cell Theory? p. 62.*

3. Ibid., *Chapter-1: An Introduction to Life on Earth – What is Evolution? p. 5.*

4. Ibid., *Chapter-15: The Origin of Species – How do new Species Form? p. 305.*

5. Ibid., *Chapter-13: Principles of Evolution – How does Natural Selection Work? p. 265.*

6. Ibid., *Chapter-16: History of Life – How did Humans Evolve? p. 332.*

7. Ibid., *Chapter-14: How Populations Evolve – What causes Evolution? p. 282.*

Chapter 2
Feudalism

1. Chomsky, Noam: *Rogue States, chapter 11: The Legacy of War* (London, Pluto Press, 2000), p. 156-159.

2. Marx, Karl. *Capital, Volume-1, Part-8: Primitive Accumulation* (1887[1867])

Chapter 3
Dynamics of Social Change

1. Audesirk, Gerald, Teresa Audesirk and Bruce Byers, Biology, Life on Earth with Physiology, Chapter-16: History of Life – Sophisticated Culture Arose Relatively Recently (Pearson.com, Pearson Education, Limited, 2016[2013]), p. 338.

Chapter 4
Basics of Economics

1. Kamerschen, David, Richard McKenzie, Clark Nardinelli, *Economics, Second Edition,Chap-1,The Economic Way of Thinking: The Economic problem*, (Boston, Houghton Mufflin Company, 1989), p. 6.

2. Krugman, Paul, Maurice Obstfeld and Marc Melitz, *International Economics, Theory and Policy, Chapter 3-6* (Pearson.com, Pearson Education, Limited, 2015), p. 56-173?

Chapter 6
Intensive Capitalism

1. Marx, Karl. *The Civil War in France* (1871), p. 3, 4.

2. Ibid. p. 29.

3. Ibid. p. 5.

4. Ibid. p. 33, 48.

5. Zinn, Howard. People's History of the United States: 1492-Present, Chapter 5: A Kind of Revolution (New York, HarperCollins Publishers, 1995[1980]), p. 76-101.

6. Marx, Karl and Friedrich Engels. *Manifesto of the Communist Party (*1888 [1848]), pp. 16.

7. Marx, Karl. *A Contribution to the Critique of Political Economy: Preface* (1859).

8. Marx, Karl and Friedrich Engels. *Manifesto of the Communist Party (*1888 [1848]), pp. 12.

9. Ibid., p.15

10. Ibid., p.21

11. Ibid., p.25

12. Marx, Karl. *Critique of the Gotha Program, Part IV* (1875).

13. Marx, Karl and Friedrich Engels. *Manifesto of the Communist Party (*1888 [1848]), pp. 27.

14. Ibid., p. 25

15. Engels, Friedrich, *Origin of the Family, Private Property and the State (1942[1884]), p. 93.*

16. Karl, Kautsky. (1918). Dictatorship of the Proletariat: Chapter-8, The Object Lesson. www.marxists.org. accessed October, 16, 2019.

17. Bernanke, Ben. S. Remarks at the H. Parker Willis Lecture in Economic Policy, Washington and Lee University, Lexington, Virginia (March 2, 2004)

18. Gorbachev, Mikhail. *Perestroika* (New York, Harper & Row Publishers, 1987), p. 192.

19. New York Times, *Our Policy Stated in Soviet-Nazi War*, June 24, 1941.

Chapter 7
Economics of Intensive Capitalism

1. Marx, Karl. *Wage, labor and capital, Relation of Wage-Labor to Capital*, (1849), p. 14-18.

2. Kamerschen, David, Richard McKenzie, Clark Nardinelli, *Economics, Second Edition, Chapter 6, Business Cycles*, (Boston, Houghton Mifflin Company, 1989), p. 152

3. Ibid., *Chapter 10, Unemployment and the Equilibrium Income Level*, p. 246-270.

4. Keynes, John Maynard, *The Economic Consequences of the Peace* (1919), Chapter VI, pp. 235-236

5. Keynes, John Maynard, *A Tract on Monetary Reform* (1923), Ch. 3, p. 80

6. Keynes, John Maynard, (1933) *National Self-sufficiency*: Section 3, republished in Collected Writings Vol. 11 (1982).

7. Keynes, John Maynard, *The End of Laissez-faire, Chapter 1* (1926).

8. Keynes, John Maynard, As quoted in *Moving Forward: Programme for a Participatory Economy* (2000) by Michael Albert, p. 128 – a recent variant of the original quote attributed by Sir George Schuster, Christianity and human relations in industry (1951), p. 109

9. Keynes, John Maynard, *The General Theory of Employment, Interest and Money* (1936), p. 155-157, 164.

Chapter 8
Benevolent Capitalism

1. Truman, Harry S. *Memoirs by Harry S. Truman, Vol-2: Years of Trial and Hope* (New York, Doubleday & Company, 1955), p. 62

2. Miller, Merle. Quoting President Truman, *in Plain Speaking: An Oral Biography of Harry S Truman: On General Douglas MacArthur* (New York, Putnam Publishing Group, 1974)

3. Rees, David. *Korea: The Limited War* (St. Martin's Press-Macmillan; 2nd printing edition, 1964), p. 418-420.

4. Gorbachev, Mikhail. *Perestroika*, p. 149.

5. Chomsky, Noam. *Rogue States*, (London, Pluto Press, 2000), p.164-169.

6. Ibid., p. 87-89.

7. Dolny, Helena. Slovo: *The Unfinished Autobiography (Melbourne, New York, Ocean Press, 1997), p. 47-48*

8. Rusk, Dean. *As I saw It: as told to Richard Rusk* (New York, W. W. Norton & Company, 1990), p. 153

9. Chomsky, Noam. *Rogue States*, (London, Pluto Press, 2000), p. 144.

10. Ibid., p. 107.

11. Ibid., p 93-97.

12. Ibid., p 17.

13. Yousaf. Mohammed and Mark Adkin. *Afghanistan: The Bear Trap: The Defeat of a Superpower* (Havertown-PA, Casemate, 2001[1992])

14. Boggs, Carl. *The Crimes of Empire* (London, Pluto Press, 2010), p. 93-97.

15. Ibid., p. 92.

Chapter 9
Economics of Benevolent Capitalism

1. Kamerschen, David, Richard McKenzie, Clark Nardinelli, *Economics, Second Edition, Chapter 19, Corporations are not Persons*, (Boston, Houghton Mufflin Company, 1989), p. 510.

Chapter 11
Globalization

1. Krugman, Paul, Maurice Obstfeld and Marc Melitz, *International Economics, Theory and Policy, Chapter 20: Financial Globalization: Opportunity and Crisis: The Structure of International Capital Market*, p. 634-636.

2. Ibid., The Shadow Banking System, p. 637.

3. Ibid., *The Global Financial Crisis of 2007-2009*, p. 647.

Chapter 12
Inter-State Trade

1. Kamerschen, David, Richard McKenzie, Clark Nardinelli, *Economics, Second Edition, Chapter 1: Introduction*, (Boston, Houghton Mufflin Company, 1989), p. 8-9.

2. Ibid., p. 30.

3. Ibid., *Chapter 34: International Trade*, p. 859-861.

4. Krugman, Paul, Maurice Obstfeld and Marc Melitz, *International Economics, Theory and Policy, Chapter 3: Labor Productivity and Comparative Advantage: The Ricardian Model* (Pearson.com, Pearson Education, Limited, 2015), p. 58.

5. Kamerschen, David, Richard McKenzie, Clark Nardinelli, *Economics, Second Edition, Chapter 34: International Trade, The Distributional Effects of Trade* (Boston, Houghton Mufflin Company, 1989), p. 863-865.

6. Ibid., p. 861.

7. Krugman, Paul, Maurice Obstfeld and Marc Melitz, *International Economics, Theory and Policy, Chapter 5: Resources and Trade: The Heckscher-Ohlin Model* (Pearson.com, Pearson Education, Limited, 2015), p. 127.

8. Ibid., p. 128.

9. Ibid., *Chapter 9: The Instruments of Trade Policy*, p. 238.

10. Ibid., *Chapter 22: Developing Countries: Growth, Crisis and Reform, The Debt Crisis of 1980s*, p. 716-720.

11. Ibid., *East Asia: Success and Crisis*, p. 720-728.

12. Ibid., *Why Have Developing Countries Accumulated Such High Levels of International Reserves?* p. 721-723.

13. Ibid., *Capital Paradoxes*, p. 736-740.

14. Ibid., *Structural Features of Developing Countries*, p. 706-709.

Chapter 13
Semi-Global Empire

1. Gorbachev, Mikhail. *Perestroika*, p. 139-140.

2. Zinn, Howard. *People's History of the United States: 1492-Present* (New York, HarperCollins Publishers, 1995[1980]), p. 620-621.

3. Boggs, Carl. *The Crimes of Empire, Chapter-5: A Tale of Broken Treaties*, p. 144-146.

Chapter 14
Threats to Existence of Mankind

1. Gorbachev, Mikhail. *Perestroika*, p. 219.

2. Quaschning, Volker. *Renewable Energy and Climate Change, Chapter 2: The Climate Before the Collapse?* (HTW Berlin - University of Applied Sciences, 2009), p. 24-42.

Chapter 16
Human Conflicts

1. Freud, Sigmund. *New Introductory Lectures on Psycho-analysis, Lecture xxxi: The Anatomy of the Mental Personality* (1933)

Bibliography

Baweja, Harinder. *Kashmir, Losing Control* ("India Today" magazine, May 31, 1993)

Berman, Morris. 2014, *Why America Failed: The Roots of Imperial Decline* (CreateSpace Independent Publishing Platform, 2014)

Berman, Morris. *Are We There Yet?* (Brattleboro, VT, Echo Point Books & Media, 2017)

Boggs, Carl. *The Crimes of Empire* (London, Pluto Press, 2010)

Carter, Jimmy. *Palestine: Peace not Apartheid* (New York, Simon & Schuster, 2007[2006])

Chomsky, Noam. And Andre Vitchek. *On Western Terrorism* (London, Pluto Press, 2017)

Chomsky, Noam. (Author), Peter Hutchison (Editor), Kelly Nyks (Editor), Jared P. Scott (Editor), *Requiem for the American Dream: The 10 Principles of Concentration of Wealth & Power* (Jacksonville, Florida, Seven Stories Press, 2017)

Chomsky, Noam. *Rogue States* (London, Pluto Press, 2000)

Chomsky, Noam. *Who Rules the World?* (New York, Picador-Macmillan Publishers, 2017)

Chomsky, Noam. Ilan Pappe and Frank Barat. *On Palestine* (Chicago, Haymarket Books, 2015)

Cohen, Stephen. *India, Emerging Power (New Delhi, Oxford University Press, 2010[2001])*

Darwin, *Charles R. Descent of Man and Selection in Relation to Sex* (1871).

Eberstadt, Nicholas. *Population Change and National Security*, (Foreign Affairs, Vol. 70, Number 3, 1991) pp. 115-131

Eberstadt, Nicholas. *The Population Implosion* (Foreign Policy, March-April 2001).

Eberstadt, Nicholas. *World Population Implosion?* (*The Public Interest Magazine, 1997*)

Engels, Friedrich. *Condition of the Working Class in* England in 1844 (1887)

Gorbachev, Mikhail. *Perestroika* (New York, Harper & Row Publishers, 1987)

Gorbachev, Mikhail. *On my Country and the World* (New York, Columbia University Press, 2000)

Gorbachev, Mikhail. The New Russia (Cambridge, UK, Polity Press, 2017[2016,2015])

Harrington, Michael. *Socialism, Past & Future* (New York, Skyhorse Publishing Inc., 2011)

Huang, Yukon. *Cracking the China Conundrum: Why Conventional Economic Wisdom Is Wrong* (New York, Oxford University Press 2017)

Bibliography

Jacques, Martin. *When China Rules the World: The End of the Western World and the Birth of a New Global Order* (New York, Penguin Books, *2012 [2009]*)

Johnson, Chalmers. *The Sorrows of Empire: Militarism, Secrecy and the End of the Republic* (New York, Metropolitan Books: Henry Holt and Company, 2005[2004])

Judis, John B. *Imperial Amnesia* (Foreign Policy magazine, July-August 2004)

Kautsky, Karl. *The Dictatorship of the Proletariat* (Vienna, National Labor Press 1919 [1918])

Leon, Abram, *The Jewish Question*: A Marxist Interpretation (New York, Pathfinder Press, 1970 [1950])

Liles, George. *The Faith of an Atheist* (MD Magazine, March 1994)

Marx, Karl and Friedrich Engels. *Manifesto of the Communist Party (*1888 [1848]), pp. 98-137.

Marx, Karl. *Value, Price and Profit* (New York, International Co. 1969[1898])

Marx, Karl. *Wage, Labor and Capital* (1847)

McCoy Alfred W. *In the Shadows of the American Century: The Rise and Decline of US Global Power* (Chicago, Haymarket Books, 2017)

McCoy, Alfred W. *Policing America's Empire: The United States, the Philippines, and the Rise of the Surveillance State* (Madison, University of Wisconsin Press, 2009)

Mill, John Stuart. *The Subjection of Women* (Cambridge Massachusetts, MIT Press, 1970[1980,1869])

Quaschning, Volker. *Renewable Energy and Climate Change*, (John Wily and Sons Ltd, Hoboken, NJ, Second Edition, 2019[2009])

Rees, David. *Korea: The Limited War* (St. Martin's Press-Macmillan; 2nd printing edition, 1964)

Rostow W. W. *The Stages of Economic Growth: A Non-communist Manifesto* (Cambridge, Cambridge University Press, 1982 [1960])

Said, Edward W. *The Question of Palestine (New York*, Vintage; Reissue edition *1992)*

Sampson, Anthony. *The Seven Sisters:* The great oil companies & the world they shaped (London, Hodder and Stoughton, 1975)

Schmitt, Richard. Introduction to Marx and Engels: A Critical Reconstruction (Boulder, Westview Press, 1987)

Stone, Oliver and Peter Kuznick. *The Untold History of the United States* ((New York, Gallery Books - Simon & Schuster, 2013 [2012])

Wise, David, *The Invisible Government (New York, Random House, 1964)*

Wolpert, Stanley. *Zulfi Bhutto of Pakistan: His Life and Times* (London, Oxford University Press,

1993)

Wright, Robert. *Feminists, Meet Mr. Darwin* (The New Republic Magazine, November 28, 1994)

Yousaf. Mohammed and Mark Adkin. *Afghanistan: The Bear Trap: The Defeat of a Superpower* (Havertown - Pennsylvania, Casemate, 2001[1992])

Zinn, Howard. *People's History of the United States*: 1492-Present (New York, HarperCollins Publishers, 1995[1980])

Index

430

Index

G

Index

Q

R

S

Index

Index

CPSIA information can be obtained
at www.ICGtesting.com
Printed in the USA
BVHW011024170321
602756BV00007B/205